IET ELECTRICAL MEASUREMENT SERIES 13

Coaxial Electrical Circuits for Interference-Free Measurements

Other volumes in this series:

Volume 4	**The current comparator** W.J.M. Moore and P.N. Miljanic
Volume 5	**Principles of microwave measurements** G.H. Bryant
Volume 7	**Radio frequency and microwave power measurement** A.E. Fantom
Volume 8	**A handbook for EMC testing and measurement** D. Morgan
Volume 9	**Microwave circuit theory and foundations of microwave metrology** G. Engen
Volume 11	**Digital and analogue instrumentation: testing and measurement** N. Kularatna
Volume 12	**Microwave measurements, 3rd edition** R. Collier and D. Skinner (Editors)

Coaxial Electrical Circuits for Interference-Free Measurements

Shakil Awan, Bryan Kibble
and Jürgen Schurr

The Institution of Engineering and Technology

Published by The Institution of Engineering and Technology, London, United Kingdom

The Institution of Engineering and Technology is registered as a Charity in England & Wales (no. 211014) and Scotland (no. SC038698).

First published 2011

The Institution of Engineering and Technology
Michael Faraday House
Six Hills Way, Stevenage
Herts, SG1 2AY, United Kingdom

www.theiet.org

British Library Cataloguing in Publication Data
A catalogue record for this product is available from the British Library

ISBN 978-1-84919-069-5 (paperback)
ISBN 978-1-84919-070-1 (PDF)

Typeset in India by MPS Ltd, a Macmillan Company
Printed in the UK by CPI Antony Rowe, Chippenham, Wiltshire

Contents

Figures

Abstract

The central thesis of this book is that it is possible to design circuitry so that the electric and magnetic fields associated with its currents and potentials are confined. If this is done, an important consequence of reciprocity is that such circuitry does not respond to external electric or magnetic fields – it is *immune to external electrical interference*. Circuitry possessing these properties ought to be much more widely used in electrical instrument, especially sensing and measurement systems within the frequency range from DC to 100 MHz.

An electrical network of coaxial cables has no significant external magnetic field if each outer conductor carries a current that is maintained equal and opposite to that flowing in the inner conductor. If the outer conductors are all at nearly the same potential, there will also be negligible electrical fields outside the cables. Consequently no cable has any means of interacting electromagnetically with any other cable, and the performance of the circuitry also is unaffected by how the cables are routed and is immune to interference.

This book describes examples of the elegance, power and utility of coaxial networks. They should be used for precise and accurate sensing and measuring circuits of all kinds. Moreover the principles involved can be applied to all circuits to prevent them generating electromagnetic pollution. It updates an earlier publication [1] by including more recent work such as AC measurements of quantum Hall resistance to provide a primary quantum impedance standard and the extension of coaxial networks to higher frequencies.

We apply first principles to common situations and then describe, as an example of the ultimate in noise matching and interference elimination, the linear networks for establishing and measuring standards of electrical impedance that are known colloquially as coaxial AC bridges.

The authors have between them more than 60 years of experience in electrical measurements carried out in national standards laboratories. These laboratories are the source of measurement standards and techniques for the rest of the technical and scientific world and are therefore dedicated to accuracy by reporting the unique correct result of a measurement within a stated uncertainty. They continually strive to reduce that uncertainty to stay ahead of the requirements of science and industry for the foreseeable future and to promote good electrical measurement practice. The principles elucidated in this book play a major part in this process.

Reference

1. Kibble B.P., Rayner G.H. *Coaxial AC Bridges*. Bristol: Adam Hilger Ltd.; 1984. (Presently available from NPL, Teddington, TW11 0LW, U.K. www.npl.co.uk)

Preface

Lord Kelvin is famously reputed to have said 'to measure is to know'. This remark might well be expanded, 'to measure is to know, to measure accurately is to understand'. The process of accurate measurement starts with setting down complete defining conditions for the measurement and devising measuring apparatus which aims to fulfil these conditions. It continues by deliberately making changes in the measurement method. If these changes still fulfil the defining conditions but different results are obtained, basic principles cannot have been obeyed. Therefore, in order to obtain a unique and accurate result, basic principles must be thoroughly understood, and defining conditions derived from them obeyed by adopting this methodical approach the very significant cost of accurate measurement will be minimised.

In particular, the basic principles of interference elimination are, in general, not well understood. Everybody who designs, constructs or uses sensitive electrical measuring circuits has to contend with issues of extraneous electrical noise and interference. A good system should be robust and insensitive to electrical and mechanical disturbances of any kind. Often these disturbing influences are accompanied by systematic errors.

The authors have gained considerable experience in combating the problem of interference through many years of work on impedance-measuring networks ('bridges'). In this specialised activity, sensitivities of nanovolts or picoamperes are common, and therefore great care has to be taken to prevent any part of the network from coupling with any other, or with electromagnetic fields in the environment. Prevention has been achieved through the elegant application of coaxial networks where connections between screened components are made with coaxial cables in which the currents in the inner conductor and the outer ('screening') conductor are constrained to be equal in magnitude but opposite in direction. The impedances to be compared are completely defined in terms of the voltage differences and currents to be established at their terminals. A network of this kind is electromagnetically isolated, and its own electromagnetic fields are contained within its components and interconnecting cables. Employing coaxial networks for accurate impedance comparisons is an esoteric activity. Standardising national laboratories obliged to acquire this capability may find that the equipment is not easily commercially available. They will be obliged to make it in-house, but then there will at least be the advantage of knowing intimately how it works. To assist this situation, we give some information on constructional techniques.

Readers not concerned with accurate impedance standardisation and measurement will find these comparatively simple systems are nevertheless well worth studying because they are an example of what constitutes good practice when designing and constructing sensitive measuring circuits. We therefore make no excuse for describing impedance-comparing circuits in some detail.

Some sections of the book (such as chapters 5 and 6) may only be relevant to these more specialised readers. The book is also intended for those in national standards laboratories who derive standards from the abstract definitions of SI units with the greatest possible accuracy.

Acknowledgements

The authors thank The Physikalisch-Technische Bundesanstalt for extensive support and encouragement; and The National Physical Laboratory for permission to include most of the content of NPL report DES 129.

We acknowledge the considerable contribution Dr. Ian A. Robinson has made to our understanding of the science of eliminating electrical interference from sensitive circuits and the technical help of Norman and Charles Lloyd (N.L. Engineering, Woodhurst, Huntingdon, Cambridgeshire, U.K.) in constructing many of the bridge components described in chapters 6–9. Jürgen Melcher has derived the topological result of section 3.2.2 and Appendix 2. We thank Dr. H. Bachmair for reading the text and making valuable suggestions.

Shakil Awan would like to express special thanks and gratitude to Amber, Gabriel and Liberty for their love, patience and support. Particular thanks to my parents for their immense support, resilience and encouragement without which parts of this book would not have been possible. Many thanks also to Sally and David Luscombe for all their help and encouragement.

Miss Janet H. Belliss and Dr. Ian A. Robinson (NPL, U.K.) are gratefully acknowledged for their help and support over the years. Thanks also to Dr. Luca Callegaro (INRIM, Italy) for useful discussions and providing Figures 9.29 and 9.30. Dr. Sze W. Chua (NMC/A*Star, Singapore) is also acknowledged for previous work on quantum Hall impedance and for providing a diagram.

Bryan Kibble thanks Anne for her support, both in general and particularly during the preparation of this book.

Chapter 1
Introduction

1.1 Interactions between circuits – eliminating electrical interference

Interference elimination is amenable to a logical approach and is most certainly not a ‘black art’. It is necessary to first discard imprecise and often incorrect notions of ‘earthing’ and ‘shielding’ before any real progress can be made. It is true that signal processing can sometimes extract the required information from a horrendous cacophony of interference, but, nevertheless, the rewards of applying correct, simple principles to produce clean, problem-free electrical circuitry are very great.

For example, by adopting the methods described in this book to confine the electric and magnetic fields associated with a circuit, whether DC or AC, to within its components and cables, the expense and inconvenience of a screened room is often unnecessary.

At the outset, we remark that it is wrong to think of a DC circuit only in DC terms. It is also an AC network and, as such, is interference prone.

Avoiding unwanted interactions between circuits is important, not only with regard to the subject matter of this book but also throughout the whole field of sensitive electrical circuits where results need to be free from inaccuracy and signal-to-noise problems caused by power line pick-up, interference from other external sources and unwanted interactions between different parts of a system. A ‘system’ could be a network of impedances and instruments or the internal workings of a single instrument. See Reference 1 for a practical example embodying the principles expounded in this book.

It is helpful to visualise the *circulating currents* and the *magnetic fields* that the currents give rise to, the *electric potentials* of conductors and the *electric fields* emanating from the conductors as the physically real entities. Unfortunately, the vast majority of electrical test and measuring instruments used to explore conditions in a circuit (e.g., voltmeters, oscilloscopes and spectrum analysers) are regarded as responding to input voltages. It must be remembered at all times that these voltages are only manifestations of currents flowing through impedances between two points in their circulating paths.

1.1.1 Basic principles

CURRENTS FLOW AROUND COMPLETE CIRCUITS

Currents start from a source – for example, a battery, power supply, signal generator, a voltage generated between the ends of a conductor in which another current is flowing – but having traversed a possibly very complicated network of conductors and impedances, subdividing and recombining on the way, they *must* recombine and return in their entirety to that source. They do *not* go to 'ground', 'earth' or a 'shield' and disappear! For example, inductance can only be properly understood if it is realised that it is only defined for a closed circuit. What is usually meant by a statement about inductance is *incremental* inductance, such as that caused by a length of wire if it were to be inserted into a closed circuit.

The concepts of 'grounding', 'earthing' and 'shielding', which are of historical origin, have caused much confusion and misery to practitioners of sensitive electrical measurement, mainly because of the woolly manner in which these concepts are frequently invoked. We suggest as an alternative that one point of a network be designated as a potential reference point to which all other potentials in the network are related and which is therefore usually given the value of zero potential. It will usually be about the same potential as the 'earth' or surroundings of the apparatus – walls, water pipes, people, etc. – bearing in mind that their potentials are rather uncertain.

It follows that the actual 'earthing' or 'grounding' conductors, which are represented by symbols like those in Figure 4.9n, have flowing in them currents returning to their source. Except for the very special case of superconductors with DC, these conductors *must* have potential differences along them. This can most certainly matter in sensitive circuits. The actual return conductors and the precise points at which they are interconnected should be drawn and considered together with the currents flowing in them. Particular care should be taken to represent the topology of the connections of the actual circuit correctly. For an example, see Figure 1.14. It is also helpful to replace the words 'earth' or 'ground' in our vocabulary with more specific terms such as 'current return conductor', 'instrument case' and 'mains safety conductor' as appropriate, and to designate a particular node of a network as a potential reference point to which all voltages are understood to be referred. The time spent thinking in this more precise way will be amply repaid, particularly in eliminating common-impedance coupling described below. A mains safety conductor is often coloured green/yellow. It is the only meaning of the Figure 4.9n symbol in our diagrams.

Currents that flow around complete circuits give rise to electric and magnetic fields, which cause interaction between circuits and between different parts of the same circuit.

The reciprocity theorem is embedded in the theory of all fields, not just electromagnetic. It connects the emanation of fields by a system to the response of that system to external fields. For our purpose it is sufficient to make use of a special case.

CIRCUITRY WHOSE LAYOUT AND SCREENING ARE SO DESIGNED THAT IT EMITS NO SIGNIFICANT ELECTRIC OR MAGNETIC FIELDS DOES NOT RESPOND TO EXTERNAL FIELDS. THEREFORE, IT IS IMMUNE TO INTERFERENCE.

We can classify these modes of interaction as follows:

(i) Electric field coupling: Either two capacitances or, more usually, one capacitance and a linking conductor (which is frequently the green/yellow or 'earth' conductor to an instrument) can couple two systems via their external electric fields, as shown in Figure 1.1 or Figure 1.2.

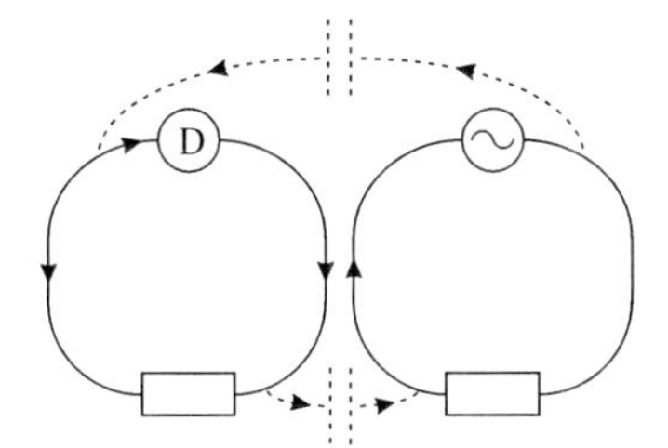

Figure 1.1 Coupling via two capacitances

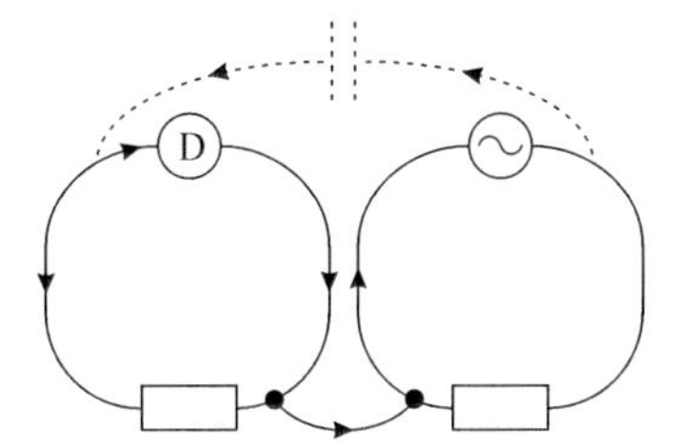

Figure 1.2 Coupling via one capacitance and a conductor

Electric field coupling is more important in high-impedance circuits having typical impedances of 10 kΩ or greater.

Capacitive coupling can be eliminated by shielding with conducting surfaces connected to appropriate points in the circuitry. Shields can intercept the capacitive currents and confine them to the interfering source, as shown in Figure 1.3, or can prevent these currents from flowing through sensitive parts of the detecting circuit.

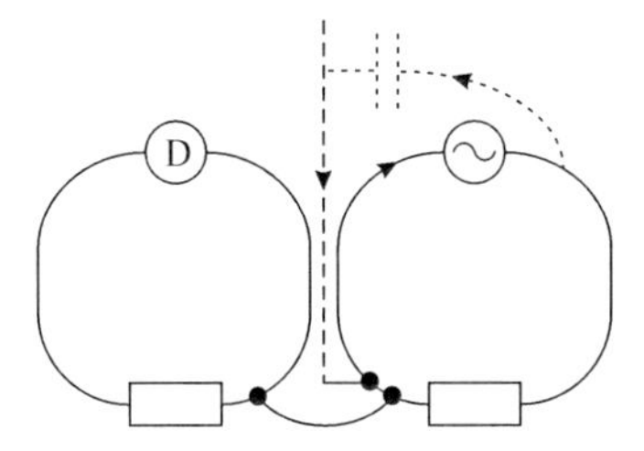

Figure 1.3 Placing a screen to eliminate capacitive coupling

Many instruments and components may appear to be encased in fairly complete conducting enclosures, but this can be illusory if the panels of these enclosures are, in fact, insulated from one another by, for example, an oxide layer or paint. Also, the shafts that mechanically couple internal controls such as switches or potentiometers to knobs or buttons outside the case are not always connected to the case and therefore provide a means of allowing currents via electric fields to enter or escape (Figure 1.4a). In Figure 1.4b, the offending shaft is connected to the case, and the effect of an oxide or paint layer is minimised by joining input and output connectors with a conductor routed close to the circuitry within the enclosure. The enclosure should then have a deliberate insulating gap so that the added conductor provides the only path for returning current. Doing this can minimise the open loop area exposed to external interfering flux, as discussed in (ii) below.

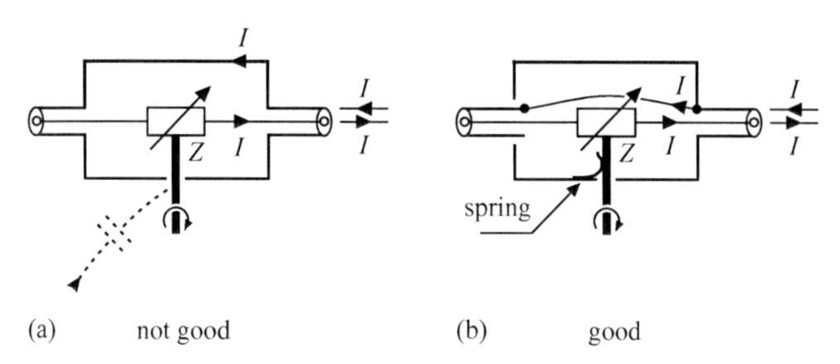

Figure 1.4 (a) A conducting control shaft penetrating a shield. (b) Connecting the shaft to the shield with a spring contact

Some aspects of capacitive coupling between various parts of a circuit, for example, between conductors on the surface of a printed circuit board, are often misunderstood. The basic principles of this situation are illustrated in Figure 1.5.

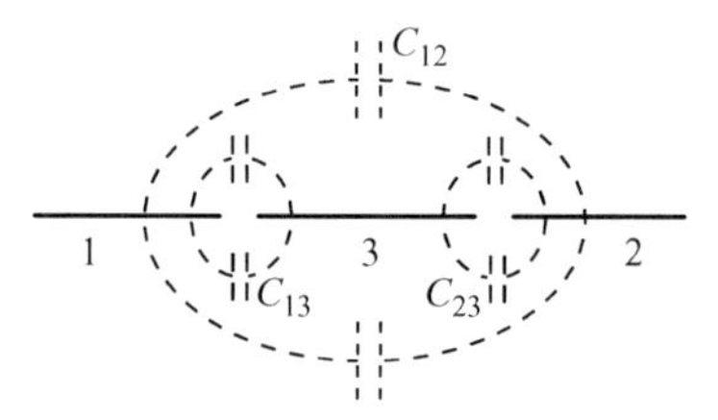

Figure 1.5 Capacitances between conductors on a printed circuit board

The capacitance C_{12} between conducting surfaces 1 and 2 is modified by the presence of a third conducting surface 3. If 3 is maintained at a *fixed* potential with respect to 1 and 2, C_{12} is decreased somewhat to, say, C'_{12}. This change is independent of the particular potential of 3, be it zero ('ground') potential or several kilovolts with respect to 1, 2 or any other nearby conducting surfaces. If the potential of 3 is indeterminate – that is, if it is 'floating', that is, not connected to circuitry which determines the potential of 1 and 2 – the current through C_{12} will be augmented by that through C_{13} and C_{23} in series.

We can apply these concepts to two conductors on a printed circuit board, which is provided with a conducting 'ground plane' in the intervening space between them, as shown in Figure 1.5, or covering the whole of the reverse side of the board, as shown in Figure 1.6. In either case, the direct capacitance

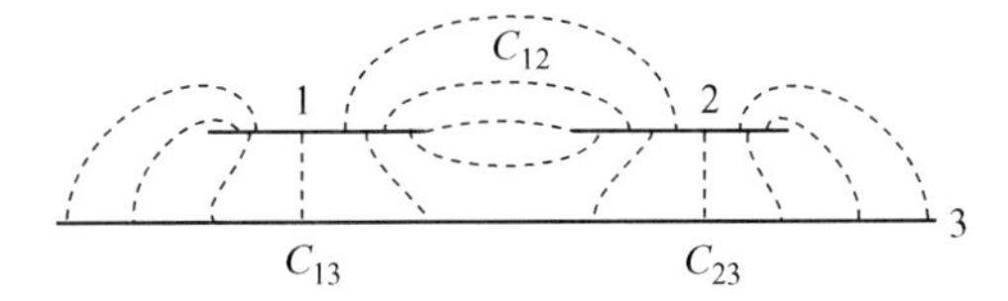

Figure 1.6 A 'ground plane' on the reverse side of a printed circuit board

C'_{12} between the two conductors is only approximately halved. In a circuit containing only impedances that are low compared with $1/(2\pi f C_{13})$, etc., the 'ground plane' has no capacitive role, but at higher frequencies (f) when its impedances $1/(2\pi f C_{13})$ etc. are lower, the potential divider formed by C'_{12} and either C_{23} or C_{13} may well be significant and useful. For more complete screening, there should be a conducting plane above, below and between 1 and 2.

(ii) Complete meshes of networks can be coupled via the magnetic field arising from current flow around one of them (Figure 1.7). That is, circuits can be coupled by mutual inductance. This coupling can be reduced, as shown in Figure 1.8, by reducing the effective area of one or, preferably, both of the meshes by, for example, using twisted wire conductor pairs or the coaxial cables, which are the subject of this book. This kind of coupling dominates in circuitry having impedances of less than 10 kΩ.

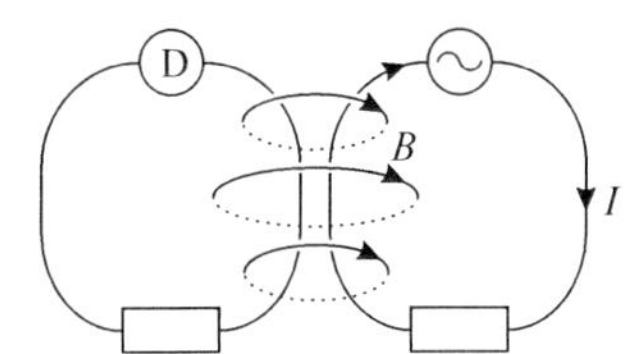

Figure 1.7 Inductive (magnetic) coupling

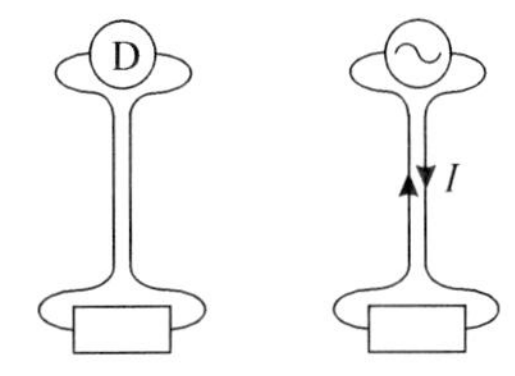

Figure 1.8 Eliminating inductive coupling

It is a fundamental property of a straight, infinitely long coaxial cable in which the current in the inner conductor is balanced by an equal and opposite current in the outer screening conductor that these currents generate no magnetic flux external to the cable (Figure 1.9). In this book we mean by 'coaxial cable' a cable having the currents in the inner and outer conductors equal and opposite. From Ampere's theorem, since there is no net current in the cable, the line integral of the flux around any circular path around the axis of the cable is zero, and since, by symmetry, the flux at every point along the path must be the same, this flux must be zero everywhere outside the cable.

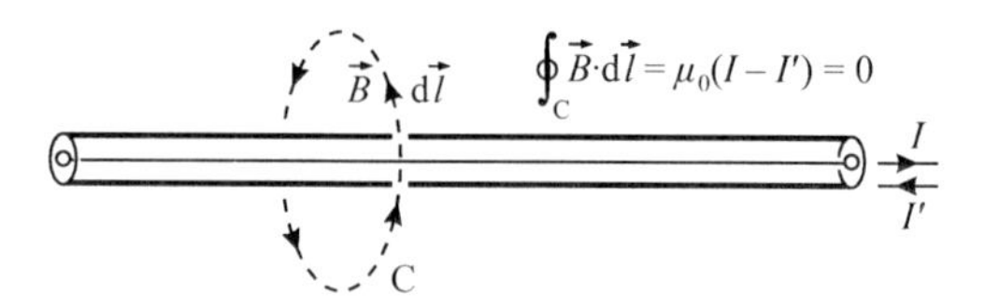

Figure 1.9 Ampere's theorem applied to a circular path concentric with the axis of a coaxial cable

In practice, actual cables are neither infinite nor straight, but if the length and radii of curvature of bends are long compared to the diameter of the cable, the external flux will be negligible.

The same is true for twisted-pair cables in which the currents are equal and opposite because the flux generated by successive twists cancels at distances larger than the pitch of the twisting (see Figure 3.10).

Components can, if necessary, be contained within enclosures of highly permeable magnetic material such as mu-metal. An unfamiliar example, which will be encountered later in the book (see section 7.2.6), is the enclosure of a toroidal transformer core, drawn in section in Figure 1.10a, in a toroidal magnetic shield. The purpose of doing this is to ensure that the magnetic flux associated with the core windings is completely confined to the interior of the shield. This technique assists in the generation of precise voltage ratios by windings threading the toroidal shield. Figure 1.10b illustrates a practical way of constructing a toroidal magnetic shield.

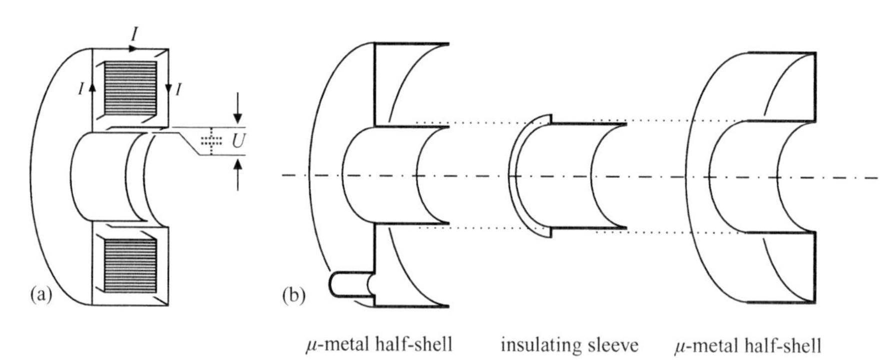

Figure 1.10 (a) A toroidal magnetic shield surrounding a toroidal magnetic core. (b) A practical construction for the shield

(iii) Figure 1.11 illustrates two circuits interacting because they share a length of conductor. The current I circulating around the right-hand circuit sets up a voltage between the ends of the common conductor, which in turn causes a current i to flow around the left-hand circuit. This common-conductor coupling can be equally important in both high- and low-impedance circuitry. In addition to this coupling mode there can be significant contributions from modes (i) and (ii), that is, via stray capacitance and magnetic flux linkage.

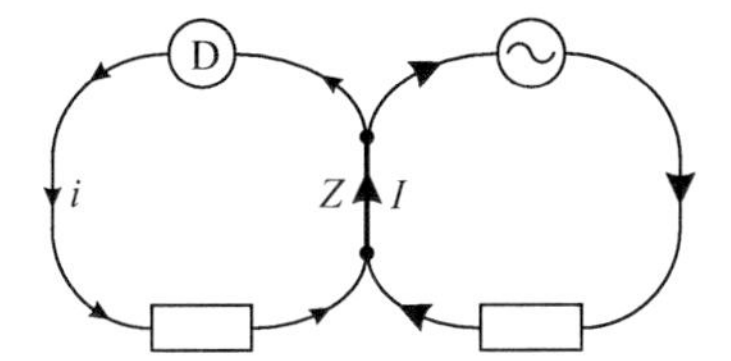

Figure 1.11 Common-conductor coupling

Assuming that there is a reason for connecting the two circuits, pure common-conductor coupling can be eliminated either by reducing the common impedance to zero by joining the two circuits with a star connection, as shown in Figure 1.12, or a branch connection, as shown in Figure 1.13. A branch connection makes it easier to also eliminate capacitance coupling with a shield, as in (i), and flux coupling, as in (ii). Another possibility is a combining network, as discussed in section 5.6.

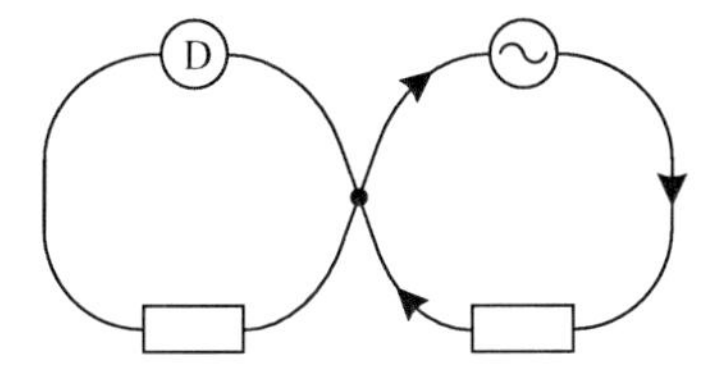

Figure 1.12 A star connection to eliminate common-conductor coupling

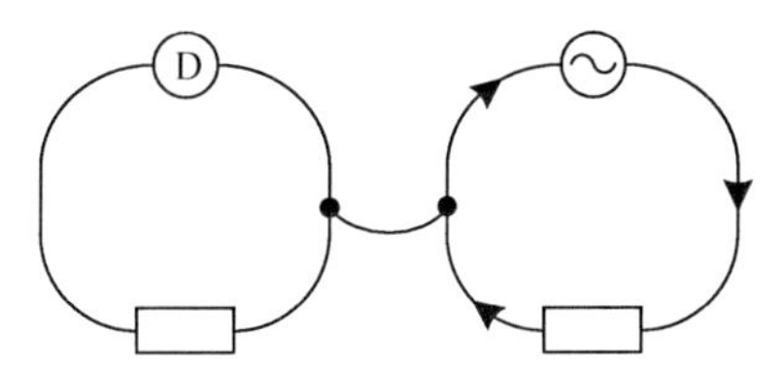

Figure 1.13 A branch connection to eliminate common-conductor coupling

As an example of eliminating common-conductor coupling, consider simple operational amplifier circuits. These could well be the first stage of matching and amplification after a sensor before conversion to a digital signal. For this purpose, we assume ideal operational amplifiers, which, when external feedback components are added, maintain $U_{in} = 0$ between their input terminals with negligible current flow into either and which ensure that neither input nor output is influenced significantly by the magnitude of either supply voltage, U_+ or U_-. Actual operational amplifiers approach this ideal rather closely.

A common application is the unsymmetrical amplifier circuit drawn in Figure 1.14. I is the total feedback current through R_f, and i_0 is the total output current flowing through the load R_L. Because of the high voltage gain of the operational amplifier, the feedback operates such that $U_{in} \approx 0$ at all times. Therefore,

$$I = \frac{U_0}{R_f} \tag{1.1}$$

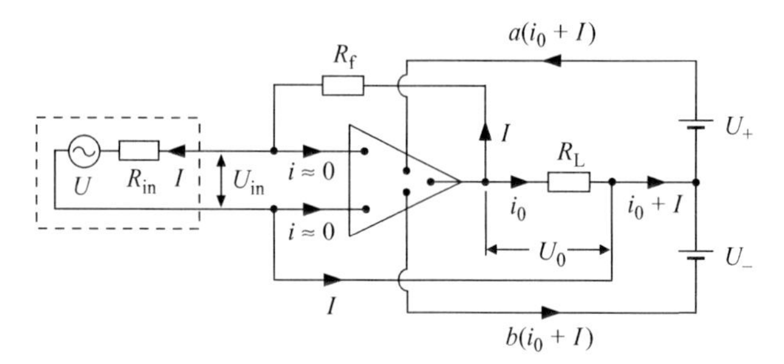

Figure 1.14 The currents flowing in an unsymmetrical amplifier circuit

and

$$U + IR_{in} = 0 \tag{1.2}$$

The voltage gain is

$$\frac{U_0}{U} = -\frac{R_f}{R_{in}} \tag{1.3}$$

Also,

$$i_0 = \frac{U_0}{R_L} = \left(\frac{R_f}{R_L}\right)I \tag{1.4}$$

Usually i_0 is considerably greater than I. As an example, for an amplifier having a voltage gain of 100, we might have $R_f = 10^5\ \Omega$, $R_{in} = 10^3\ \Omega$ and $R_L = 10^3\ \Omega$, so that $i_0 = 100I$. If $U = 10$ mV, $I = 10\ \mu$A and $i_0 = 1$ mA. Note that I and i_0 complete their circuits through the power supplies, which supply fractions a and b, respectively, of the total current, plus the currents (not shown) needed to operate the internal circuitry of the amplifier. Because i_0 and therefore the output voltage across the load have been made immune, by the design of the operational amplifier, to variations in either supply voltage, it is highly immune to noise voltages in either supply, and the noise current i_n from any such noise voltage flows only around the supply loop.

This simple situation becomes much more complicated when two amplifier circuits are connected in cascade and fed, as is usual, from the same supply rails. The circuit is drawn in Figure 1.15.

Common-conductor coupling at C provides additional input to the second operational amplifier, and if the operational amplifiers are, in practice, not completely immune to changes in the supply voltages, common-conductor coupling along the supply rails and current return conductor can cause problems. In severe cases, these common-conductor couplings can cause feedback and oscillations in high-gain circuits.

Some of these problems are avoided if the circuit connections are remade as shown in Figure 1.16. Star connections avoid some common-conductor couplings,

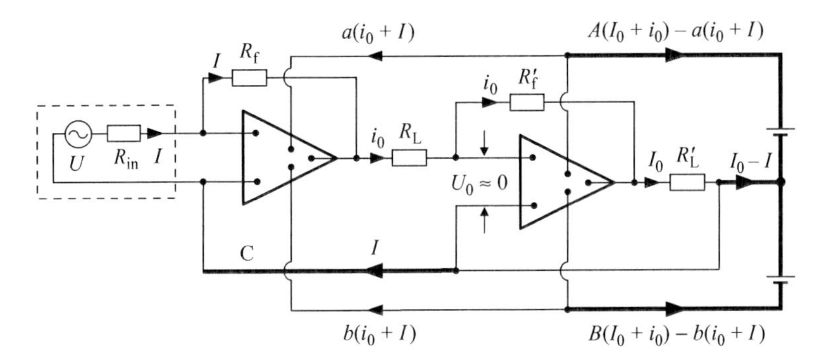

Figure 1.15 The currents in a two-stage operational amplifier circuit. Heavy lines indicate occurrences of common-conductor coupling

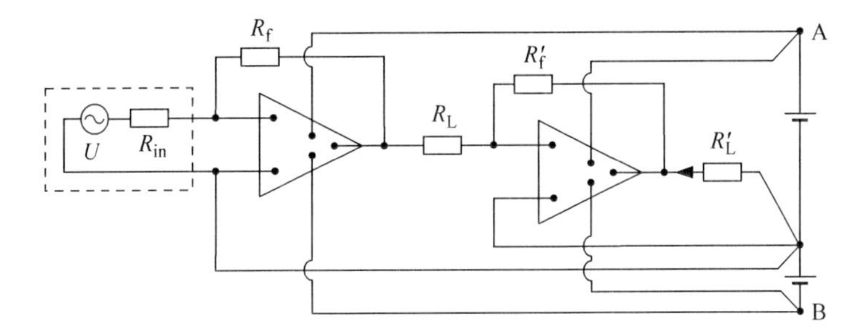

Figure 1.16 Removing common-conductor couplings from the circuit of Figure 1.15

and the differential-input properties of both operational amplifiers sense the remote voltage source. Mutual inductive coupling (see (ii) earlier) should be minimised.

The whole subject of interference elimination is often viewed as something of a 'black art' with ad hoc 'cures' such as ferrite beads or mains interference filters being used on a trial-and-error basis. One cause of the adoption of this straw-grasping approach lies in the *vector* addition of induced voltages A, B, C, etc., which come from a common primary source. For example, consider each single-frequency component of mains and mains-borne interference entering circuitry by the routes identified earlier in this section. The voltage induced at each frequency by each route will have a different magnitude and phase associated with it, and the resultant is the vector sum S, as illustrated in Figure 1.17. S is observed between two points of the circuit by some test instrument such as an oscilloscope or voltmeter, and as a further complication, some additional voltage components may arise from the loops added to the network by the very act of connecting the test instrument.

Suppose that one suspected cause of entry of interference is identified and removed so that one of the contributing voltage vectors, say C, is eliminated. The remaining vectors, represented by the broken lines, then sum to give a new resultant S',

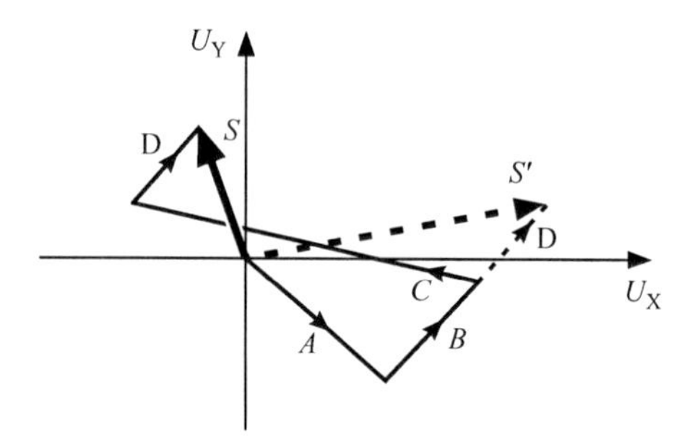

Figure 1.17 S is the vector sum of several (A, B, C, etc.) phase-related interferences from a common source. S′ is the new vector sum if C is eliminated

which will very possibly *have a greater magnitude than S*. There is now a strong tendency to think that the action taken to eliminate *C* was incorrect; to restore the status quo, abandon rational thought and try some other remedy at random instead.

In these circumstances, one must revert to sound basic principles and have the courage of one's convictions to persist with correct remedial steps, even though some appear to make the total observed interference worse. When all causes are dealt with, the interference will disappear totally. To stop short of this ideal is to court disaster because a small resultant can be obtained by the accidental partial cancellation of many large contributions. Indeed, one terrible approach to eliminating interference is to deliberately introduce a further large component equal and opposite to *S*. The result will only be temporarily satisfactory because any change in the circuit layout, the nature of the interference and so on could alter the magnitude or phase of one or more of the vectors and bring back the problem, probably unsuspectedly.

The good news in this difficult situation is that if one frequency component, for example, the fundamental harmonic of 50-Hz mains interference, is properly eliminated, all other frequency components including those of mains-borne interference from other sources will also be eliminated because the underlying causes, common-conductor, capacitive and mutual inductive coupling, have been reduced to insignificance.

In the literature on interference rejection, reference is often made to problems caused by 'earth loops'. This phrase usually means the accidental or deliberate provision of more than one current return path in the interconnections between circuits. Quite large interfering voltages can be propagated amongst these circuits by common-conductor or mutual inductive coupling. These voltages can be eliminated by removing unnecessary return conductors and making appropriate star connections between the remainder.

Finally, by applying the reciprocity theorem to balanced-current coaxial or twisted-pair networks, which generate no significant external magnetic or electric fields, we reiterate the vital conclusion from this section that these networks *do not respond to external fields*. If common-conductor coupling has also been eliminated, these networks are *interference free*.

1.1.2 An illustrative example – using a phase-sensitive detector

Consider a phase-sensitive detector used as a null detector to detect the absence of a signal. Measuring a wanted signal involves exactly the same considerations in that the object is to ensure that there is no unwanted component added to this signal. Phase-sensitive detectors most acutely exemplify this problem because they need an auxiliary reference input, which must be phase-locked to the signal to be detected. Incorrect wiring can cause coupling of this reference signal into the measurement input, and this causes consequent erroneous results, particularly where a signal source is used to both stimulate the measurement circuitry and provide the reference signal, as illustrated in Figure 1.18. Currents at the source frequency circulate in the outer conductors of coaxial cables, which link the source, measurement circuit and detector, and this results in mutual inductive and common-conductor coupling in the detector. The important point is that there should be no significant current in the inner or outer conductor of the coaxial cable connecting the measurement circuitry to the signal input of the detector. This is because a null indication of a properly constructed detector means that the voltage difference across its input port is zero and that there is no current in the inner conductor. Therefore, a necessary and sufficient condition for an accurate null indication, when connected by a cable to a remote port in the measuring system, is that there should be no net current in this cable. Current equalisation, by incorporating a current equaliser (see section 3.1.1) into each mesh of the network, can achieve this condition. A simple test for the effect of any current in the outer conductor is to temporarily insert a resistance in series with the outer conductor. Doing this, if all is well, should have no effect on the indication of the detector.

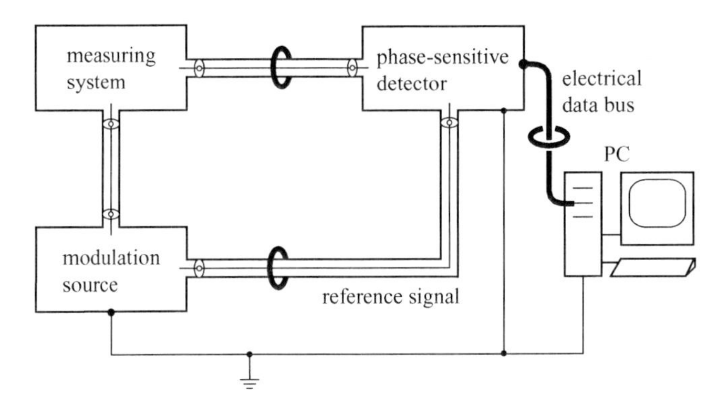

Figure 1.18 Connecting a phase-sensitive detector to a network

An alternative approach is to use isolating transformers. The relevant parts of the network, including the green/yellow safety conductor of the mains, which also connects sources and detectors, are shown in Figure 1.19. There are two undesired currents. First, if the two screens of the isolating transformer A (see section 1.1.5) are not quite at the same potential, a current will flow across the capacitance between them, through the network of outer conductors in the bridge, through the outer of the cable to the detector input and return to the transformer primary and its

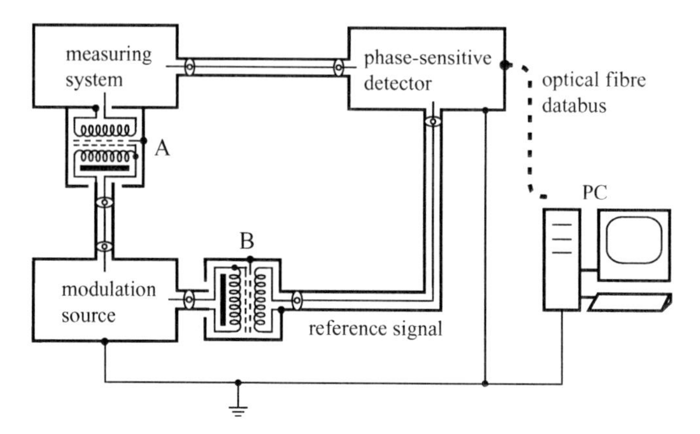

Figure 1.19 An isolating transformer can break the loop

screen via the green/yellow mains safety conductor and the reference supply cable outer conductor (see Figure 1.19). Second, a proportion of the reference signal current will return to the source via the green/yellow mains safety conductor and a small proportion via the detector output lead outer conductor, the bridge output outer conductors and the inter-screen capacitance of the transformer. The currents from both of these causes flow down the outer conductor, but not the inner, of the input lead to the detector and cause a potential drop down the outer conductor. Hence, the condition $U = 0$ between the inner and outer conductors, which holds at the detector input when it registers a null, does not hold where this input cable joins the network, and this leads to an incorrect indication.

The net current in the detector lead can be detected by the total current detection transformer described in section 7.3.5 and may be reduced to an acceptably low value in either of the following two ways.

(i) An isolating transformer B in the reference signal cable will prevent current flowing in the outer conductor of this cable.
(ii) An additional single conductor may be used to join the secondary screen of transformer A to the outer of the detector case, and an equaliser put in the cable to the detector input (Figure 1.20). Alternatively, an auxiliary small voltage source comprising an injection transformer and associated circuit (see section 7.3.4) may be connected between the primary supply outer and the primary screen to modify the potential of the latter until it is equal to that of the secondary screen, so that no current flows between them, as registered by total current detection transformer surrounding the detector input lead.

1.1.3 Diagnostic equipment

There are some devices that, when used in conjunction with an oscilloscope or spectrum analyser of the requisite bandwidth and sensitivity, are particularly useful in revealing the causes of problems with interference.

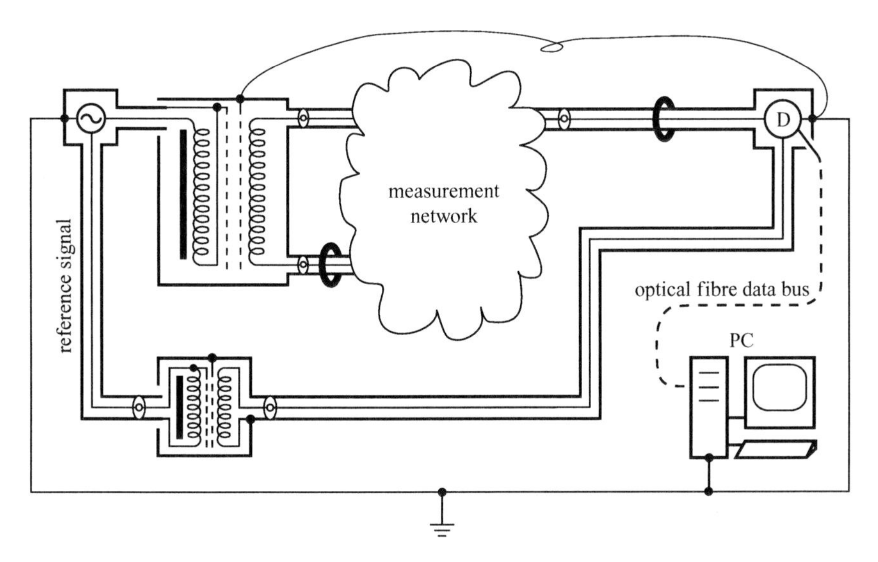

Figure 1.20 An added conductor and an equaliser reduce the current in the outer conductor of the input cable to the detector

The ubiquitous hand-held multimeter is a very useful diagnostic tool whose limited accuracy of the order of 0.1% is usually adequate. Being battery powered confers on it a fair degree of isolation, but nevertheless, simple precautions are necessary in order not to get misleading results. The metal tips of probes should not be held in the hands. Apart from the obvious safety implications, the shunt leakage currents through the human body will cause errors when measuring high resistances or voltages of sources having high internal impedance. When measuring alternating voltages having associated high internal impedance, capacitance between the leads or via the operator's body can be significant, and separating the leads from each other and placing the instrument on an insulating surface might reveal the existence of this effect. Conversely, when sensing alternating voltages across very low impedances, the inductance of the loop of separated leads could be significant and can be reduced by twisting the leads together.

A battery-powered oscilloscope having a wide bandwidth is an extremely useful diagnostic tool for eliminating interference. Its degree of isolation is very helpful in not altering circuit conditions by the very act of connecting it into the circuit. Mains-powered oscilloscopes with effective differential inputs can also be useful, but the consequences of connecting the mains safety conductor to the circuit being examined need consideration. A spectrum analyser is equally useful, but these instruments are not usually completely isolated, and again, this may have consequences.

A differential-input preamplifier, to a large extent, avoids the problem of additional significant current paths with concomitant interference being introduced by the very act of connecting the diagnostic equipment itself. A properly designed differential input is a three-terminal device, as illustrated in Figure 1.21, which has

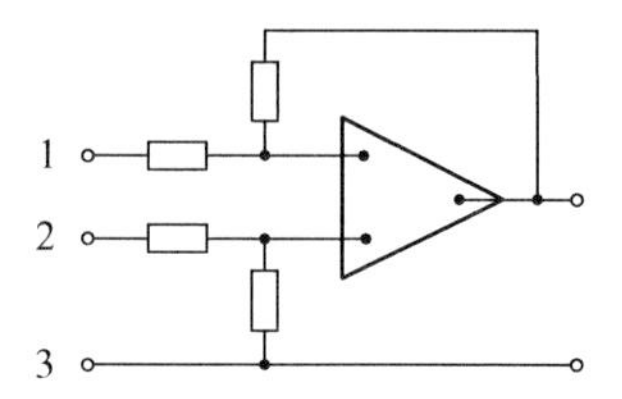

Figure 1.21 A differential-input preamplifier

the property of responding to a voltage input between terminals 1 and 2 while having a very much smaller (common-mode) response to voltages between 1 or 2 and 3. Therefore, there will be a true indication of circuit conditions if terminals 1 and 2 are connected to the nodes of the circuit between which we want to observe the voltage. Careful thought must be given, however, as to where to connect terminal 3. If there is no other connection between the circuit and the test instrument, via, for example, the green/yellow safety conductor, terminal 3 should be connected to a point on the current return path of the circuit. If there is already a low-impedance connection of this kind between the circuit and the preamplifier, terminal 3 should be left unconnected to avoid introducing an extra current return path. The existence or otherwise of a connection can quickly be determined by a hand-held resistance-measuring meter connected between a point on the current return path of the circuit and terminal 3.

A sensitive differential-input preamplifier correctly used in this manner can detect common-conductor problems by connecting inputs 1 and 2 to the two points on the conductor. An observed voltage indicates a common-conductor current in the conductor, which may cause a problem.

We have emphasised that it is best to view the *currents* in circuitry as having primary physical significance, but that, unfortunately, detection and measurement of *voltage* differences are almost universal. But if it is too difficult to unravel the problems of troublesome circuitry by making voltage difference measurements alone, direct detection and measurement of currents can prove very enlightening. This is possible if the conductor carrying the alternating current can be threaded through a high-permeability magnetic core, which is provided with a winding connected to a voltage-detecting instrument. This operation can be made very easy if the core can be temporarily split so that it can be clipped around the conductor. Direct currents can also be detected in the same way if the core has a transverse gap in which a Hall effect sensor of magnetic flux is mounted. Clip-on ammeters are familiar instruments based on this principle, and by using a high-permeability core and a multi-turn winding, sensitivity can be extended up to the milliampere or even the microampere range. The diagnostic power of this approach in difficult situations cannot be overemphasised, but unfortunately, suitable commercial equipment is rare.

A short solenoidal coil wound on an insulating former and covered with a conducting screen, as shown in Figure 1.22, can be connected to an oscilloscope. The screen should have an insulated overlap so that it does not form a shorted turn.

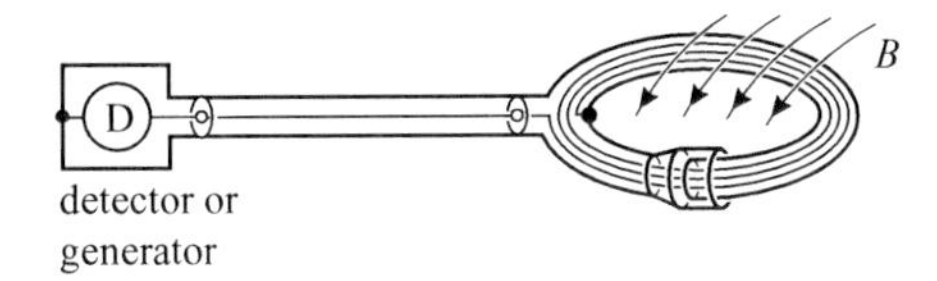

Figure 1.22 A pick-up coil for detecting magnetically coupled interference

The output of the coil is an induced voltage proportional to any AC magnetic flux threading the coil. An induced voltage reveals the presence of unwanted magnetic flux threading the circuit under examination. Employing equalised-current circuits and enclosing components in magnetic screens will eliminate problems caused by magnetic flux.

If the coil is deliberately supplied with current from a source in the circuit, its magnetic flux can test the sensitivity of the circuit to other unwanted flux couplings by positioning the coil at various positions adjacent to the circuit. This can be a very sensitive test to be interpreted with care as the flux the coil produces is likely to be much greater than the actual unwanted fluxes present.

A capacitive probe is simply a sheet of conductor, preferably covered with an insulating layer to protect circuitry from short-circuits caused by accidental contact, which is connected via a length of screened cable to the input of an oscilloscope or some other detecting instrument (Figure 1.23). When held by the screened cable (to connect the handler to the screen) and placed in the vicinity of the circuit being examined, it will reveal the presence of any unwanted electric fields, which might be part of a capacitive current flow. In a similar manner to the magnetic pick-up coil, it can be connected to a generator to create electric fields to which circuitry might be sensitive. Because of the high-impedance nature of this useful device, it is somewhat harder to interpret the results in comparison with those obtained with a magnetic pick-up coil.

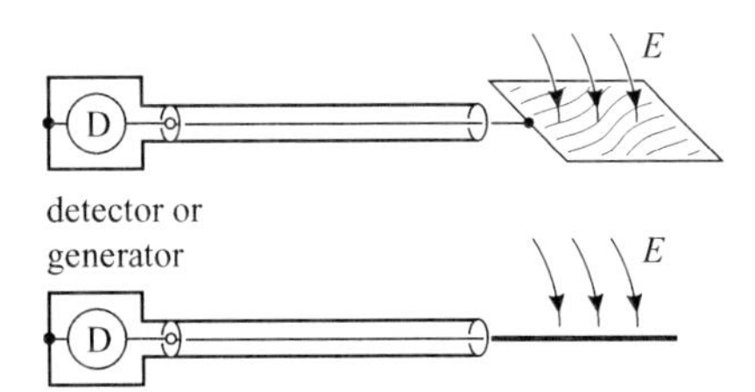

Figure 1.23 Probes to detect or produce electric fields

Coaxial networks can be used over a very wide frequency range, from DC to several GHz. In principle, the frequency range over which coaxial cables not terminated by their characteristic impedance operate in networks is limited only by the requirement to keep the length of the cables rather less than the wavelength of the electromagnetic field in the cable. In some applications, the phase change down a cable, which occurs if its length becomes comparable with the propagation wavelength in the cable, is a serious drawback. To overcome this, cables can be kept shorter if the components of the network are physically smaller. Typical figures are less than 1 m at 1 MHz, 100 mm at 10 MHz and so on. This need for increasing miniaturisation will no doubt reach the ultimate in micromachined circuitry as this technology develops.

For planar circuits on circuit boards, at still higher frequencies, a continuous sheet of conductor on one side of the board, termed a 'ground plane', is often employed. Currents in circuitry just above this plane prefer to flow in routes of minimum inductance by creating return paths in the plane immediately below themselves. There may still remain capacitive couplings between parts of the circuitry above the plane.

The concept of a *linear network* is very powerful in designing and understanding networks of *linear circuit elements*, that is, those for which the voltage across them is strictly proportional to the current through them under normal operating conditions.

The superposition theorem provides an extremely powerful tool for unravelling complicated network problems. It should be borne in mind at all times when considering interference elimination. It states that in a network of linear circuit elements – for example, impedances, sub-circuits and sources – in which the voltage across them is proportional to the current through them, the current from each individual source *behaves as if no other currents are present*. We can conceive of a current setting out from a source, dividing and subdividing as it goes into the network, and then these separate components of the current combine and recombine with the original current value just before returning to the source. Superposition is much more than an abstraction – it has immediate and practical application in many circumstances.

In a linear network, many currents, each existing independently and originating in separate sources, can sum to give the total current in any one path of the network.

A 'source' can be the voltage drop across a component. If one were to replace the component by a voltage source equal to the voltage across it, the rest of the network would have no means of registering the replacement.

There are many theorems relating to linear networks – Kirchoff's laws, Thevenin's theorem, etc. – but many useful conclusions can be reached directly from the above basic principles without recourse to mathematics.

But beware if some circuit components are not sufficiently linear, as, for example, semiconductor junctions and poor joints. The result will be generation of harmonics from an original sinusoidal source and, in general, voltage-dependent responses.

The reciprocity principle is also vital for understanding some aspects of circuitry. It is deeply embedded in the theory of classical fields, including electromagnetic ones. As a simple example, there is reciprocity of source and detector in a Wheatstone's bridge network. That is, if the bridge is balanced so that the current through the detector is zero, on interchanging the positions of detector and source, the detector current is also zero – that is, the bridge is still balanced. On a more sophisticated level, the directional radiation pattern of a transmitting antenna is identical with the sensitivity receiving pattern of the same antenna.

For our present purpose, we have noted that we can more easily visualise how a circuit will emit electromagnetic fields than how it will respond to external ones. By arranging its layout and screening, we can minimise its emission and therefore ensure that it responds less to external interference. Coaxial techniques, properly applied, can reduce sensitivity to interference to negligible proportions. The most powerful tool of all in the battle with interference is to develop the imaginative powers of the brain for visualising the electric and magnetic fields generated by circuitry and then take steps to minimise these.

Interference is usually, but not invariably, conveyed electrically. Earlier we have assumed that all interference coupling modes between circuits are electrical. In fact, significant energy can sometimes be coupled acoustically – that is, the interfering circuit can have some element that behaves like a loudspeaker, and the measuring circuit, an element that behaves like a microphone. Examples of the former include mechanical movement of power transformer cores and conductors, vibration of cooling fans, piezoelectric elements and electric field or mechanically induced movement of shielding conductors. Examples of the latter are acoustic modulation of the reluctance of matching and isolation transformer cores, triboelectric effects in dielectric insulators and capacitance modulation of shields. Coupling modes can often be identified simply by altering the damping of the emitter, transmitting medium or receiver by clamping or grasping in the hand, or by altering the disposition of components. Once identified, it is usually a simple matter to provide the necessary acoustic isolation by using non-microphonic cables for critical connections and acoustically lossy mountings for troublesome components.

To summarise, eliminating electrical interference depends on applying elementary principles concerning circuit coupling. Some persistence is often necessary in real, complicated situations to avoid ad hoc unsatisfactory partial solutions.

1.1.4 Isolation

Eliminating all possible sources of extraneous interference is vital for successful sensitive analogue electrical measurements. If interference is present, not only is the resolution of the measurement impaired through degradation of the signal-to-noise ratio but also the rectification of the interference by any of the myriad semiconductor rectifying junctions in modern electrical circuitry may well result in systematic error. The ideal, to be attained as closely as possible, is to *isolate* the low-level circuitry. That is, to achieve that no significant extraneous currents enter or leave the isolated circuits.

When the necessary isolation has been accomplished, there remains the problem of conveying of the analogue information gained by the low-level circuitry from the isolated region, either in analogue form or, more commonly and satisfactorily these days, in digital form, without destroying the integrity of the isolation. We discuss the principal ways of accomplishing this.

Once in high-level analogue or in digital form, the information should be reasonably safe from further significant corruption by interference. This is particularly true for on/off pulsed digital information, on which interference below a certain maximum level has no effect at all. The information can then be safely further processed by computers.

By proper design, it is possible to totally isolate a circuit so that it can only communicate electrically with anything outside its screened boundary by desired couplings. This concept, which is central to interference rejection, can be appreciated by referring to Figure 1.24. The processing circuit may well contain one or more noise sources, and there must be a filter to short the output of these sources to its local screen so that the noise currents are constrained to follow paths that are totally within the local screen. The circuit should be constructed by either using the conductor-pair techniques of section 1.1.10 or making the local screen so that magnetic or electric fields do not escape it. The local screen is connected to the overall screen, which isolates the measurement system as a whole from external electric fields, at just one point A, as shown. The measurement shield may be connected to the green/yellow mains safety conductor for safety reasons at just one point B.

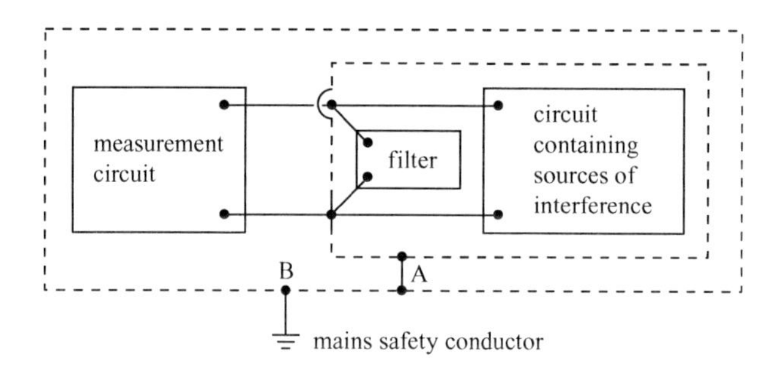

Figure 1.24 An isolating enclosure with a filtered output

The two examples in section 1.1.5 and 1.1.6 should help to clarify how to achieve isolation in practice.

1.1.5 Totally isolating transformers and power supplies

Consider the mains power supply circuit outlined in Figure 1.25.

The 'filter' consists of storage capacitors and voltage regulation circuitry. These ensure that only a DC potential is allowed to exist between the output terminals. The transformer is uniformly wound on a toroidal magnetic core, and the winding has an anti-progression turn (see section 7.2.1) to cancel the single-turn effect of the winding as it advances around the core. This form of transformer winding does not significantly propagate magnetic flux into its surroundings.

The two toroidal shields over the primary winding need to be constructed with their circumferential overlaps as shown in Figure 1.26, which shows a cross section through the axis of a toroidal core and these nested shields. For perfect capacitive

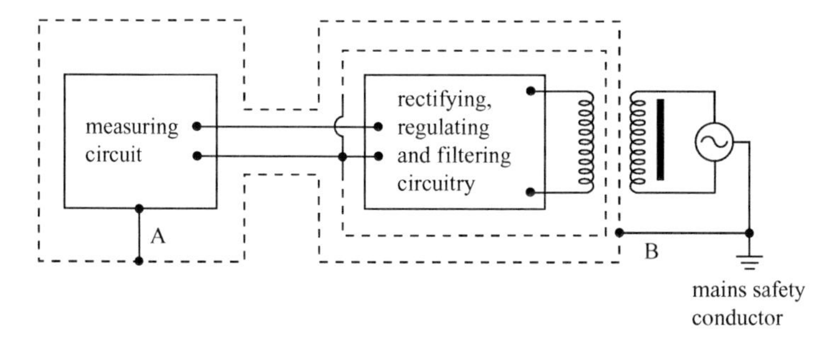

Figure 1.25 A totally isolated mains-driven power supply

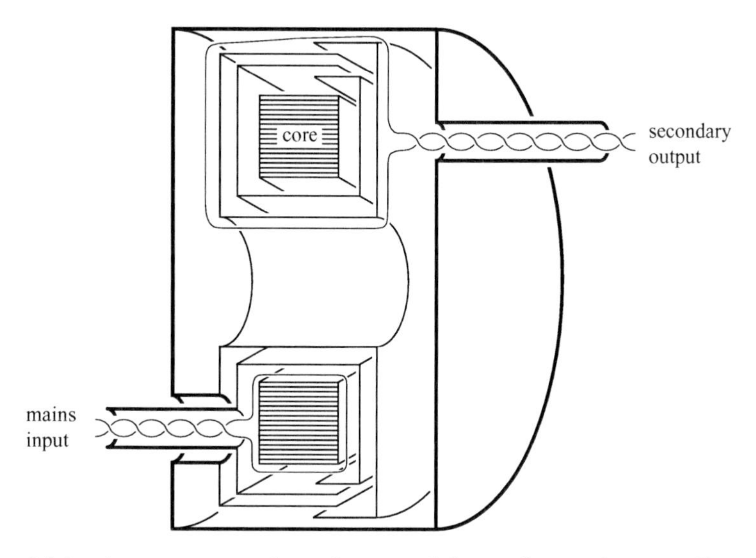

Figure 1.26 A cross section through a toroidal transformer for a totally isolated supply

isolation, the cross-capacitances between them at their overlapping gaps, which are of the order of a few picofarads, need to be equal, and they are best made as shown in Figure 1.27d. If necessary, this can be done by connecting a trimmer capacitor in parallel with the lesser cross-capacitance. This adjustment can be carried out by connecting a highly sensitive oscilloscope between the inner and outer shields and adjusting the capacitor for minimum observed 50-Hz signal.

It is also possible to construct screens of a poorly conducting material such as a loaded plastic. The conductivity of the sheet material should be sufficient to act as a screen for electric fields but low enough so that a complete screen having a conducting joint rather than an overlap does not constitute a significant shorted turn of the transformer.

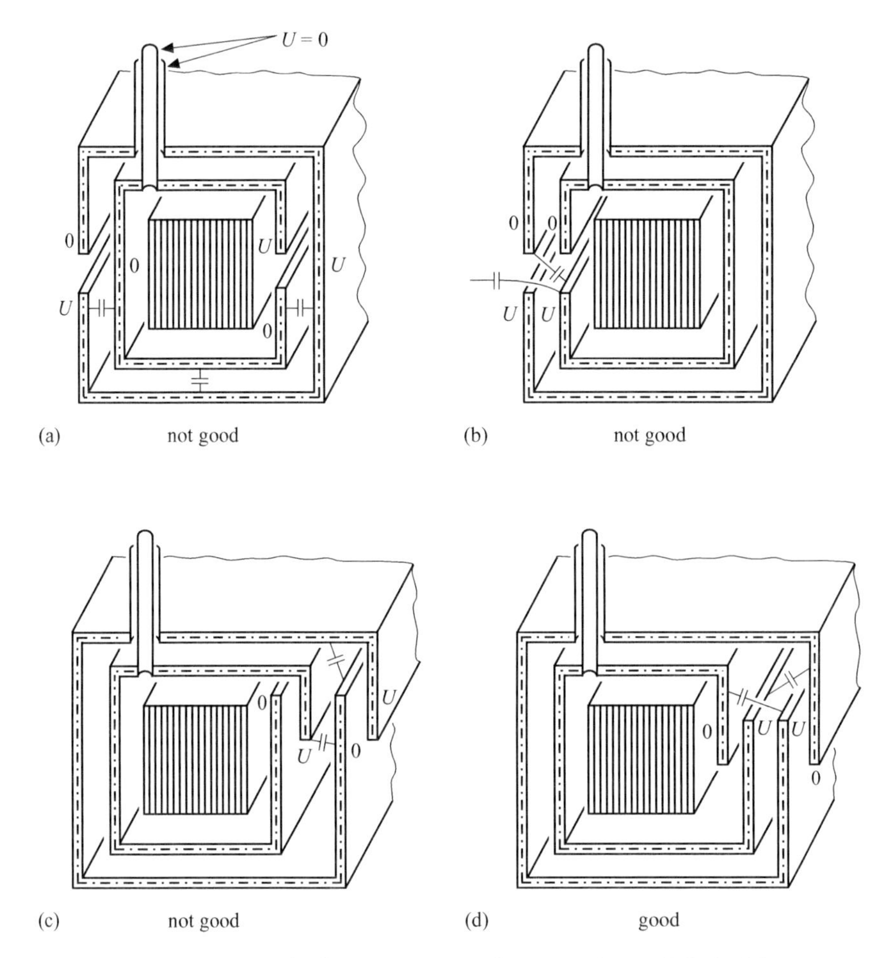

Figure 1.27 The best way to overlap nested toroidal shields

The isolation of a supply constructed along these lines can be such that only a fraction of a microampere of current will flow at mains frequencies from the primary to the measurement system, and thence to the measurement shield and back to the primary winding.

1.1.6 Isolating a noisy instrument

Many commercial instruments, for example, digital voltmeters, are insufficiently isolated from the mains by their internal power supplies. They also often generate noise from their internal digital and analogue workings, which appears as noise sources *between their input terminals* and between these terminals and their case or internal shield. The instrument is designed to be sufficiently immune to this self-generated noise, but the rest of the measurement system connected to the instrument may well not be so, particularly if it involves other insufficiently isolated components which provide return paths for interference currents. Figures for this noise are not usually given in the manufacturer's specifications. Possible noise sources and couplings in a typical instrument are illustrated in Figure 1.28.

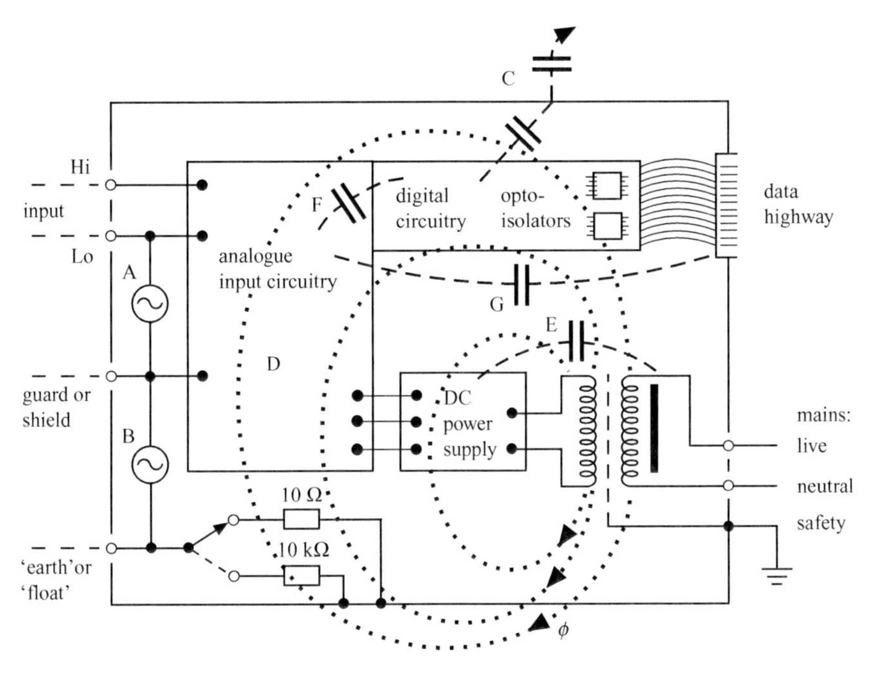

Figure 1.28 A typical commercial instrument as a noise source

The problem is amenable to the general approach given in this section, as illustrated in Figure 1.29. A mains voltage to mains voltage isolation transformer must be constructed according to the method given in section 1.1.5, and the instrument needs to be placed in two nested shields, as shown. In practice, the conducting case of the instrument will usually be, or can be made to be adequate to serve as the inner shield. The most significant noise source is often that between the 'low' input terminal and the instrument case or shield. Its effects can be confined to

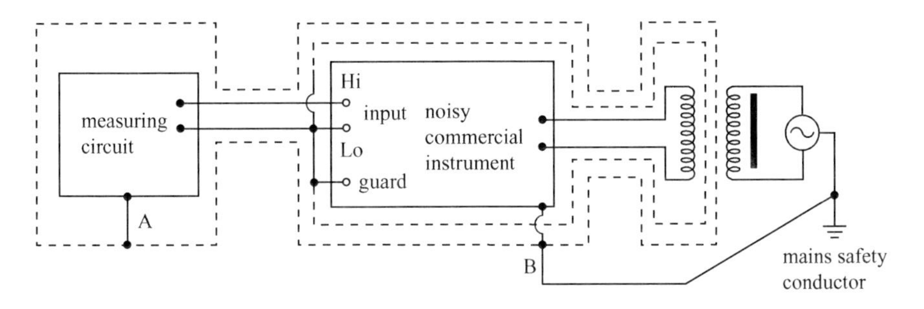

Figure 1.29 Isolating a noisy instrument

the interior of the inner shield by connecting this shield to the ‘low’ terminal. The noise source between the ‘low’ and ‘high’ input terminals may well be insignificant, so that a filter is unnecessary and degradation of the performance of the instrument can be avoided.

1.1.7 The available methods for isolating outputs

This section addresses the problem of replacing direct connections via conductors between circuits by indirect coupling methods involving negligible capacitance between the circuits. Common-conductor coupling, inductive coupling and, with appropriate shields, capacitive coupling can be eliminated.

Many integrated-circuit packages are available whose internal circuitry is divided into two separate halves. Information, usually digital, is passed between them via light-emitting diodes in one circuit and light-sensitive receivers on the other circuit. The only direct electrical connection between the circuits is via the high insulation resistance of the substrate and package and the capacitance between the two circuits. Because of the small dimensions of the circuits, this capacitance is only of the order of a picofarad. These opto-isolators are valuable, though not perfect, isolation devices. The circuit on the measurement side may produce unconfined interference of its own. To gain full benefit from the smallness of the capacitance between the circuits, the integrated-circuit package ought to be arranged to be threaded through a hole of minimal size in the measurement screen so that the plane of the screen coincides with the gap between the two sides of the opto-isolator. Otherwise, interference generated at high potential on the receiving side may be carried via capacitance to and from the measurement side. Unfortunately, opto-isolators are rarely used with this degree of care.

There are also several commercial examples of these devices where the light between the two circuits to be isolated from one another is transmitted via optical fibres. Since the circuits can now be as far apart as desired and the fibres are easily threaded through small holes in the measurement shield, capacitive coupling can easily be made to be completely negligible. But in the commercial versions, the circuit within the measurement screen can create interference as it generates and receives the light pulses from the other circuit.

This interference has been greatly reduced in an optical fibre data highway devised by Robinson [2] by two measures.

1. The 20-MHz clock needed in the circuitry within the measurement screen to synchronise the transfer of information with the other circuits on the highway is totally enclosed within its own nested screen and has no output until data transfer is requested.
2. The transfer of data on the highway is so arranged and controlled that a given outstation circuit within a measurement shield is not activated unless the circuitry within that shield is ready to report the results of a measurement. Therefore, it can often be so arranged that the outstation is totally electrically quiet, while the circuitry within its measurement shield that it reports on is actually making measurements.

Isolating transformers (see section 1.1.5) have internal screens constructed so that there is only magnetic coupling between windings. Their purpose is to prevent any current flowing through capacitance or leakage resistance from one portion of a network to another.

Electronic operational amplifiers can provide isolation if connection via conductors, which carry negligible current, is permissible. Figure 1.30 is a diagram of a circuit whose input is a symmetrical pair of FET input operational amplifiers coupled to an output operational amplifier. By choice of the resistance ratios R_2/R_1 and R_4/R_3, the voltage gain of the circuit can have any desired value less than, equal to or greater than unity. The circuit is differential, that is, the output voltage U_{OC} is the amplified difference voltage U_{AB} and contains no component of the 'common-mode' voltages U_{AC} and U_{BC}. Because the operational amplifiers have FET input devices, the input currents need to be only a few picoamperes flowing through a high input impedance of about 1 pF in parallel with $10^{13}\Omega$ slope resistance. In addition, there will be the DC input bias currents of the devices, which will be constant and of the order of 1 pA. The sum of these currents will flow back to the measurement circuit through the current return connection CC′, but will usually be so small that they will not perturb the measurement circuit significantly.

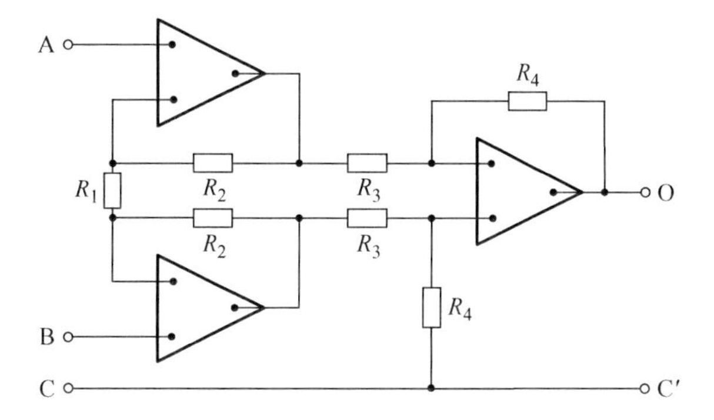

Figure 1.30 An isolation amplifier where one direct connection is permissible

If no direct conducting connection is permissible, isolating amplifiers perform the same function for analogue information as do opto-isolators for digital information. The circuitry within the measurement shield is based on a conventional operational amplifier. Its output is converted to a form suitable for reception by the circuitry outside the measurement shield without any direct electrical connections by conductors, with one of a variety of techniques including optical coupling, conversion of the information to a high frequency followed by transformer coupling or coupling via a capacitor of only a few picofarads. Similar feedback techniques across the isolation can ensure that the analogue voltages or currents in the measurement circuit transmitter are accurately represented by those generated to match them in the receiving circuitry. Conversion of the output of the measurement circuitry to digital form renders this linearisation unnecessary. An optical fibre data link can enable the isolation to withstand kilovolts. The isolation is compromised only by the impedance between transmitter and receiver, and this is typically only a few picofarads in parallel with a few gigaohms.

Each of the isolation devices described above needs power to be supplied to the circuit within the measurement shield in a manner that does not significantly degrade its isolation. Disposable or rechargeable batteries contained entirely within the measurement shield can do this, but at the expense and inconvenience of replacing or recharging them when exhausted. The isolated power supply described in section 1.1.5 was devised to fulfil this need, but the commercially available solution provided by isolated DC–DC converters may be adequate. These converters are similar to the isolation amplifiers described above in that DC power outside the measurement shield is converted to high-frequency AC and conveyed to the circuit inside the measurement shield by a high-frequency transformer having physically separated windings. The AC is reconverted to DC power within the measurement shield and any residual AC is filtered and returned to the measurement shield to prevent its passage back into the measurement circuit. The isolation is incomplete, being compromised by the winding-to-winding capacitance and leakage resistance of the transformer, which might amount to a few picofarads in parallel with a few gigaohms, but this is often adequate in practice.

1.1.8 Balancing

An alternative approach to current equalisation, which applies to both DC and AC as well as when the return current conductors cannot be of low impedance, is to employ symmetrical balanced circuit design. Consider the example illustrated in Figure 1.31. Information is sent down a symmetrical conductor pair by sources A and B, which are of equal amplitude and in anti-phase. These sources could, for example, be the secondary windings of a transformer or a balun device. The information is received by detecting circuits D_1 and D_2, which are of equal sensitivity and input impedance. The outputs of these detectors are subtracted (because A and B are in anti-phase) in circuit D.

Because of the symmetry of the system, any interfering magnetic flux ϕ_{ext}, which threads the loop formed by the conductor that conveys residual imbalance current back to the sources, causes equal interference voltages to be registered by D_1 and D_2. Subtraction of these voltages by Deliminates them. Any capacitive interference such as the interfering source C can be conveyed to a screen that is symmetrical with respect to the conductor pair and will then also be balanced out.

1.1.9 Minimising the effects of insufficiently isolated commercial instruments

Sections 1.1.5 and 1.1.6 are a counsel of perfection. In very particular circumstances, when systematic errors caused by rectification of noise by measuring instruments (a common problem) are significant or when very noise-sensitive instrumentation is involved, the construction of the correctly designed isolating

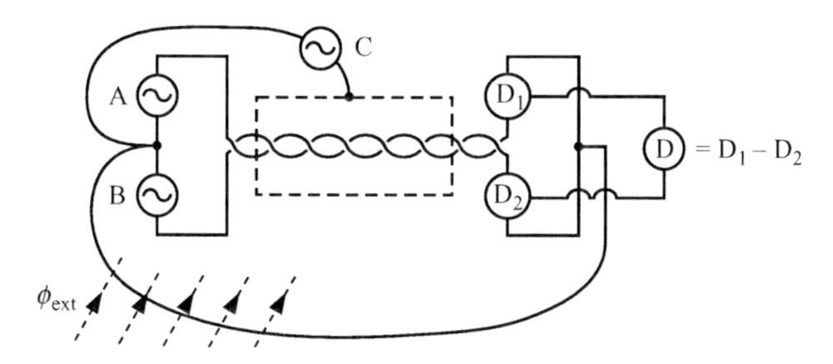

Figure 1.31 Balancing a conductor-pair circuit

systems described in these sections is very worthwhile, if not mandatory. An example of the compulsory use of these techniques is the automated operation of measuring systems involving a SQUID [1,3].

But these specialised transformers and supplies are not yet readily available. By making use of the fundamental principles of section 1.1.1, we attempt in this section to suggest practical guidelines for getting the best interference-rejecting performance from the usual commercial equipment. We can only give general guidelines because the effort instrument designers make for easy interference rejection varies considerably and is often not apparent from the instrument specifications. Each case must be considered separately and is usually an exercise in damage limitation. Nevertheless, simple general principles applied with a view to the relative importance of the effects that should be minimised can often achieve acceptable results in practice.

Typical instruments (all of which, we assume, are mains powered) are

1. power supplies,
2. signal sources (which can be regarded as AC power supplies) and
3. low-level measuring instruments such as sensitive digital voltmeters, phase-sensitive detectors and preamplifiers.

A diagrammatic representation of any of these can, for the present purpose, be drawn as shown in Figure 1.28. The 'Hi' and 'Lo' terminals may very well be the inner and outer connections made by a coaxial socket.

Some guidelines for obtaining optimum noise rejection are as follows:

1. Eliminate common-conductor couplings in the external wiring. Remember that data highway cables plugged in later might violate a carefully thought-out circuit.
2. Consider whether the circuit connected to the input of a low-level measuring instrument is of high- or low-output impedance.
3. The mutual inductive coupling, which causes problems in low-impedance circuits, is perhaps easier to eliminate than capacitive coupling. Use impedance-transforming devices such as operational amplifiers having a low-output impedance, or impedance-matching transformers or isolation amplifiers to

transform to low-impedance interconnection networks. Then pay particular attention to eliminating outer conductor loops (see sections 1.1.1 and 2.1.3).

4. Only the minimum number of connections necessary for the functioning of a circuit should be made. In particular, the connection of everything in sight to 'earth' is a council of despair. Instead, draw a *complete* circuit diagram, remembering that the circuit that exists in reality may not correspond with the working circuit diagram in small, but vital, details. For example, there may be conducting or capacitive routes for currents through ostensibly non-electrical apparatus such as optical tables and pressure sensors. Routes of this kind should be drawn explicitly on the circuit diagram, and their effects considered.

1.1.10 The 'traditional' approach to DC and low-frequency circuitry versus the current-balanced conductor-pair coaxial approach

The techniques used in DC sensing and measurement circuits have traditionally been viewed as completely different from those employed for AC. This is a serious error. From the point of view of, at least, interference elimination, there is no difference whatsoever, and DC practice would greatly benefit from being considered from an AC viewpoint. In particular, coaxial circuitry could be relevant, and although the current equalisation technique described in section 3.1.1 is effective only at frequencies above a few tens of hertz, this is all that is required to eliminate most interference. Current balancing in a screened, twisted-pair network is even more appropriate for DC circuits. Great benefit can be obtained by attention to the concept of isolation and to minimal and correct connection to the mains safety conductors.

There are problems in DC measurements in addition to those encountered in AC measurements. One problem is that of thermoelectric emfs, and fairly rapid source polarity reversal is needed to eliminate drifts in them. A second problem, mostly encountered in high-impedance circuitry, is caused by the phenomenon of dielectric storage of charge in capacitors including inter-conductor and inter-component insulation. Dielectric materials can store charge within their volume if subjected to a unidirectional voltage. The discharge current can persist, exponentially decaying away in seconds, minutes or even hours once the voltage source is removed or reversed. This can create a problem in sensitive DC circuitry. Two instances are (i) if insulated high- and low-potential leads are in close contact without an intervening screen and (ii) if an analogue operational amplifier is configured as an integrator with a capacitor in its feedback loop.

It is often possible to design the circuit so that the discharge current takes a route that does not affect the measurement, usually by providing proper shielding or a coaxial approach. The concept of direct impedance is very useful in this regard (see section 4.1.2).

The magnitude of the phenomenon depends on the particular dielectric. Polysulphone and poly-tetraflouride-ethelene (PTFE) plastics exhibit a much lower dielectric storage than the more common polythene insulation.

1.1.11 Thermoelectric emfs

The combination of temperature gradients and circuits composed of different metals produces emfs in the microvolt range. These emfs can cause errors when making low-level DC measurements, and if the temperature gradients are time dependent, the consequent variation of thermoelectric voltage is a source of low-frequency noise.

In principle, the problem can be eliminated from low-level DC measurements by reversing the polarity of the source of the desired signal. If the measuring circuit is linear, from the superposition theorem, only that part of the circuit response due to the source will reverse. Taking the mean of the difference between the response with the source 'forward' and the source 'reversed' then eliminates unwanted stray voltages. A similar result can be obtained from the difference between the 'source forward' and 'source off' conditions. If the source cannot be turned off or reversed rapidly, thermoelectric voltages must be minimised by reducing temperature gradients and ensuring that dissimilar metals are eliminated as much as possible from the circuit. Input terminals to instruments can be lagged and protected from draughts to reduce temperature differences between them, and copper-to-copper connections used elsewhere in the circuit.

For applications where reduction of thermoelectric emfs to the nanovolt level is important, it may be necessary to use oxygen-free high-purity copper, or materials with proven low thermal emf with respect to it, everywhere in the circuit. The unstressed crystalline structure of the conductors must not be compromised by excessive bending, stretching or work hardening. Permanent or temporary connections around the circuit need special attention. Switch contacts used for this kind of work should be made from special alloys or silver- or gold-plated pure copper. If a rubbing action accompanies making a contact, several seconds may be needed for induced thermoelectric emfs to die away.

Soft-soldered joints need careful treatment. Ordinary soft solder as used for electrical connections has a high thermoelectric emf with respect to copper. It can nevertheless be used if the copper conductors to be joined are first brought into intimate contact by applying pressure while being soldered together. If there is no great excess of solder surrounding the resulting joint, thermoelectric voltages will be minimal because the solder-copper voltages are shorted by the lower-resistance copper–copper contact. This approach is, in our experience, preferable to using 'low-thermal' solder containing cadmium as it is difficult to obtain an electrically or mechanically sound joint with this material. Pure copper crimped joints are an alternative to soldering as the stress in the metal is only localised.

Thermoelectric effects in the internal defining point junctions of standard resistors of value of the order of 1 Ω or less can be unexpectedly troublesome. Peltier effects resulting from the flow of measuring currents through them can cause local heating or cooling. The concomitant thermoelectric emfs generated appear at the potential terminals so that the apparent value of the resistor can be altered by several parts in 10^7 for a 1-Ω resistor, an amount that increases in proportion to the decrease in value of smaller resistances. The alteration decays away in time Δt between the previous application or reversal of the measuring current and when the measurement is made. Δt depends on the thermal capacity of the defining point junctions and is typically of the order of a second.

1.1.12 Designing temperature-controlled enclosures

In high-accuracy work voltage, current and impedance components, which are as stable as possible, are a prime necessity. Unstable components hamper investigation by making changes to the type B (systematic) errors of the measurement system. Changes should produce an anticipated alteration of the result, and obtaining a different result from that anticipated should indicate the presence of a systematic error rather than an alteration of the value of an unstable component.

Temperature is usually the main environmental effect that changes the value of a component. A temperature coefficient producing a relative change of about 10^{-5}/°C is typical. Therefore, if measurements having a relative accuracy of 10^{-7} to 10^{-8} are to be made, temperature control of the order of millikelvins is needed. The controlled temperature should be a few degrees above any likely fluctuations in the ambient laboratory temperature. The combination of less precise control and a thermometer to sense the temperature of the standard is seldom a completely satisfactory solution. The existence of thermal transport delay times within the standard, thermometer response times and possible hysteresis of the standard in response to temperature changes means that the thermometer reading at any given instant will correspond not to the present value of the standard but to its value some time previously.

Successful design of a temperature-controlled enclosure depends at least as much on thermal design to ensure close coupling of the controlling temperature sensor to the heater of the enclosure as on the stability and adequate gain of the sensor signal – amplifier – heater system. This section aims to describe the principles of good thermal design. Details will depend on the thermal properties of the enclosure. Tighter temperature control can be attained by a 'two-staged' design with an inner enclosure totally surrounded by a separately controlled outer enclosure. Critical parts of the control circuitry can be located between the enclosures to maintain them at a more constant temperature.

The main principles are as follows:

(i) Tight thermal coupling of the heater to the controlling sensor or sensors to minimise thermal delay between them and to facilitate the design of a stable, high-gain control amplifier.
(ii) Tight thermal coupling of cables, etc. to the enclosure, the component within the enclosure and the temperature sensor will minimise the effect of heat flow between the environment and the enclosure along the cables. Thermal coupling can be attained by thermally anchoring to the enclosure a length of each cable before egress to the environment. This will minimise heat flow along the cable length within the enclosure.
(iii) Thermal anchoring to an outer thermally shielding enclosure in a similar manner will reduce temperature gradients in the walls of the inner enclosure.

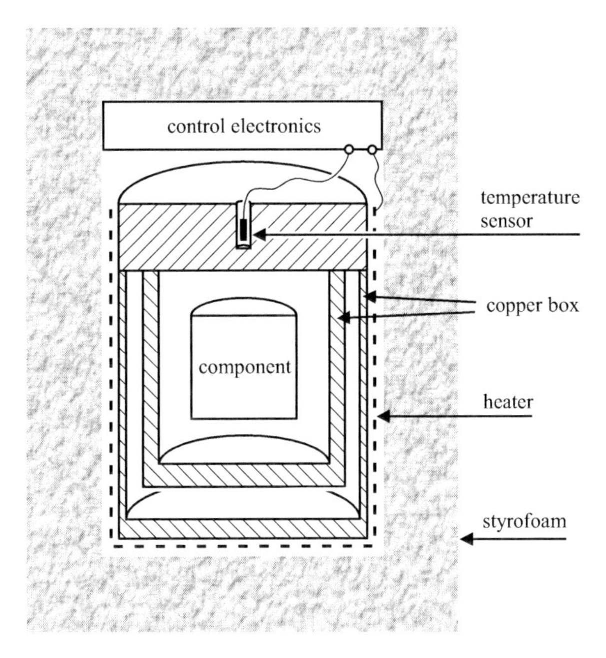

Figure 1.32 An example of a small temperature-controlled enclosure

Figure 1.32 illustrates a design of a temperature-controlled enclosure for small components based on these principles. The design philosophy is to maintain one point of the enclosure, where the temperature sensor of the heat-controlling circuit is located, at a constant temperature by heaters uniformly distributed over the surface of the enclosure. If there are two or more sensors in series, they should be adjacent, and the controlled point is then at the mean position between them. By constructing a double-walled enclosure, the outer, controlled wall will act as a guard to shield the inner wall from the effects of temperature gradients.

For larger enclosures, immunity from temperature gradients induced by ambient conditions can be obtained by heating uniformly all the outside surfaces of an enclosure and sensing its temperature as the mean of several sensors located at the centres of each surface of the enclosure.

A thermometer should be mounted in such a way that it represents an average temperature for the whole enclosure. The effect of its self-heating on the temperature of the enclosure should be considered. There is no effect if the thermometer is kept continuously energised at a constant level, because this produces only a constant offset. Proper thermal anchoring of the thermometer leads in a similar way to other connections to the device will prevent ambient temperature affecting the thermometer reading.

A single-stage (sensor + amplifier + heater) system is likely to achieve a temperature control ratio against ambient temperature of less than one hundred – that is, worse than 10 mK/K. Tighter control requires a two-stage design, which can potentially achieve 0.1 mK/K.

Useful thermally conducting materials for constructing temperature-controlled enclosures include copper, aluminium or aluminium alloys. A jointing technique such as soft-soldering is needed to ensure good thermal contact between parts of the enclosure. An alternative technique for demountable enclosures is to fill close-fitting joints with a high thermal-conductivity paste. These pastes are also useful for efficient thermal anchoring of cables, etc. Aluminium oxide or beryllium oxide (use the latter with great care – health hazard) or mica are good electrical insulators having high thermal conductivity for washers, plates and other components.

Two high-stability negative temperature coefficient (NTC) thermistors can be combined in opposite arms of a DC or AC Wheatstone bridge for control or for sensing the measurement temperature. Connecting these as shown in Figure 1.33 both doubles sensitivity and enables the mean temperature of two separated points of the enclosure to be sensed.

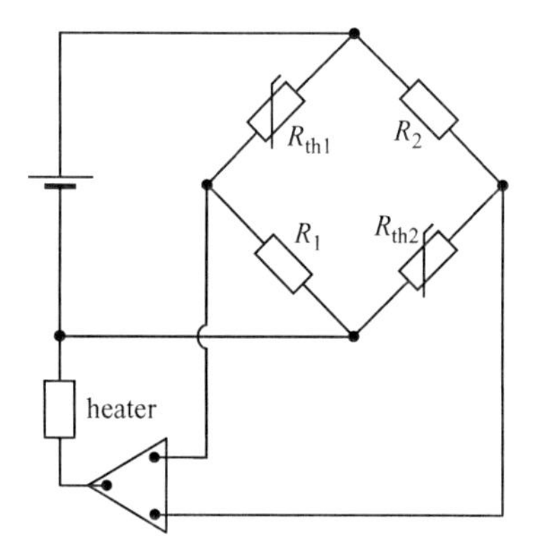

Figure 1.33 Two thermistors in a Wheatstone bridge for temperature sensing

Heaters can be either commercially-made mat or flat ribbon types, or purpose-constructed by winding a flat ribbon resistive conductor over adhesive Mylar or Kapton tape layer applied directly to the surface of the enclosure. Care should be taken to ensure that the heating conductor has no sharp edges. Further layers of Mylar or Kapton can be applied to keep the heater in place to ensure good thermal contact between it and the enclosure. Alternatively, an aluminium enclosure can have an anodised layer added for direct application of a heating conductor, again kept in good thermal contact by an overlayer of Mylar or Kapton tape.

1.1.13 Ionising radiation (cosmic rays, etc.)

This is mentioned more for the sake of completeness than for being a likely practical nuisance, although it can significantly affect certain high-impedance instrumentation, such as photomultipliers and electrometers. Shielding is usually impractical, and elimination by suitable signal conditioning must be resorted to.

1.1.14 Final remarks

To a large extent, the correct design of an interference-free electrical measurement system is a matter of circuit topology – that is, the correct arrangement of the connectivity of conductors and conducting surfaces, and the layout of magnetic circuits. The exact geometrical disposition of the system, other than ensuring contiguous paths of conductor pairs, should then be unimportant. Our experience is that there is one, and only one, correct topological arrangement of a circuit and that the search for it is very worthwhile.

We reiterate the most fundamental principle of all – that the *paths of the circulating currents* are what matter. It pays to develop the ability to imagine the resulting voltage drops and electric and magnetic fields they create. Every interference problem has a rational, if sometimes obscure, explanation, and often considerable persistence is needed.

References

1. Williams J.M., Smith D.R., Georgakopoulos D., Patel P.D., Pickering J.R. 'Design and metrological applications of a low noise, high electrical isolation measurement unit'. *IET Sci. Meas. Technol.* 2009;**3**(2):165–74
2. Robinson I.A. 'An optical-fibre ring interface bus for precise electrical measurements'. *Meas. Sci. Technol.* 1991;**2**:949–56
3. Williams J.M., Hartland A. 'An automated cryogenic current comparator resistance ratio bridge'. *IEEE Trans. Instrum. Meas.* 1991;**40**:267–70

Chapter 2
Sources, detectors, cables and connectors

In this chapter, we discuss the miscellaneous but nonetheless important properties of sources and detectors and of coaxial cables and connectors so that the practical examples of network design described in chapter 8 can be successfully implemented.

2.1 General principles

The voltage and current waveforms in the bridge networks discussed in this book are usually assumed to be closely sinusoidal and of a single angular frequency $\omega_0 = 2\pi f_0$, but we can consider any periodic waveform of arbitrary shape by representing it as a Fourier series, that is, as a superposition of phase-related harmonics (multiples) of the fundamental frequency whose amplitudes are such that the added result is the original waveform.

We need to consider departures from the ideal situation, where

(i) the bridge source generates a pure sine wave at ω_0
(ii) the network is strictly linear with respect to voltages and currents
(iii) the detector responds to only a single frequency, which can be adjusted to be ω_0

In principle, if (iii) is true, either condition (i) or (ii) can be relaxed without incurring error, but if (iii) is not true, then both (i) and (ii) must hold if a correct result is to be obtained.

2.1.1 The output impedance of a network affects detector sensitivity

In Figure 2.1, Z_{out} is the output impedance of a network as would be measured at the detector port in the absence of detector and input cable and with all network sources set to zero. In the special case of a bridge measuring the ratio of two impedances Z_1 and Z_2 connected to voltage sources of negligible output impedance, $Z_{out} = (1/Z_1 + 1/Z_2)^{-1}$. If Z_1 and Z_2 are resistances R_1 and R_2, $R_{out} = (1/R_1 + 1/R_2)^{-1}$.

The total shunt admittance presented to the input of the detector is $Y_{total} = Y + j\omega C_D + 1/R_D$, where Y is the sum admittance of all cables connected between the network components comprising Z_{out} and the detector input, C_D is the input capacitance of the detector and R_D is its shunt input resistance.

The sensitivity of the detector is reduced by the sensitivity factor s, defined as the ratio of the voltage actually detected to that detected by an ideal detector of

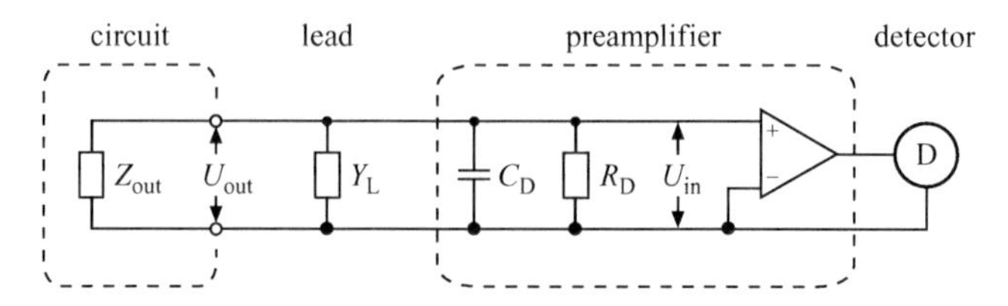

Figure 2.1 The output impedance of a network and the shunt admittance of an input cable and detector affect sensitivity

negligible shunt admittance, connected to the network by cables of negligible shunt capacitance. Hence, $s = 1/(1 + Y_{total} Z_{out})$, where s is a complex quantity whose components govern the in-phase and quadrature response of the detector. If the output impedance of the network is a resistance R_{out} and the cable and detector admittances are purely a capacitance C_{total}, $s = 1/(1 + R_{out}/R_D + j\omega C_{total}R_{out})$. If the network is a capacitance ratio bridge as described in section 5.1.2, the corresponding expression is $s = 1/(1 + C_{total}/C_{out})$.

2.1.2 The sensitivity of detectors to harmonic content

Bridge networks can be divided into two types for the purpose of analysing the departure of detectors from the ideal. Some bridge networks have a balance condition that is independent of frequency, apart from minor departures arising from small frequency dependencies of components, but others have an inherently frequency-dependent balance condition. The 'twin-T' bridge network drawn in Figure 2.2 is an example of the latter, as its balance condition is $\omega RC = 1$. The quadrature bridge described in sections 9.2.1–9.2.3 is another example.

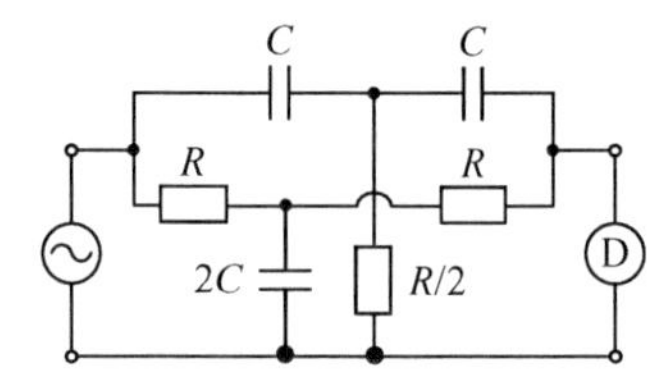

Figure 2.2 A twin-T bridge

If a bridge is nearly frequency independent, there will be only a small amount of higher harmonics appearing at the detector terminals when the balance condition is satisfied for the fundamental frequency ω_0. This amount will be proportional to the product of the harmonic content of the source and the residual frequency dependence of the bridge balance. Configuring the detector to detect $2\omega_0$, $3\omega_0$, etc. will indicate the presence of these higher harmonics, but the only consequence of them is a possible increase in the background indication when the detector is configured to respond to ω_0. The presence of this background increases the difficulty of judging the minimum indication of a simple tuned detector, but this condition still corresponds to the true bridge balance. The effect is usually negligible if a phase-sensitive detector is used rather than a tuned detector.

A more subtle problem arises if the amplification stages of the detector, before any frequency selection takes place, are non-linear. If there are successive harmonics present at the input to the detector whose frequencies differ by ω_0, then the non-linearity will produce signals at the various sum and difference frequencies, and in particular, a signal actually at ω_0. An erroneous balance will result. Fortunately, good present-day designs of amplifiers are such that any non-linearity is rarely large enough to cause any trouble when used with frequency-independent bridges.

Formally, we can see how the fundamental frequency arises from mixing harmonics if we take the detector response r to contain higher terms than linear in the applied voltage:

$$r = a_1 U + a_2 U^2 + a_3 U^3 + \cdots + a_n U^n + \cdots \tag{2.1}$$

where the coefficients a_n usually decrease rapidly with increasing n so that only the first two terms of (2.1) need be considered.

Then, if the applied voltage contains harmonics,

$$U = \sum_{k=0}^{\infty} \alpha_k \sin[(k+1)\omega_0 t] \tag{2.2}$$

On substituting (2.2) into (2.1), we find terms like

$$a_2\, \alpha_k \sin[(k+1)\omega_0 t]\alpha_{k+1} \sin[(k+2)\omega_0 t]$$

which can be rewritten as

$$a_2\, \alpha_k \alpha_{k+1} \frac{\cos[\omega_0 t] - \cos[(2k+3)\omega_0 t]}{2} \tag{2.3}$$

The $\cos[\omega_0 t]$ terms represent the spurious signal at the fundamental frequency.

The phenomenon is called intermodulation distortion, and the magnitude of the effect is proportional to the product of amplitudes of two successive harmonics. A simple test for its presence is to observe whether there is any change in the bridge balance when the size of the input signal to the detector is altered with a linear attenuator. Network components whose values are slightly voltage dependent will give a similar effect if the network source voltage is changed, and so the test of deliberately increasing the harmonic content of the source while maintaining its amplitude unchanged is also valuable.

When the balance condition of a bridge network is frequency dependent, the network will pass the harmonics present in the source through to the detector relatively unattenuated, while at balance, the amount of fundamental frequency should be zero. The problems of insufficient frequency selectivity and non-linearity of the detector are then far worse, and it is necessary in work of the highest accuracy to precede the detector with a frequency-selective circuit of sufficient linearity so that this circuit itself does not cause intermodulation. The magnitudes of harmonics in a source usually decrease as the harmonic order increases, so that it is often sufficient to eliminate the second and third harmonics to a high degree of perfection by tuned filters and to reduce higher harmonics with a passive low-pass filter. Figure 2.3a is such a filter. The 0.1 H inductors together with the adjustable 100 Ω resistors and adjustable capacitors constitute one arm of two nested Wheatstone bridges whose other arms are the 6800 and 3300 pF capacitors and the 100 Ω fixed-value resistor. When adjusted, these bridges, in principle, reject both phases of the second and third harmonics completely. Higher harmonics are rejected by the input transformer and the inductor–capacitor (L–C) circuit at the output, both of which are tuned to pass ω_0. The component values given are suitable for matching the 100 kΩ output impedance of a room temperature quadrature bridge to a detector. This filter adds some noise but it is only of the same order of that inherent in the bridge.

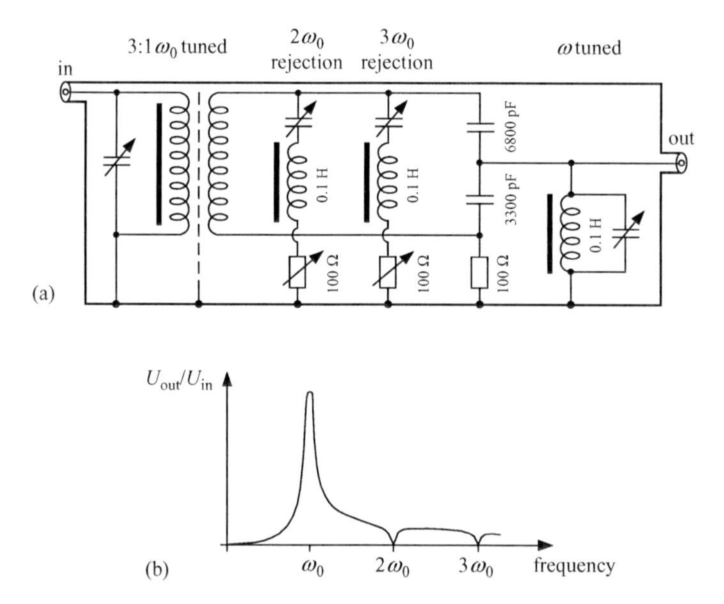

Figure 2.3 (a) A harmonic-rejecting filter. All cores are high-Q amorphous metal material. (b) Its frequency response

The quadrature bridge described in section 9.3.7 in connection with the quantum Hall effect has components at a cryogenic temperature and a consequent much lower noise. To take full advantage of this, a different filter is needed in which the most troublesome harmonics are greatly attenuated by simple series-tuned circuits, as shown in Figure 2.4. To achieve low-noise performance, the inductors must be wound on low-loss toroidal amorphous metal cores capable of yielding a Q-factor of several hundreds, and the capacitors must have a dielectric loss factor of less than 10^{-3}. The detector following the filter must reject other harmonics adequately.

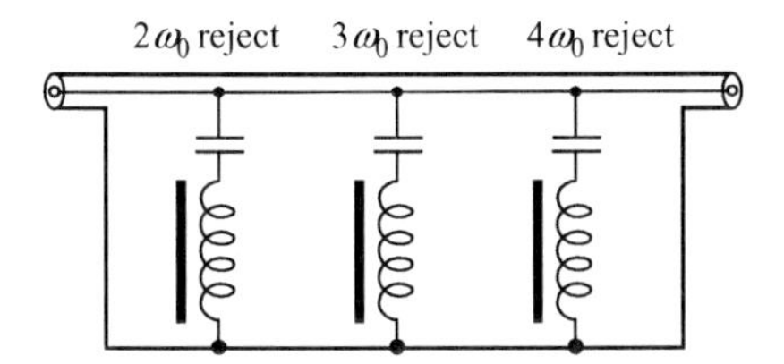

Figure 2.4 An ultra-low noise harmonic-rejecting filter for use with a bridge having components at cryogenic temperatures

2.1.3 Noise and noise matching a detector to a network

There is an extensive literature concerned with noise in electrical circuits (see e.g. Reference 1 or a more recent Reference 2). Here, we discuss only some simple considerations, which are often sufficient for our purpose.

It is usually easy to amplify a small output signal from a network sufficiently so that the sole criterion to be investigated is the optimisation of the signal-to-noise ratio of the amplifier. When we are concerned in this book with networks containing only passive components, we need consider only Johnson noise. (We will assume that troublesome, but in principle avoidable, noise from interference, microphony, etc. has been eliminated.) Johnson noise generates a power

$$\frac{U_n^2}{R} = i_n^2 R = 4kT\Delta f \tag{2.4}$$

in a resistance R. k is Boltzmann constant ($\approx 1.38 \times 10^{-23}$ J/K), T is the absolute temperature (≈ 300 K for room temperature) and Δf (Hz) is the bandwidth of the detecting device, so the noise power is about 1.6×10^{-23} W for a 1-Hz bandwidth at room temperature. For a network connected to a detector, the resistance R concerned represents the total shunt loss; that is, *all* sources of energy loss in the network as well as in the detector are included in it. The noise performance of most AC detectors now approaches the Johnson noise limit over a range of source impedances presented to their inputs. Optimising their performance is therefore a matter of impedance matching of the network output impedance to an impedance that is within the optimum input impedance of the detector.

The noise situation can be analysed by replacing the actual resistors and other loss sources with equivalent noise sources in series with ideal noise-free resistors when we are considering a voltage noise source or in parallel when we are considering a current noise source. The superposition theorem then allows us to consider the effect on a detector of the signal source and the various noise sources separately.

Noise arising from components that have an outer coating of lossy dielectric, which forms a lossy capacitance between the inner and outer conductors, can be eliminated by surrounding the components with an intermediate conducting casing supported by low-loss dielectric, which is connected to the inner conductor, as shown in Figure 2.5.

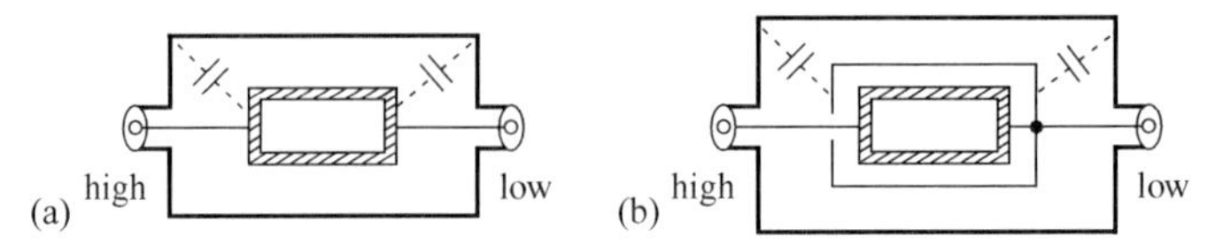

Figure 2.5 (a) A component coated with a lossy dielectric. (b) Removing the loss contribution to the noise

In this section and the following section, we assume that all impedances are pure resistances in order to bring out the point we wish to make as simply as possible. We also take as a simple model of a real detector the equivalent circuit of Figure 2.6 where D is an ideal noise-free detector of high input impedance, that is, of negligible input admittance. Its actual noise resistance and input impedance are represented by a shunt conductance G_D. We suppose that the total loss of the network can be represented by a single conductance G_S, as would be the case for a network having only resistive components and negligible shunt admittance.

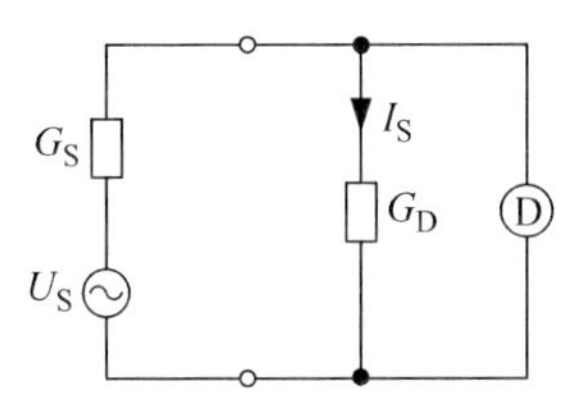

Figure 2.6 The equivalent circuit of a detector connected to a noise source

The total noise admittance presented to D is $G = G_S + G_D$. The noise power is

$$U_n^2 G = 4kT\Delta f \tag{2.5}$$

The signal current in the loop is

$$I_S = \frac{U_S G_S G_D}{G_S + G_D} = \frac{U_S G_S G_D}{G} \tag{2.6}$$

The voltage across the detector is

$$U_D = \frac{I_S}{G_D} = \frac{U_S G_S}{G} \tag{2.7}$$

Hence the signal-to-noise power ratio

$$\begin{aligned} S &= \frac{U_D^2}{U_n^2} \\ &= \frac{U_S^2 G_S (G_S/G)}{4kT\Delta f} \end{aligned} \tag{2.8}$$

Another source of noise in addition to Johnson noise occurs at low frequencies in non-metallic conductors and semiconductor devices. This is termed 'flicker noise' or '$1/f$ noise' because its magnitude is approximately inversely proportional to the frequency f for unit bandwidth; it is unlikely to be detectable at 1 kHz but will increase in relative importance as the frequency is reduced.

2.1.4 The concept of a noise figure

$P = U_S^2 G_S/4$ is the maximum possible signal power that can be transmitted into the detector terminals by varying G_D until it equals G_S. The signal-to-noise ratio obtained when the value of P equals the total noise power $4kT\Delta f$ is called the noise figure or factor n; it is a measure of by how much the system fails to meet the ideal signal-to-noise ratio. In the above instance,

$$n = \frac{G}{G_S} = \frac{G_S + G_D}{G_S} \tag{2.9}$$

Often $N = 10 \log_{10} n$ is used instead of n and the condition of optimum power transfer into the detector ($G_D = G_S$) gives $n = 2$ or $N = 3.01$ dB.

In the present example, it is instructive to note that

$$n \rightarrow 1 \text{ or } N \rightarrow 0 \text{ dB as } G_D \rightarrow 0 \text{ or } G_S \rightarrow \infty \tag{2.10}$$

that is, the best signal-to-noise ratio is *not* obtained under conditions of maximum signal power transfer. Thus, in general, it is important not to confuse power matching with noise matching; it is noise matching that is usually relevant.

For a real detector, G_D is often fixed by the instrument design, but by interposing a matching network between the source and the detector, provided that it itself does not introduce further significant noise, the apparent value of G_S presented to the detector terminals can be altered. Two such matching devices that can be used either separately or in combination are a tuned L–C circuit and a ratio transformer.

In Figure 2.7a, we illustrate a typical problem of matching a network modelled as a source having an output impedance Z_n, the real component of which generates noise, to a preamplifier having both a noise voltage U_N and a noise current I_N.

A narrowband tuned L–C circuit (Figure 2.7b) can simultaneously perform the useful task of rejecting harmonic components of U_S. A ratio autotransformer (Figure 2.7c) is a broadband simple matching device from low to high impedance as shown, or, by reversing input and output connections, from high to low impedance (Figure 2.7d). A ratio transformer (Figure 2.7e; section 1.1.5) could also provide isolation of the detector from the network. Harmonic rejection can be accomplished by tuning the transformer with a capacitor that shunts either the primary or secondary winding. In a practical transformer, it is the inductance of each winding that makes tuning possible. The inductor or transformer core should have a low loss so as not to introduce further noise.

From the discussion so far, it might be thought that using a matching device to make the network impedance presented to the detector look very low would be

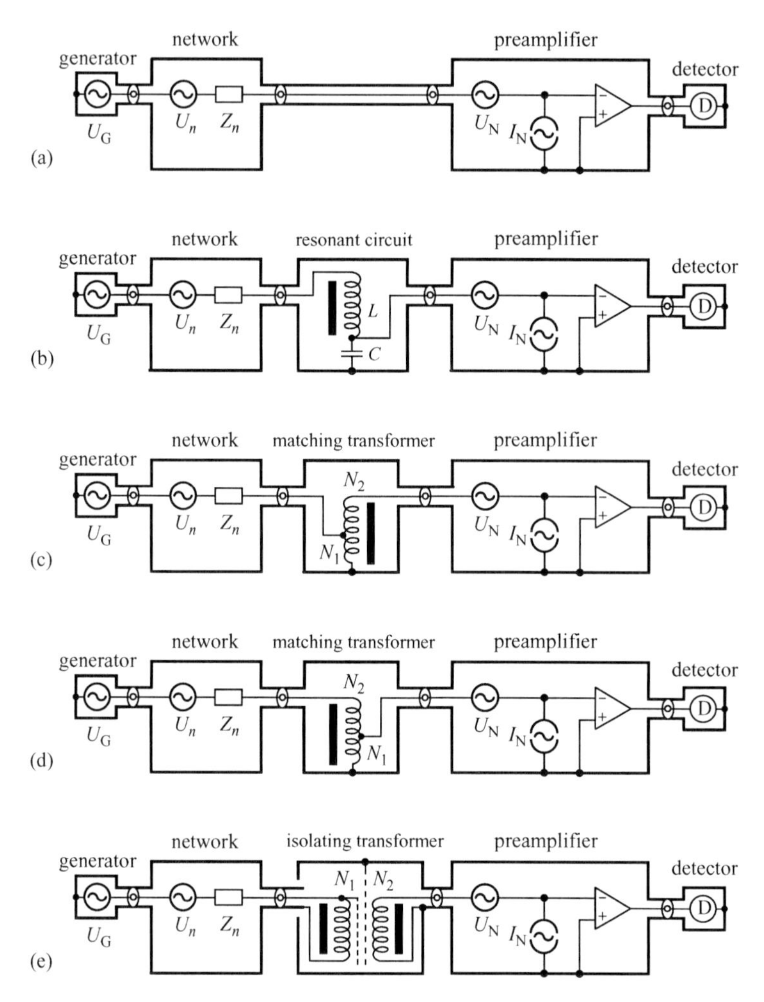

Figure 2.7 Circuits to match a detector to a network for better sensitivity: (a) a source, the equivalent circuit of a network, the equivalent noise representation of a preamplifier and a final detector; (b) and (c) matching a low-impedance source and network to a high-impedance detector; (d) matching a high-impedance source and network to a low-impedance detector; (e) an isolating matching transformer

sufficient to resolve all signal-to-noise problems. Unfortunately, this is not so because we have adopted too simple model for a real detector; its noise attributes can rarely be represented by a simple conductance G_D.

Figure 2.8 is a better representation of a real detector. An 'ideal' detector (noiseless and of infinite input impedance) is shunted by a noiseless admittance Y_D in series with a voltage noise U_D, the whole being shunted by a current noise source I_D.

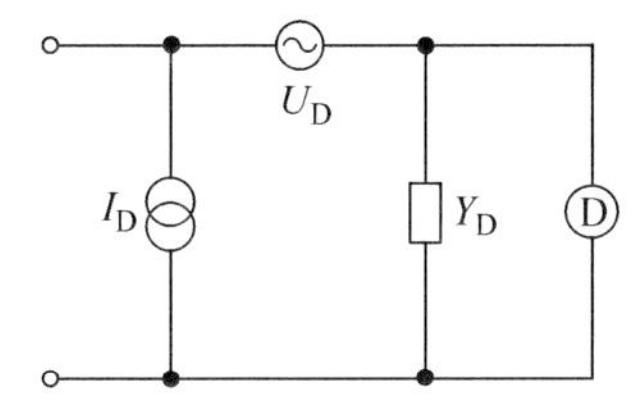

Figure 2.8 The equivalent circuit of a detector

If, therefore, a detector for which this representation is valid is connected to a network by a matching circuit of low output impedance and low signal voltage but high signal current so that I_D and Y_D are shorted out, the detector voltage noise U_D is still registered and can swamp a small signal voltage. If the matching circuit is designed to have high output impedance, high signal voltage and low signal current, the effect of U_D may well be negligible but the signal current will be swamped by I_D in parallel with it. There is an optimum between these two extremes.

U_D and I_D are easily measured for a given detector. Having calibrated the sensitivity of the detector for both voltage and current by connecting it to standard signal generators and attenuators, U_D can be measured as the voltage indicated when the input terminals are short-circuited. I_D is the voltage indicated times the input admittance of the detector Y_D when the input terminals are open-circuited and screened to prevent electric field pick-up.

The optimum signal-to-noise situation arises when the voltage generated by I_D flowing in the source impedance Y_S added to Y_D equals the voltage noise U_D.

Manufacturers of detectors often supply 'noise contours', that is, loci of constant noise figures on a graph of source frequency against source impedance. From these, the optimum impedance to be presented to the detector at a given frequency can be deduced and a suitable matching network designed.

2.2 Attributes of sources

A network source will usually consist of a signal generator, possibly followed by a power amplifier and a power transformer if the output power of the generator is insufficient. It is useful to combine the functions of matching and isolation (by means of properly constructed screens – see section 1.1.5) in one device. The arrangement must be capable of delivering the necessary power without excessive distortion, but if the output goes straight into a transformer forming part of a measurement network, there is no point in striving to achieve excessively low distortion because the transformer will add some distortion on its own account. This can be minimised for a voltage transformer if the output impedance of the source is made very low compared to the input impedance of the transformer. A transformer adds distortion because the instantaneous input impedance of a transformer depends on the relative permeability of the core, which in turn depends on the flux level and therefore on the excitation level. This variation of its instantaneous input impedance causes a varying voltage drop across the output impedance of the source, and this distorts the waveform. Attaining a total harmonic distortion of 0.1% or less is reasonable.

If a bridge balance is frequency dependent, the source must have the necessary frequency stability. This is easily attained if the source is a frequency synthesiser locked to a precision frequency standard. The frequency can be measured by an electronic counter also locked to a precision frequency. For low-frequency measurement, better resolution is obtained by operating the counter in the period mode.

2.3 Properties of different detectors

Phase-sensitive detectors (also called lock-in detectors or vector voltmeters) are most commonly employed for observing small signals and for balancing bridge networks but wideband and simple tuned detectors are also useful, particularly for diagnostic purposes. To obtain an optimum signal-to-noise ratio an appropriate preamplifier is needed.

2.3.1 Preamplifiers

Preamplifiers are needed for all kinds of electrical measuring circuits. They can be isolated by being powered by batteries contained within the conducting case of the preamplifier or have a mains-powered supply. A mains-powered supply must be properly isolated (see section 1.1.5) to eliminate any net current into the input of the preamplifier.

The preamplifier shown in the circuit of Figure 2.9 is especially useful for high source impedances, but it also performs well for other values of source impedance. The input transistor is an *n*-junction field-effect transistor (JFET), for example, type 2SK170 made by Toshiba. The input resistor R_{in} is of the order of a gigaohm, so, being in parallel with the source impedance, it does not usually significantly increase the noise. Also, the noise current of the transistor must be small enough not to generate significant noise when it flows through the source impedance.

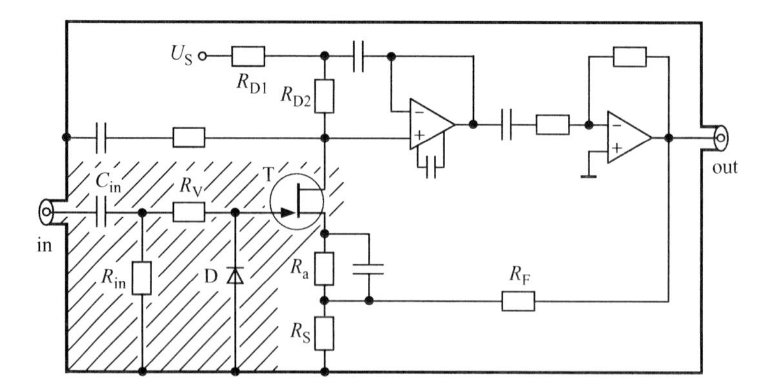

Figure 2.9 A low-noise preamplifier suitable for high-impedance sources

Typical values for the equivalent input noise voltage and noise current are 1.5 nV/$\sqrt{}$Hz and a few fA/$\sqrt{}$Hz, respectively. Preamplifiers for optimum noise matching to high output impedances need special attention to avoid including lossy dielectric materials in all capacitances to the screen in the region of their input circuitry. In this example, this applies to all materials within the shaded area. A two-stage feedback amplifier sets a stable gain as the ratio of the feedback resistor R_F and the transistor source resistor R_S. A gain of 100 is typical.

Figure 2.10 is a low-noise preamplifier circuit [3] optimised for cryogenic source impedances of the order of 10 kΩ such as that of the quadrature bridge involving two cryogenic Hall resistances described in section 9.3.7. This bridge network has an output impedance of 12.9 kΩ (at a temperature of the order of 1 K) in parallel with a virtually noise-free capacitance of 10 nF so that the noise amplitude is of the order of 1 nV/$\sqrt{}$Hz. Using a matching transformer (see section 2.1.3) in conjunction with a low-noise high-impedance preamplifier is not practicable because of the excess noise generated by transformer core losses.

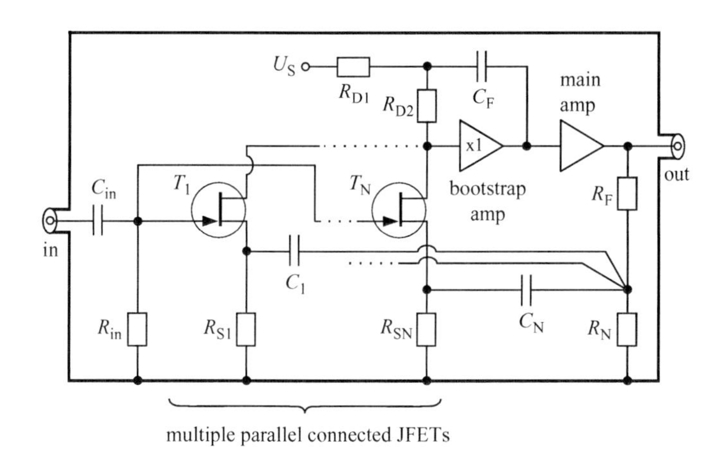

Figure 2.10 A low-noise preamplifier suitable for source impedances of the order of 10 kΩ

The preamplifier achieves low voltage noise by using multiple parallel-connected JFETs having a common drain resistor and individual source resistors. The bootstrap action of the amplifier increases the drain resistance virtually by means of the feedback capacitor C_F. Optimum performance is obtained by selection of the individual JFETs for low noise and, for example, four parallel JFETs can produce a noise voltage of only 0.5 nV/$\sqrt{}$Hz.

2.3.2 Wideband (untuned) detectors

The most useful example of these is the oscilloscope, which can reveal the presence of unwanted interference. A newly constructed network should always be scrutinised in this way, and it is prudent to re-examine it at intervals also.

2.3.3 *Narrowband (tuned) detectors*

These often incorporate a frequency-selective network (usually a twin-T bridge, Figure 2.2) in the feedback loop of an electronic amplifier. The resulting frequency response is similar to that of an L–C tuned circuit; that is, it attenuates at frequencies away from the resonant frequency by a rate governed by an effective Q of the circuit. Successive harmonics are attenuated with respect to the preceding one by a factor of the order of Q. It is difficult to design stable circuits with Q-values much greater than 10^2–10^3, so the detector will have for the second harmonic $2\omega_0$ not much less than 1% of the sensitivity that it has for the fundamental frequency ω_0. As we have noted in section 2.1.2, this situation can cause errors in detecting a bridge balance condition.

Further, a simple tuned detector responds to both in-phase and quadrature signals alike; the indication corresponds to the square root of the sum-of-squares of the in-phase and quadrature components. This can make the balancing of a bridge network where the supposed in-phase and quadrature adjustments are not in fact orthogonal somewhat difficult. A degree of phase dependence can be obtained by deliberately introducing a signal of the desired phase directly into the detector from the network source. Smaller quadrature signals will then produce very little additional indication, but in-phase signals will be fully displayed as an addition to the standing deflection.

A simple extension of this technique can improve the sensitivity in balancing a bridge network to almost the limit imposed by detector noise by taking as the balance point the mean of two adjustments, which produce equal relatively large indications on either side of the correct balance condition. This technique is superior to simply trying to adjust for a minimum indication because it gives the detector a modicum of phase sensitivity against both a small quadrature component and noise.

2.3.4 *Phase-sensitive detectors that employ a switching technique*

All phase-sensitive detectors need a reference signal; in bridge networks it is usually derived from the bridge source. In the past, many phase-sensitive detectors applied the reference signal to switching circuitry to rectify the input signal. Since they are phase related, any harmonics present in the input signal will also be rectified. The rectification of even harmonics sums to zero but the last of the n cycles of the nth harmonic in a rectification cycle does not cancel so that the odd nth harmonic is detected with an efficiency of $1/n$ relative to that of the fundamental frequency. Therefore, switching phase-sensitive detectors need to be used with caution, particularly in bridge networks. If the bridge balance condition is frequency dependent (see section 2.1.2), they must be preceded by a tuned filter to reject harmonics sufficiently.

2.3.5 *Phase-sensitive detectors employing a modulating technique*

Newer phase-sensitive detectors in effect multiply the incoming signal and the reference signal, and display the result after passage through a low-pass filter. The effect on all signals other than that at the reference frequency is a time-varying output at the difference frequency, and this is rejected by the low-pass filter. Therefore, there is, in principle, nearly complete rejection of all harmonics except the one desired. Naturally, the reference signal must not contain any significant harmonic content. Also, non-linearity at small signal levels and internal harmonic generation as saturation levels are approached lead to some likelihood of significant intermodulation distortion, so checks for all these effects should be made.

The design philosophy of most signal-detecting instruments now is to convert an analogue input to digital form as soon as is practicable after some amplification at the input stage. Any signal recovery, manipulation, filtering, etc., including multiplication with a similarly digitised reference signal, is then carried out by mathematical operations on the digital representation of the signal. This approach has the potential problem of aliasing [4] if the variation rate of the signal approaches the digitisation rate, but it results in better linearity.

2.4 Cables and connectors

The properties of coaxial cable used for high-accuracy work need careful consideration. Where networks containing small direct admittances (such as small-valued capacitors) are concerned, any loss in the cable dielectric, particularly of cables making connections to the detector and any associated sub-network, will add noise. Polythene and PTFE are good low-loss dielectrics. If the capacitance per unit length of the cable is unduly large, the available out-of-balance signal will be shunted and the balance sensitivity will be decreased (see section 2.1.1).

In some cables, the central conductor is of steel (to increase its mechanical strength) plated with copper. This increases its series resistance but has no other consequence.

Coaxial cables suffer from microphony at audio frequencies, and, in an extreme case, acoustic oscillations in a cable or a component carrying a relatively large current or voltage can be transmitted to a touching cable associated with a detector where they can be re-converted into an electrical signal. Therefore, it is better if all cables are of anti-microphonic construction. This is usually accomplished by putting a layer of conducting plastic between the dielectric and the outer conductor where microphonic signals are generated or re-converted by the triboelectric effect. Such a layer also prevents leakage of electric field through holes in the woven outer conductor. In some examples, the resistivity of the plastic layer can be unduly high and will thus cause some loss and constitute another source of noise, particularly at higher frequencies.

Therefore, the choice of cable at audio frequencies involves a compromise between low shunt capacitance, low series inductance, low series resistance, reasonable physical size, flexibility and cost. The authors have made extensive use of PTFE-dielectric cable of about 6 mm overall diameter that has a characteristic impedance of 75 Ω, a capacitance of 70 pF/m having a loss angle of 35×10^{-6} radians and a series inductance and resistance of 0.4 μH/m and 0.07 Ω/m, respectively. Microphonic effects are reduced by a coating of conducting varnish applied to the dielectric. This varnish is easily removed locally when terminating the cable with a connector.

The principal requirements for coaxial connectors are that the series resistance they introduce between their mating surfaces should be low and reproducible, and they should have low loss in any dielectric surrounding the inner conductor. The inner conductor should be completely screened by the surrounding outer conductor when the connectors are mated, and there should be available T-connectors of low impedance. Laboratories undertaking work of the highest accuracy at audio frequencies usually use the British Post Office (BPO) MUSA connectors that, besides the above desirable properties, are simple to attach to cable ends and components, are relatively inexpensive, and mate and unmate with a simple push or pull. The more readily available Bayonet Neill–Concelman (BNC) connectors suffer from a higher and more variable contact resistance. Ordinary ready-made cables with BNC connectors do not usually have an anti-microphonic conducting layer. General Radio GR900 connectors are unnecessarily elaborate and expensive for lower-frequency applications. The hermaphrodite GR874 connectors are suitable but are somewhat bulky. Care must be taken to ensure that the central mating conductors do not become partially unscrewed and provide a source of unwelcome variable resistance. At higher frequencies, the sub-miniature SMA, B and C connector series and ready-made cables are good and convenient because they are impedance-matched to a cable. Care must be taken when mating and unmating them to provide only axial force or else the life of the connectors will be very limited.

References

1. Robinson F.N.H. *Noise and fluctuations in electronic devices and circuits*. Oxford: Clarendon; 1974
2. Ramm G., Bachmair H. 'Optimisation of the signal-to-noise ratio of AC bridges'. *Tech. Mess.* 1982;**49**:321–24
3. Schurr J., Moser H., Pierz K., Ramm G., Kibble B.P. Johnson-Nyquist noise of the quantised Hall resistance. To be published in *IEEE Trans. Instrum. Meas.* 2011
4. *A Guide to Measuring Direct and Alternating Current and Voltage Below 1 MHz*. London: The Institute of Measurement and Control; 2003

Chapter 3

The concept of a low-frequency coaxial network

Interference-free balanced-current circuitry evolved from the need to develop bridge measurements of standard impedances to obtain ever higher accuracies to satisfy demands for better characterisation of passive electronic and electrical circuit components.

Early comparisons of the value of impedances were confined to resistance measurement because only direct current sources and detectors were available. A little later, some form of commutator provided intermittent or reversed current, and this allowed capacitances and mutual and self-inductances (which are only defined for the passage of time-varying currents) to be compared. Bridge design continued to evolve around DC techniques for which the spatial disposition of conductors, impedances, detectors and sources is immaterial (interference considerations apart). The basic standard of impedance was a DC resistor, assumed or designed to have nearly the same value with AC.

Even after sinusoidal sources and detectors responding to a single frequency became available and some of the advantages of AC bridge comparisons became apparent, the past history of the subject lingered on in the form of single-conductor networks, but the advent of the calculable mutual inductor (see section 6.4.12) and the even more accurate calculable capacitor (see section 6.2) necessitated the development of better techniques.

Electrical impedances store or dissipate the energy of electric and magnetic fields and currents, and so for a reproducible, well-defined standard of impedance, the location and strength of its fields must be defined by fixing the boundary of the space they occupy. A bridge network relates the energy flow through two or more sets of impedance standards in order to compare their values accurately; to do this, the standards need to be interconnected by conductors to form a network. Therefore, the electric and magnetic fields surrounding these conductors must also be controlled and accounted for. If the impedance standards have exposed terminals and single-wire conductors interconnect them, fields surround both conductors and standards. The strength and distribution of these fields are altered by changes in the relative position of the standards, conductors and external objects, including the experimenter. The use of such networks inevitably leads to imprecise relationships between the values of the standards because the stray fields associated with the standards and networks are altered by changes in their environment.

Therefore, the subject of accurate impedance measurement was developed by using coaxial cables which carry equal and opposite currents in their inner and concentric outer shielding conductors. Components and standards were encased in electric and magnetic screens. Fields are thereby confined to the inside of cables and standards, and a bridge network becomes well defined. Meaningful measurements can then be made with relative uncertainties of as little as 1 in 10^9, if required.

This approach may be viewed simply as the application of the coaxial techniques common in radiofrequency work to low-frequency networks, combined with more careful control of the fields associated with circuit components. The ideas have been developed by many people over many years, but principally by R.D. Cutkosky, at the National Institute of Standards and Technology, USA, and A.M. Thompson, at the National Measurement Laboratory, Australia.

One way of looking at coaxial networks is to see them as two superimposed networks. The first network consists of the meshes of components and the conductors connecting them. The second network comprises the screens of the components and the outer coaxial screen of the interconnecting cables. The configurations of the two networks are identical, and, by providing each mesh with an equalising device such as a current equaliser (see section 3.1.1) or a differential-input amplifier (section 1.1.7), the current in the outer screen of each cable and component is constrained to be equal and opposite to the current in the components and central conductor. The current in any cable as a whole is then zero over the frequency range of equalisation and no external magnetic field is created. The second network of screens and cable outer conductors has only a small impedance and consequently is all at nearly the same potential, so that there is no significant external electric field either. Consequently, the layout of cables and components has no effect on the performance of the circuits. Also, the reciprocity theorem states that such networks also do not respond to fields from external sources; that is, they are interference free. Because of these desirable properties, the purpose of this book is to encourage coaxial techniques for DC and AC circuitry up to at least 1 MHz. Unexpected interactions between apparently unrelated parts of a circuit are a frequent source of frustration, if not outright design failure. They can be avoided by visualising and minimising the electric and magnetic fields associated with the various parts of the circuit. The role of power supply leads and their associated current return conductors ('earth' or 'ground' conductors) to and from parts of circuits needs careful consideration in relation to the network as a whole.

The tendency now in designing electronic instruments is to convert the output of very simple analogue input circuitry to digital form by periodic sampling of its amplitude as early as possible in the chain from input to final presentation of the result. Any manipulation of the signal, for example, for bandwidth or band-pass filtering or for combining with another input for phase-sensitive detection is carried out by arithmetic algorithms applied to the numbers representing the signal. This approach has frequency limitations and other difficulties because the bandwidth is inherently limited to half the sampling frequency. At greater frequencies, false results are produced (see Reference 1). These considerations are outside the scope of this book, which is concerned only with the correct gathering of the input signal, and, in particular, in avoiding its corruption by noise, interference and crosstalk from other input channels. It is true that some of these effects can be minimised in the signal-processing chain by filtering, etc., but it is a much better and cost-effective practice to eliminate these problems from DC to 100s of MHz DC and AC circuitry in the first place (see Reference 2).

3.1 The coaxial conductor

In section 1.1.1, we proved that for an infinitely long straight coaxial cable in which, by some means, the current through the outer conductor is maintained at all times equal and opposite to the current in the inner conductor, there is no external magnetic field at all. In practice, cables are neither infinitely long nor straight or perfectly coaxial, but the external magnetic field of a finite and curved length is still usually negligible, provided the length and radius of curvature are large compared with the diameter of the cable – conditions usually fulfilled by virtue of its mechanical stiffness.

In a network of components interconnected by balanced-current coaxial cables, the outer conductors should be electrically continuous and without any appreciable impedances in their network. Therefore, despite the currents flowing in them, they form a nearly equipotential surface and the electric field outside the cables is also negligibly small. Therefore, we have a system of conductors outside which there is no appreciable electromagnetic field. That is, the network of cables cannot radiate interference, and according to the reciprocity theorem, it cannot respond to external interference either, and we have an interference-immune system.

It cannot be too strongly re-emphasised that there is nothing different about similarly constructed DC networks in this regard; constructed according to these principles, they will also be interference-immune.

Equality of the opposing currents in the inner and outer conductors of a coaxial cable in a mesh, where the outer conductor has low impedance, occurs automatically at frequencies higher than a MHz or so, because the go-and-return impedance comprising the resistance and inductance of the coaxial cable is much lower than that of the impedance of the loop traversed by the outer conductor. At lower frequencies, the current equaliser technique described in section 3.1.1 is necessary. This technique does not equalise direct currents or currents whose frequencies are below 50 Hz or so, but it suffices for eliminating interference which usually is of higher frequency.

3.1.1 *Achieving current equalisation*

When a single coaxial cable goes to an isolated component, which has no other external electrical connection to it, the current flowing in the inner conductor of the cable must return via the outer conductor, and current equalisation is automatic.

But if the component has other connecting conductors so that it is a part of a mesh in a network, positive steps must be taken to achieve current equalisation.

A current-equalising technique was developed for kHz frequencies in conjunction with the coaxial AC bridges described later in this book because it ensures that the electrical definition of the standards measured by these bridges is complete. This technique also eliminates from circuit networks in general interference induced by external magnetic fluxes within the working frequency range of the technique.

Consider, as illustrated in Figure 3.1a, just one mesh of a network of conductor pairs (usually coaxial cables). The circuit of the inner conductor usually involves a voltage source in series with a comparatively high impedance Z that determines the value of the current I_P flowing around it. The low-impedance circuit of the outer shield conductors has no additional impedances or voltage sources. If the cable is threaded through a high-permeability magnetic transformer core as shown, the core and mesh become a current transformer. The inner conductor is its high-impedance primary winding and the outer conductor is the secondary winding, which is approximately a short-circuit across which very little voltage can be induced. The circuit is re-drawn in Figure 3.1b to illustrate this. Because there is very little voltage across the secondary winding, there can be very little flux in the core, and this can only be so if there is an almost equal and opposite current I_S in the secondary (i.e. the circuit of the outer conductors) to I_P in the primary windings (i.e. the circuit of the inner conductors). The total current in the cable as a whole, which is the sum of the current in the inner and the current in the outer, is therefore approximately zero, as required. The effectiveness of the device is greatly enhanced if the cable is threaded through the core a few times rather than just once, and this device can achieve an approximate equalised-current condition for frequencies from about 10 Hz to 1 MHz.

Let ϕ_P and ϕ_S be the magnetic fluxes in the core caused by the primary and secondary currents I_P and I_S, respectively, which thread the core n times. The total flux in the core is

$$\phi = \phi_P - \phi_S = n(I_P - I_S)$$

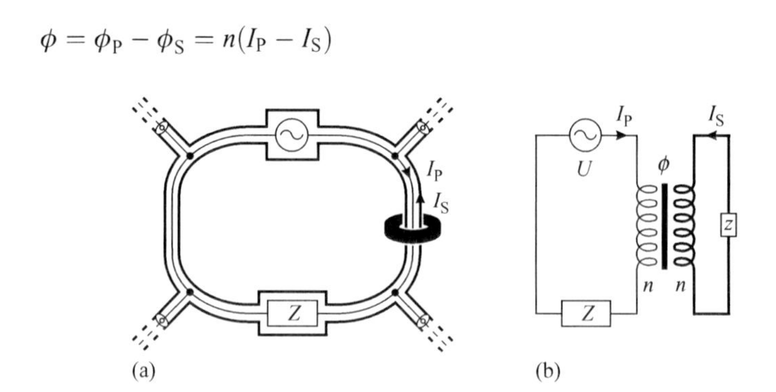

Figure 3.1 (a) One mesh of a coaxial network threaded through a core and (b) the equivalent circuit

Then

$$n\frac{d\phi}{dt} = U - I_P Z = I_S z = n^2\left(\frac{dI_P}{dt} - \frac{dI_S}{dt}\right) = j\omega L(I_P - I_S)$$

for a sinusoidal current. L is the inductance of the wound magnetic core, Z is the impedance of the circuit of inner conductors and $z = r + j\omega L$ is the impedance of the circuit of outer conductors.

Hence,

$$I_P = I_S + \frac{I_S z}{j\omega L}$$

and

$$U = I_P Z + I_S z = I_P\left\{Z + \frac{z}{[1 + (z/j\omega L)]}\right\} \approx I_P\left(Z + z - \frac{z^2}{j\omega L}\right)$$

If $|z| \ll |j\omega L|$

$$U \approx I_P(Z + z) \tag{3.1}$$

and

$$I_P \approx I_S \tag{3.2}$$

Equation (3.1) shows that the apparent impedance of the primary circuit has been increased by z, whilst (3.2) shows that equality of currents has been achieved to a good approximation.

The above analysis can be refined by including the energy loss in the core. This loss can be represented by a resistance R in series with the inductive impedance ωL of the wound core. That is, its Q-factor is equal to $\omega L/R$. Typical values for ωL and R are 100 and 10 Ω, respectively. The fractional current inequality is

$$\frac{I_P - I_S}{I_P} = \frac{z}{j\omega L + R} \tag{3.3}$$

The apparent impedance of the primary circuit has been increased by

$$Z + z - \frac{z^2}{j\omega L + R} \tag{3.4}$$

If the circuit consists of several meshes, each independent mesh ought to be provided with an equaliser, but it is irrelevant from the point of view of equalisation at which point around the mesh it is placed.

For frequencies in the range above 100 kHz, the resistance of the meshes of outer conductors becomes small compared with the impedance of the inductance of their open loops, and this latter impedance is small compared to the impedance of a go-and-return path through inner and outer conductors of cables. Then current equalisation occurs without the need for a transformer core.

So to achieve successful equalisation, two conditions must be met:

1. The circuit of one of the two conductors (the outer of coaxial cable conductors or one of a twisted pair of conductors) must have an impedance that is mainly its resistance, which is low compared with the inductive impedance of the wound equaliser.
2. The circuits of the two conductors must nearly coincide in space, that is, the open area between them must be small.

If a set of equalisers are a component of a balanced coaxial bridge network, their imperfect operation caused by the finite impedance of the meshes of the outer conductors means that the currents in the inner and outer conductors of the cables are not perfectly equal and opposite. This has a small effect on the balance condition of bridge networks for the following reasons:

(i) The cable corrections (see section 5.4) assume perfect current equalisation, and if equalisation is imperfect a fraction of the open-loop inductance contributes to the series inductance of the cables.
(ii) The contribution of the outer impedance of terminal-pair components is altered. For four-terminal-pair components, the relevant impedance is the four-terminal impedance of the outer conductor that is usually very small so that the alteration of the four-terminal-pair impedance is usually negligible.
(iii) The output voltages between inner and outer conductors of the coaxial connectors of transformer taps are slightly changed.
(iv) There are residual magnetic couplings between meshes of the network.

A single total correction for all these small effects can be obtained from the simple procedure of equaliser evaluation described in section 3.2.2. Equaliser evaluation generates a small correction that can be applied to the measurements to give results that would be obtained with perfect current equalisation everywhere. Note that the effect of residual magnetic couplings implies that the relative orientations and positions of equalisers and meshes should not be altered once evaluation has been carried out. The need for attention to this point can easily be tested by deliberately temporarily altering positions and orientations of cables and equalisers to observe whether the bridge balance is significantly affected.

As noted earlier, the interference rejection properties of a current-equalised coaxial network follow from the fact that it generates no significant external magnetic flux. The reciprocity theorem then ensures that the network does not respond to interference from externally generated fluxes.

A current equaliser is represented symbolically by a toroidal core surrounding a coaxial cable, as shown in Figure 3.2. The magnetic transformer cores used for this purpose are usually, but not necessarily, toroidal.

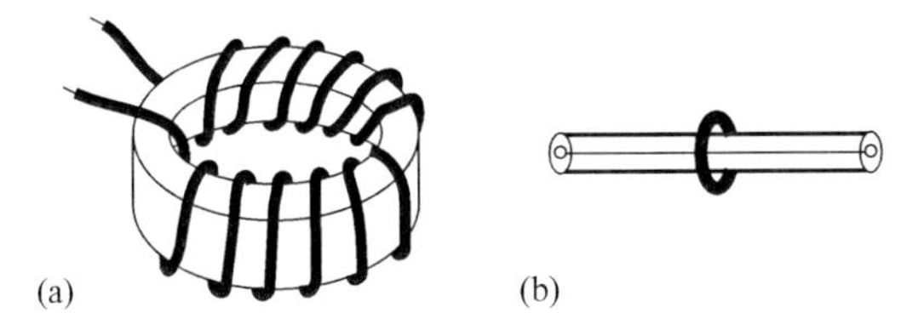

Figure 3.2 *(a) A high-permeability toroidal transformer core used as a current equaliser. (b) Its symbolic representation*

3.1.2 The concept of a coaxial network

In this book, we are concerned with conductor-pair networks where one network of the pair has a low impedance compared to the other, and where the current in any one conductor is matched by an equal and opposite current in the other conductor of the pair. The potential differences between corresponding nodes of the pair of networks are quantities of interest because, for example, detectors or sources are to be connected between a pair of corresponding nodes.

This concept can be generalised to any current-equalised network. With reference to Figure 3.3, which shows one mesh of an equalised-current conductor-pair network, where the potentials at its nodes, U_A, U_B, etc., are referenced to some arbitrary zero,

$$(U_A - U_{A'}) - (U_B - U_{B'}) = (U_A - U_B) - (U_{A'} - U_{B'}) \approx I(Z + z)$$

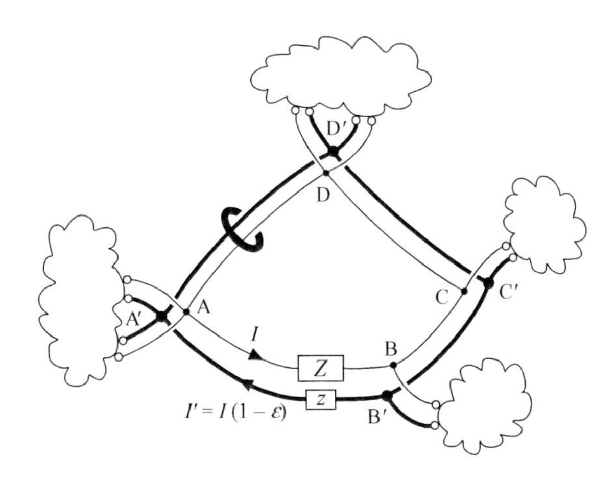

Figure 3.3 *An equalised-current conductor-pair mesh*

The small inequality caused by imperfect equaliser action, denoted by ε in the figure, can be removed by the procedure of equaliser evaluation described in section 3.2.2.

Usually, the impedances like Z, which represent the impedances of components in the circuit, have much greater values than those like z, which represent the impedances of the current return conductors.

If we replace the small impedances z of the primed mesh with ones of zero impedance, $U_{A'} = U_{B'}$, and if we then enhance every impedance between two nodes of the unprimed mesh by the corresponding impedance of the old primed mesh, an equivalent single mesh with the same nodal equations

$$U_A - U_B = I(Z + z'), \text{etc.} \tag{3.5}$$

as the conductor-pair mesh is obtained.

If the conductor-pair mesh is, in fact, coaxial circuitry as drawn in Figure 3.4, the entire primed mesh constitutes an equipotential surface and there are no external electric fields. Hence, the behaviour of a conductor-pair net only differs from that of the corresponding single-conductor net in the vital respect that a coaxial conductor-pair net has no interactions between its constituent meshes or between itself and external electromagnetic fields.

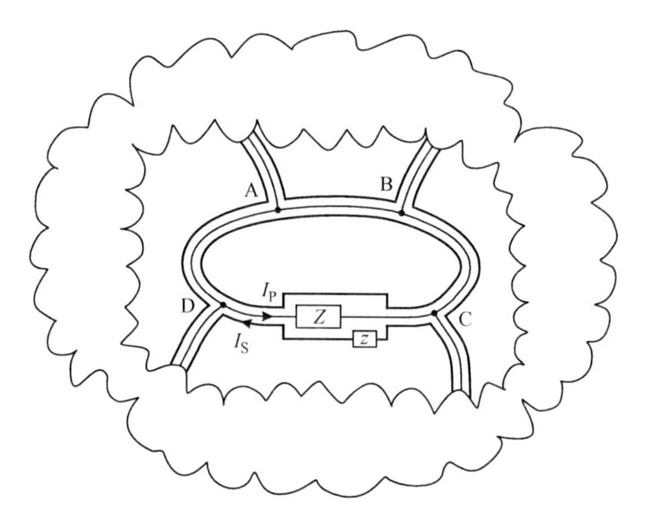

Figure 3.4 A coaxial version of Figure 3.3

At frequencies of the order of, or greater than, 1 MHz, current equalisation occurs without the need for equalisers because the open-loop inductance of a mesh is much greater than the go-and-return inductance of a conductor pair.

3.2 Construction and properties of coaxial networks

3.2.1 Equalisers in bridge or other measuring networks

It must be understood that the set of current equalisers in a network is at least as important as any of the other components (such as ratio transformers, impedances to be compared, combining networks, etc.). The equalisers need very careful and detailed attention if the highest possible accuracy and degree of interference elimination are to be achieved. Note that imbalanced mesh currents result from imperfectly efficient equalisers and *affect the main outcome of the measurement network* (see section 3.1.1). The process of assessment of all the equalisers acting together will quantify this effect (detailed in section 3.2.2). Simply measuring the unbalanced mesh currents will not enable the total effect of equalisers to be assessed.

There are at present three ways in which equalisers can be constructed:

(i) The simplest and most familiar way is to wind 10–20 turns of coaxial cable through a high-permeability toroidal core (see section 3.1.1 and Figure 3.2). The equality of the opposing currents in the inner and outer conductors can be improved by increasing the number of turns, n, on the core because the inductance of the winding, L, increases as n^2, whereas the resistive component of z increases only as n. Enhanced efficiency can therefore be obtained by winding 20–40 or more turns until the window of the toroid is full. Threading a greater number of turns through the usual toroidal core may require a thinner coaxial cable, and thinner and longer coaxial cable will further increase the series resistance of the mesh of outer conductors. This will reduce the efficiency somewhat, but it can be regained, if necessary, by further increasing the number of turns. The longer coaxial lead of such an equaliser has a greater parallel capacitance and series inductance, and this might be undesirable in some leads. For example, it would increase the cable correction of a defining lead (see section 5.4). In other meshes of a network, such increased-efficiency equalisers are often a good choice.

(ii) It is possible to achieve near-perfect current equalisation by manually balancing the flux in an equaliser core to zero, using the circuit and procedure described in testing equaliser efficiency in section 3.2.2.

(iii) As an extension of this technique, the balancing can be done automatically, as shown in Figure 3.5, by applying feedback with an operational amplifier to create a so-called *active* current equaliser [3]. This device employs two cores, one mounted on the top of the other, and the coaxial lead is wound about ten times around both cores. To simplify the figure, the coaxial leads and the single conductors are shown threaded only once through each core. A single-conductor winding on one of the cores senses the induced voltage resulting from the magnetic flux caused by the net current in the coaxial lead. An operational amplifier amplifies and converts this voltage to a current that is passed through a single-conductor winding on the second core in such a sense as to null the net current together with the voltage it causes in the first core (see Figure 3.6). This feedback circuit increases the equaliser efficiency by the voltage-to-current gain. This gain is the ratio of the feedback resistor R_2 to the input resistor R_1 multiplied by the ratio of turns of the voltage-detection and the current-injection windings:

$$\text{Gain} = \left(\frac{R_2}{R_1}\right)\left(\frac{N_{\text{det}}}{N_{\text{inj}}}\right) \tag{3.6}$$

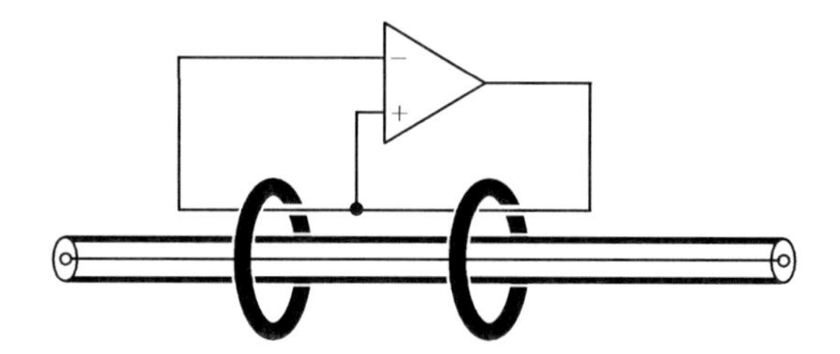

Figure 3.5 An active current equaliser

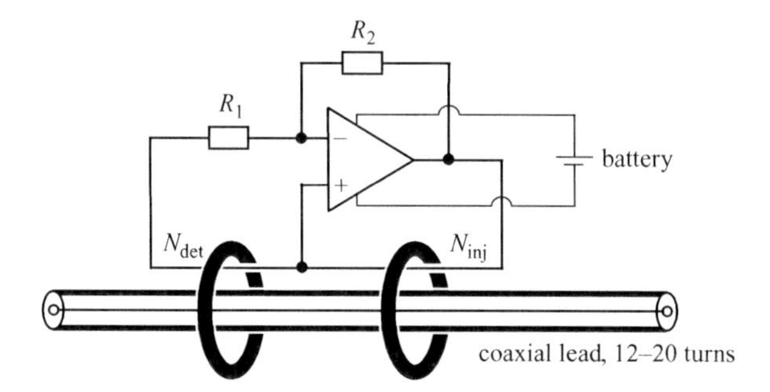

Figure 3.6 Circuit details of an active current equaliser

Typical values are $R_2 = 1\ \mathrm{M\Omega}$, $R_1 = 1\ \mathrm{k\Omega}$, $N_{det} = 10$ and $N_{inj} = 50$, corresponding to a gain of 200. To derive this equation, it has been assumed that the impedance of the detection winding is small compared to the input resistor, and then the gain is practically frequency independent. Above a certain limit frequency, this assumption is no longer valid and the gain decreases with frequency. Furthermore, the gain should not be too large (in practice not larger than approximately 1000), otherwise the active equalisers in a bridge network begin to oscillate and generate excess noise. The more active equalisers that are used in a bridge network, the smaller is the maximum gain allowing stable operation. A suitable gain is of the order of 200–500.

Care must be taken that stray capacitive currents are not coupled from the outside into the network of inner conductors via the power supply leads to the operational amplifier. To avoid this, active equalisers are often powered by batteries contained within the shielding box housing the device, but it must then be ascertained that the device has been switched on and that the batteries remain in good order for the duration of critical measurements. A better solution is to use a totally isolating power supply as described in section 1.1.5.

Often, passive equalisers of enhanced efficiency are adequate if the operation of the equalisers is assessed and a correction made for their imperfect efficiency, as described in section 3.2.2. A core wound with a bifilar winding of ordinary conductors where one winding is connected into the inner circuit and the other into the circuit of the outer conductors can give good efficiency at frequencies below a kilohertz and is therefore good for rejection of low-frequency interference in DC circuitry. Unfortunately, the efficiency rapidly drops off at higher frequencies because some of the current in the inner conductor is shunted through the device by the inter-winding capacitances without being equalised.

The efficiency of particular equalisers can sometimes be enhanced by reducing the potential between nodes of the network of outer conductors forming part of their mesh. These potential differences can be regarded, as in linear network theory, as sources driving unwanted currents round other meshes. A potential difference can be reduced by connecting a low-impedance single conductor in parallel with the cable producing it. If there is an equaliser in this cable, naturally the single conductor must be wound through it together with the cable, and in any case the conductor must be routed alongside the cable (ideally, twisted with it) as its role is simply to reduce the impedance of its outer conductor.

3.2.2 Assessing the efficiency of current equalisers

The efficiency of an equaliser in isolation is *meaningless* and cannot be assessed because equalisers all act together as a single component of the network of outer conductors.

A simple qualitative way of experimentally assessing whether the presence of a given equaliser is necessary to obtain the desired accuracy from a measurement network is to short some added single-conductor turns around the core. If the transformed impedance of these turns is low compared to the mesh circuit of the outer coaxial conductor, the equalised current in the outer conductor will flow in these turns instead, thus eliminating the effect of the equaliser almost completely. Note, however, that one possible reason for no significant effect being observed could be that there is another, redundant, equaliser in the same mesh.

The method of making a quantitative evaluation is as follows (see Figure 3.7):

(i) Each cable which incorporates an equaliser is threaded in turn through a net current detection transformer (see section 7.3.3). The voltage registered by a detector connected to this transformer is a measure of the unbalanced current in the cable.

(ii) By also threading the cable through a net current injection transformer supplied from an adjustable two-phase quasi-current source, both phases of the detector signal can be nulled. If desired, near-perfect equalisation will be preserved in the cable so long as the current supply is maintained. This constitutes a manual balance to enhance equaliser efficiency.

(iii) Both phases of the current are then increased over that required to null the detector by a suitable ratio such as 100 times, and the effect on the phases detected by the *main* detector of the network is observed. Evidently, from the linearity of the network, the effect of the actual current in the equaliser will be 100 times less, and a correction to both phases of these amounts of the opposite sign can be applied.

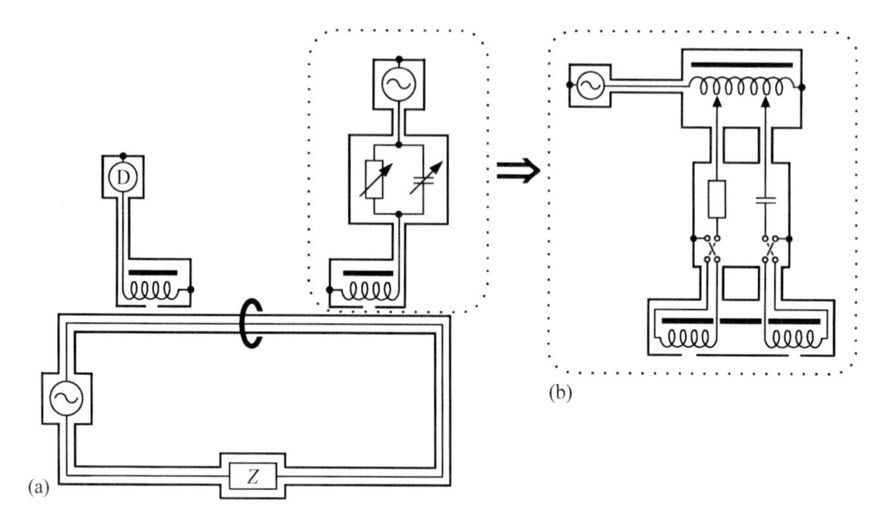

Figure 3.7 Evaluating the effectiveness of a current equaliser: (a) the injection and detection circuits; (b) a practical in-phase and quadrature quasi-current source

In this way, each of the equalisers in a network can be assessed, and from the superposition theorem, the net effect of the simultaneous imbalanced currents in all the equalisers and network meshes will be the algebraic sum of the separate contributions from each equaliser. This correction to the result obtained from a measurement network is completely separate from the correction required for the effect of current and voltage loss down cables (the so-called cable corrections) analysed in section 5.4.

The evaluation of active equalisers can be more difficult than the evaluation of passive ones. First, the net currents might be close to, or even smaller, than the resolution of the net current detector. Second, the evaluation procedure requires measuring the effect of injecting a greatly enhanced net current, but the active equaliser almost compensates for this injection (or it may go into saturation if the injected current is too large). A simple solution is to temporarily switch off the feedback amplifier of the particular equaliser while the other ones are active, to evaluate this equaliser in passive mode and to assume that its efficiency in active mode is enhanced by its gain. (For this purpose, the measured gain should be used because the equation to calculate the gain is an approximation that, for example, holds only at frequencies below an upper limit.)

It is essential that each mesh of a measurement network contains an equaliser, unless equaliser assessment shows that the effects of particular ones are negligible. Even then, an equaliser may be desirable for interference rejection. For equaliser assessment to be meaningful, each mesh should contain only one equaliser. That is, the number of equalisers in a measurement network should be equal to the number of meshes, and the equalisers should be properly located. If too many equalisers were to be employed, equaliser assessment will be compromised. This is because if a mesh incorporates two or more equalisers, then when attempting to assess one of them, the others will take over the function of opposing the unbalance current.

Proper placement of equalisers is a topological problem, which can be solved by drawing the mesh of outer conductors and network nodes, as shown in Figure 3.8b, which is an example of this process being applied to the measurement network of section 9.1.4 and its accompanying Figure 9.4. The process consists of conceptually providing cables with equalisers and then deleting each cable having an equaliser in succession to remove the mesh that they equalise.

A topological network theorem states that if n is the total number of nodes, c the total number of cables and m the number of independent meshes, $m = c - n + 1$. This theorem is proved in Appendix 1, and it provides a useful way of checking correct equaliser placement. A component that has more than two cables connected to it is a node for this purpose.

Because there is not more than one equaliser per cable and each of the m meshes has just one cable incorporating an equaliser, $c = c_0 + m$, where c_0 is the number of cables without an equaliser. Consequently, $m = c - c_0 = c - n + 1$ and hence $c_0 = n - 1$, that is, the number of cables without an equaliser must be one less than the number of nodes. Therefore, there must be exactly one cable without an equaliser joining up all the nodes, for if any node were to have more than one cable incorporating an equaliser going to it, there would be a mesh without an equaliser, contradicting the original requirement. If all the cables going to a node have equalisers, either the condition $c_0 = n - 1$ cannot be met or there is a node elsewhere having two or more leads without an equaliser, which is not allowed. The final solution can therefore be checked by drawing the tree graph of conductors that do *not* incorporate an equaliser, as shown in Figure 3.8c. This graph should contain no loops or isolated branches.

In the bridge networks discussed later in this book, it is important that the placement of equalisers is such that there is an equaliser in the input cable to the detector. Proper operation of this equaliser will ensure that the balance condition $U = 0$, $I = 0$ across the input port of this detector is also true, as required, at the network node at the end of this cable remote from the detector.

3.2.3 Single conductors added to an equalised network

The currents flowing in the network of low-impedance coaxial outer conductors of an equalised conductor-pair net cause small but finite potential differences between parts of these conductors. The currents flowing because of these potential differences through the small capacitances between different regions of the outer conductors are usually negligible. An exception arises from the relatively large inter-screen capacitance of an isolating transformer when the output of the network it supplies is detected by a mains-energised detector joined by a common single conductor (the green-yellow safety conductor of the mains supply) to a mains-energised source.

In general, two points on the outer conductors of a coaxial equalised conductor-pair network can be made to be at the same potential if we add an equaliser between

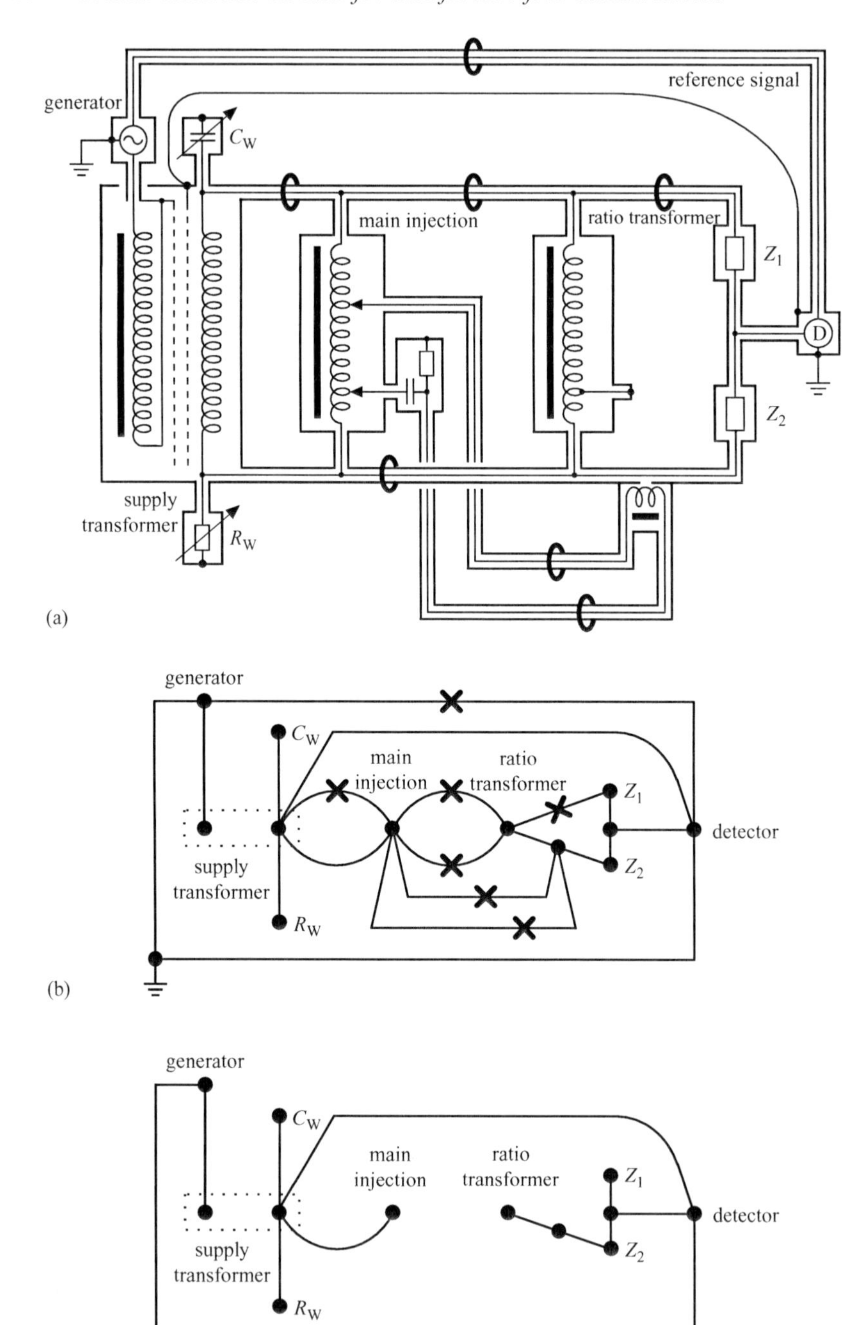

Figure 3.8 (a) The network and (b) the nodes of the circuit of outer conductors. (c) In the network of cable without equalisers

the points and connect them with a single low-impedance conductor. The unbalance current in the cable is diverted into the added conductor, and the equalisation of the currents in the original network is improved. A single unpaired conductor of this kind can be brought into the philosophy of equalised conductor-pair networks by regarding it as being the low-impedance member of an equalised pair whose (non-existent) partner has infinite impedance.

In Figure 3.9, the points A and B have been joined by a single conductor C and an equaliser added to the original mesh.

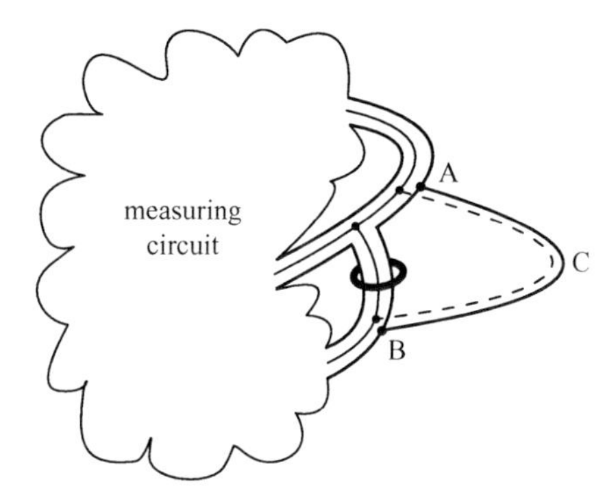

Figure 3.9 A single conductor added to a current-equalised network

3.2.4 Other conductor systems having similar properties

Although the content of this book mainly relates to coaxial circuitry because it enables interference to be eliminated from audio- and higher-frequency measuring circuits, other configurations such as twisted-pair conductors having either separate shields for each conductor or a single shield surrounding both have similar properties if the current in the second conductor is equal and opposite to the first, and there is no significant current in the shield. This configuration can be simpler and more relevant for some applications, particularly DC and low-frequency circuitry. Shielding each conductor separately can eliminate leakage currents between the conductors, and the circuit can usually be arranged in such a way that leakage currents between conductors and their surrounding shields are routed so that they do not affect the measurement (see section 4.1.2, direct impedance).

In Figure 3.10, a tightly and uniformly twisted conductor pair carry equal 'go' and 'return' currents. At a small distance from these conductors, the magnetic fields from each current tends to cancel and at distances greater than the twist length this cancellation is further enhanced by the opposing directions of the field produced by successive twists. The magnetic field is then as small as that produced by a practical coaxial cable. The electric field is dealt with by surrounding the twisted conductors with a continuous shielding foil or close-woven mesh, but the interconnections between the shields of different cables in a network must be arranged

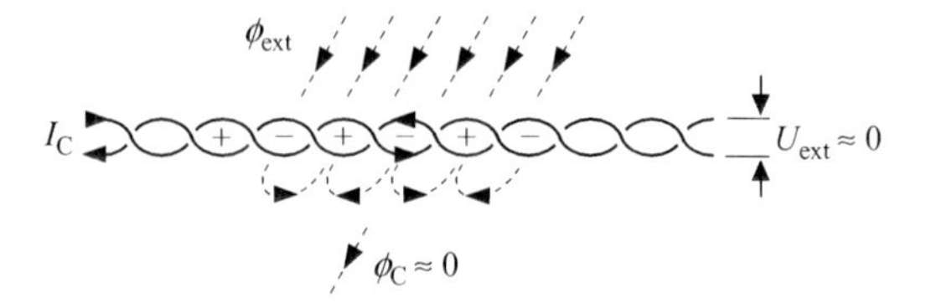

Figure 3.10 The magnetic field surrounding a twisted-pair cable

differently from a coaxial network. The shield of each cable ought to be connected to the rest, using single-conductor extensions if necessary, without forming any loops, as illustrated in the example of Figure 3.11. In practice, many of the shield connections will be made directly to the outer shielding case of an instrument. This shield should be connected to the twisted-pair shields at one point, and the routes taken by capacitive currents from the twisted-pair conductors to the shield should be considered.

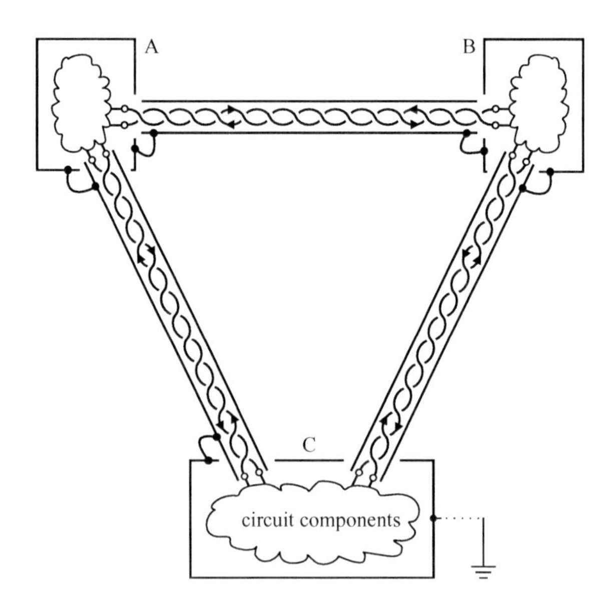

Figure 3.11 Interconnection of the shields of twisted-pair cables

It is fundamental that the 'go' and 'return' currents of a pair must be constrained to be equal for *all* current values and frequencies within the range over which interference is to be eliminated – adjusting equality for one current or frequency value only will not ensure that the circuitry is immune from interference. Equality is automatic for a twisted-pair or a coaxial cable which constitutes the only connection to an isolated instrument, but measures such as symmetrical loads (see section 1.1.8) or current equalisers are needed for loop circuits.

3.2.5 DC networks

It is important to appreciate that there should be no difference in approach when designing either AC or DC measurement circuitry. The principles given earlier for the elimination of interference, isolation and elimination of shunting admittances (e.g. poor insulation) from AC networks apply equally to DC networks. It is not sound practice to rely on extensive filtering to eliminate interference from DC systems because interference can readily be rectified by the many semiconductor junctions in instruments to create systematic errors.

Because current equalisers are not very efficient at frequencies less than the order of 100 Hz, other conductor and shield arrangements are more appropriate to avoid very low-frequency noise being coupled in and disturbing the measurements. The usual solution is to employ overall shielded or individually shielded twisted wire pairs (see section 3.2.4) or bundles, with the circuitry so arranged that the total current in any pair or bundle is zero.

Using individually shielded conductors has the advantage that the effect of shunting insulation leakage resistances can be eliminated in the same way as small capacitances and loss admittances in an AC network. For example, a Wheatstone bridge network designed to compare high resistances of value greater than the order of 10 kΩ, for which MΩ and GΩ leakage resistances are significant, can be provided with a DC Wagner balance auxiliary network (see section 8.1.3).

3.2.6 The effect of a length of cable on a measured value

In some accurate measurements it is necessary to take into account the effect that a connecting coaxial cable might have on the measured value. As a very simple example, suppose a cable connects a DC voltage standard of appreciable internal resistance to a voltmeter that has very high input resistance so that it draws negligible current in order to make the measurement. If the cable has leakage resistance between inner and outer per unit length, a current will be drawn from the voltage standard and will cause a measurement error whose magnitude depends on the length of the cable.

More realistic examples occur in making AC measurements, especially at higher frequencies, because the leakage resistance is augmented by the much greater leakage through the capacitance between inner and outer conductors, and the inductive impedance of the cable is added to the ‘go’ plus ‘return’ series resistance of the inner and outer conductors. It is, therefore, necessary to be able to calculate the effect of cables on currents and voltage differences conveyed from one end of a cable to the other.

In this book, we are predominately concerned with coaxial cables used to make such connections, and so we will not consider any other type of connection, such as, for example, twisted wire pairs, but if the ‘go’ and ‘return’ currents they carry are equalised, the conclusions are the same.

A coaxial cable will have a shunt admittance because of the capacitance and conductance between the inner and outer conductors, and these conductors will also possess finite series impedances because of their resistances and the ‘go’ and ‘return’ self-inductance. Strictly, a cable should be modelled in terms of distributed

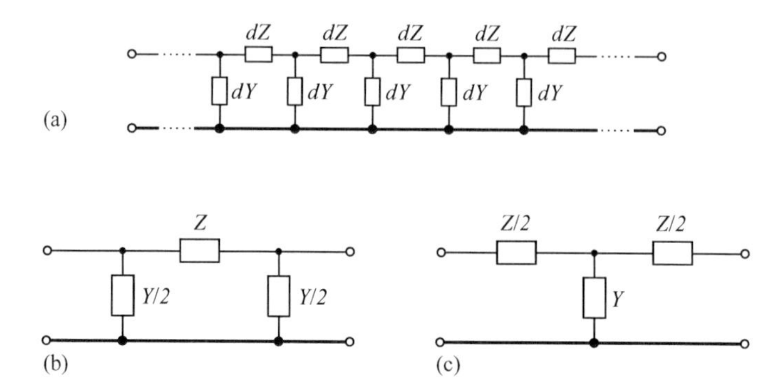

Figure 3.12 The equivalent circuit of a coaxial cable

series impedances and shunt conductances, as shown in Figure 3.12a, but since we are considering only the approximation where the lengths of cable are negligibly short compared with the wavelength of the electromagnetic field in the cable, the distributed admittance and impedance can be considered as 'lumped', that is, as if the admittance and impedance of the cable act at a single point along the cable. Further, we can analyse the effect of cables as if the impedance of the outer conductor is zero and that of the inner is increased to take account of it. This is because, in the circuits we are concerned with, the currents in the inner and outer conductors are constrained to be equal and opposite so that the change of the voltage difference between inner and outer conductors between two points along a cable can be ascribed to an impedance in which the resistive component is the algebraic sum of the resistance of inner and outer conductors. The inductive component of the impedance is the small inductance of the circuit of the 'go' (in the inner) and 'return' (in the outer) currents. Therefore, in this section, the impedance ascribed to the inner conductor is in reality the combined impedance of inner and outer conductors (see section 3.1.2).

By the admittance Y of a cable, we mean the capacitance and shunt conductance measured at either end of the cable with the other end open-circuited. Strictly speaking, to eliminate end effects when measuring Y, a difference measurement should be made with the cable inserted between the measuring instrument and a dummy lead of identical cable, but such end-effects are usually negligible compared to the accuracy with which we need to know Y.

Similarly, by the impedance Z, we mean the resistance and series inductance measured at one end with the other end short-circuited. End effects could again be eliminated, if required, by taking the difference between inserting the cable (not shorted) between the measuring instrument and a dummy shorted length that is measured separately.

We take the circuit shown in Figure 3.12c as the equivalent circuit of a cable for analytical purposes. The alternative of Figure 3.12b is also possible. We shall find that even if we regard the quantity to be measured, for example, the impedance of a standard, as having been redefined with reference to new exit ports at the ends of the added cables, uncertainties still arise because of variable connector admittance and impedance at these and the original ports. The successive elaboration of the definitions of the impedance of a standard described in sections 5.3.2–5.3.8 is largely to minimise this effect.

Two special cases of conditions obtained at one end of a cable with specified conditions at the other are of interest to us.

(i) The relation between the current at one end of a cable and the current at the other when the potential difference is zero. Note, in Figure 3.13, that the network in the right-hand 'cloud' must contain sources and be capable of setting up the specified conditions and the left-hand termination could be a short-circuit, but is not necessarily so.

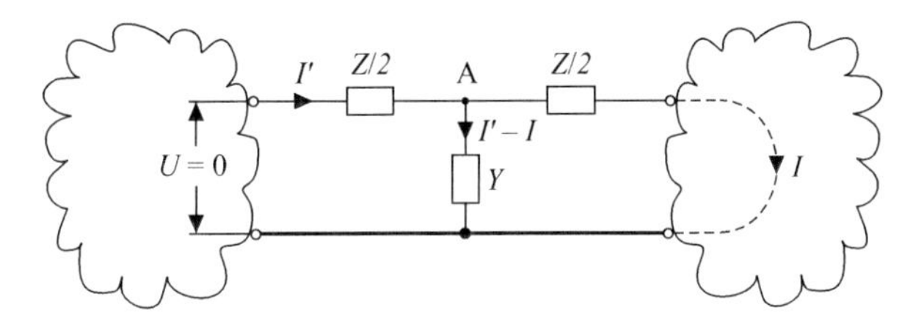

Figure 3.13 The relationship between the currents at the ends of a cable

The sum of the currents at the network junction A is zero, hence the current through Y is $I - I'$. Traversing the left-hand mesh,

$$\frac{I'Z}{2} + \frac{I' - I}{Y} = 0$$

that is,

$$I = I'\left(1 + \frac{YZ}{2}\right) \tag{3.7}$$

(ii) The dual situation of Figure 3.14 illustrates how the voltage at one end of the cable is related to that at the other under conditions of zero current there. Again, as shown in Figure 3.13, the networks in the clouds set up the required conditions; the termination on the right could be an open circuit, but is not necessarily so.

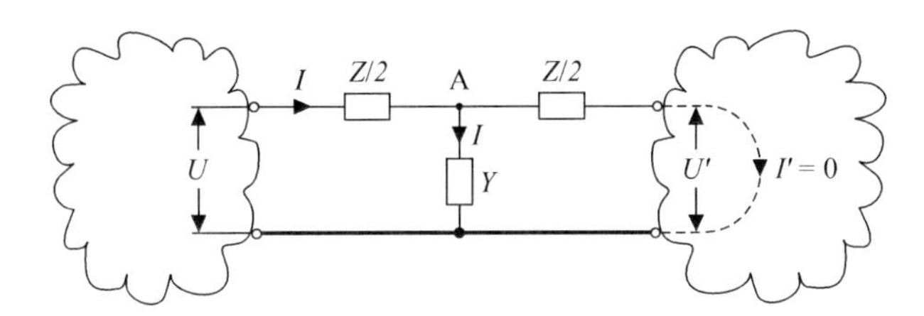

Figure 3.14 The relationship between the voltages at the ends of an open-circuited cable

Traversing the left-hand mesh,

$$U = \frac{IZ}{2} + \frac{I}{Y} = I\left(\frac{Z}{2} + \frac{1}{Y}\right)$$

and the right-hand mesh, $U' = I/Y$, hence

$$U = U'\left(1 + \frac{YZ}{2}\right) \tag{3.8}$$

In contrast to the original paper of Cutkosky [4], our representation of a cable has the impedance Z divided in equal halves at the internal point A, as drawn in Figure 3.12c. This gives rise to the factor of 2 in our expression where Cutkosky has unity. We adopt our representation to simplify the cases when the clouds are connected with two cables in series. The cables have admittances Y_1, Y_2 and impedances Z_1, Z_2, respectively, either because the lengths of similar cables are different or because the cables have different characteristic impedances (see Figure 3.15).

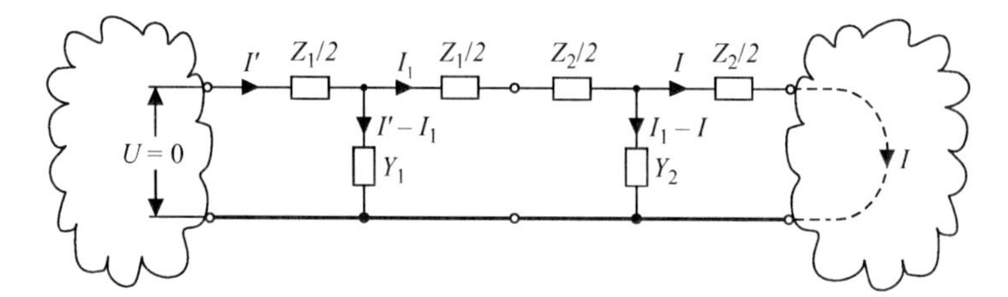

Figure 3.15 The combined effect of two cables in series

Traversing the left-hand mesh,

$$\frac{I'Z_1}{2} + \frac{I' - I_1}{Y_1} = 0$$

and traversing the central mesh,

$$-\frac{I' - I_1}{Y_1} + \frac{I_1(Z_1 + Z_2)}{2} + \frac{I_1 - I}{Y_2} = 0$$

Since $YZ \leq 10^{-6}$ for moderate lengths of the order of a metre of cable at audio frequencies, we may neglect the second-order terms, giving, on eliminating I_1,

$$I = I'\left(1 + \frac{Z_1Y_1 + 2Z_1Y_2 + Z_2Y_2}{2}\right) \tag{3.9}$$

an expression which is equivalent to (3.7) for a single cable.

The correction for potential-transferring cables in series is identical.

If the two cables have the same characteristic impedance, then $Z_1/Y_1 = Z_2/Y_2$, and (3.9) can be written as

$$I = I'\left[1 + \frac{(Z_1 + Z_2)(Y_1 + Y_2)}{2}\right] \tag{3.10}$$

On comparing this result with (3.9) above, as expected, we see that the combined cable behaves as if it had total lumped impedance $Z_1 + Z_2$ and admittance $Y_1 + Y_2$. These total lumped impedances are what, to a sufficient approximation, would be obtained by a direct measurement. Therefore, if two cables having differing inductances and capacitances per metre have to be used, as for instance in making connections inside and outside a cryostat, it is nevertheless a considerable simplification if they have the same characteristic impedance and can be treated as one cable for purposes of calculating a cable correction.

It also follows from the above that the effect of two cables in series, even if they have the same admittance and impedance per unit length, cannot be directly added. Combination is non-linear because the effect of each cable is proportional to the *square* of its length.

3.2.7 Tri-axial cable

The results of the previous section can be modified by replacing a coaxial cable with a tri-axial cable where the intermediate shield conductor can be driven at a remote end B by a voltage source U_{guard} adjusted to raise the potential of the intermediate shield to that of the inner conductor (see Figure 3.16).

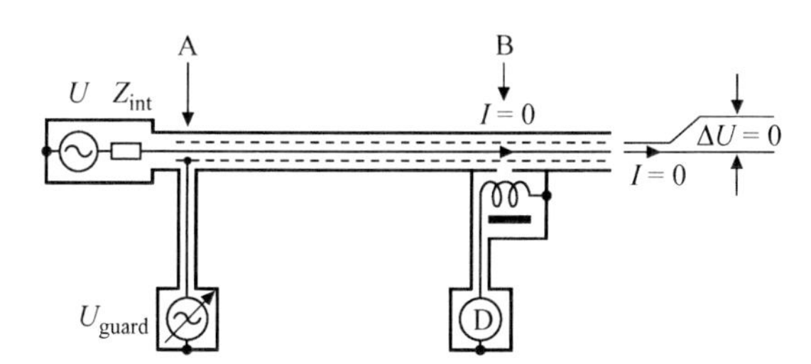

Figure 3.16 A tri-axial cable with the intermediate shield raised to the same potential as the inner conductor

The immediate obvious advantage is that the loading of the cable on a voltage source U of high internal impedance Z_{int} is removed. Furthermore, if the cable is used for remote voltage sensing and U_{guard} finely adjusted to null a current detection transformer (see section 7.3.5) at the remote end B, the cable correction for shunt admittance between inner and intermediate shield acting with the series impedance of the cable is the same as calculated in the previous section, but the sign is reversed. A similar conclusion follows if the cable is used to measure the current *I* in the inner at the remote end when a voltage detector between inner and shield is nulled there. In either case, the effect is to create a defining cable that has moved the plane where a voltage difference or a current is defined from A to a remote point B.

The technique of employing tri-axial cable to make four-terminal-pair AC measurements of quantum Hall impedance has been described by Ricketts [6]. This technique is an alternative to multiple-series connections (see section 6.7.1) but involves extra auxiliary balances and is hence not so elegant. Cabiati and D'Emilio [5] have extended this treatment to multiple-shield cables.

References

1. *A Guide to Measuring Direct and Alternating Current and Voltage Below 1 MHz.* London: The Institute of Measurement and Control; 2003. pp. 79–84
2. Kibble B.P., Robinson I.A. *Guidance on Eliminating Interference from Sensitive Electrical Circuits.* NPL report DES 129. Available from http://publications.npl.co.uk/ (search under publication)
3. Homan D.N. 'Applications of coaxial chokes to AC bridge circuits'. *J. Res. NBS – C.* 1968;**72C**:161–65
4. Cutkosky R.D. 'Four terminal-pair networks as precision admittance and impedance standards'. *Trans. IEEE Commun. Electron.* 1964;**83**:19–22
5. Cabiati F., D'Emilio S. 'Low frequency transmission errors in multi-coaxial cables and four-port admittance standard definition'. *Alta Freq.* 1975;**44**:609–16
6. Ricketts B.W., Fiander J.R., Johnson H.L., Small G.W. *IEEE Trans. Instrum. Meas.* 2003;**52**(2):579–83

Chapter 4
Impedance measurement

4.1 Improvements in defining what is to be observed or measured

The standards compared by coaxial networks must be provided with terminal pairs. The physical form of a terminal pair is a coaxial connector. Its inner contact is the terminal connected to the component, and the outer contact is the terminal connected to its enclosing conducting case. A coaxial connector, which is a terminal pair, is often called a 'port'.

Unfortunately, the majority of low-frequency reactance standards in laboratories, and which are still being manufactured, are not provided with coaxial connections. It is usually a simple matter, however, to replace the existing terminals with coaxial connectors and, where necessary, to provide a new outer conducting container.

It is very much the purpose of this book to encourage coaxial techniques within the formalism of coaxial networks, to show that these techniques have the great advantage of obtaining accurate and certain results, and to show how to implement each concept with practical, constructable apparatus.

The bridge networks with which we are concerned will therefore consist of impedances provided with, and defined in terms of, terminal-pair coaxial terminals and will be connected with conductor-pair coaxial cables. Equalising the current in the inner conductors and screening outer conductors of cables so that the currents are equal and opposite will ensure that the cables have negligible external fields (see section 1.1.1), and therefore, the cables do not interact with each other so that their routing is immaterial. The network will also have little response to external interference.

In general, the techniques we describe are appropriate for frequencies from 10 Hz to at least 1 MHz. Highest accuracy is attained for frequencies from about 400 Hz to 2 kHz.

At first sight, passive conductor-pair networks seem unrelated to the design of electronic circuits and equipment, but because of its emphasis on a complete understanding of circuits and their interactions, AC bridges are a valuable subject to study. The equalised conductor-pair concept is also a powerful design tool for electronic circuits, although somewhat different means such as the use of differential input amplifiers are employed to achieve equalisation or minimisation of loop currents. The role of power supply conductors and their associated current return conductor ('earth' or 'ground') to and from a packaged circuit needs careful consideration in relation to the circuitry as a whole.

Magnetically operated switches (relays) are often required in the inner circuit of a coaxial system, for example, in designing a system for automated operation. The switch contacts of these should be properly isolated by screens from the energising coil, but inexpensive commercial components having this property are rare.

4.1.1 Ratio devices

Ratio devices which are voltage or current ratio transformers, are so constructed that almost all of the magnetic flux threading one winding also threads another. Their use produces more accurate and convenient bridge networks than the four-arm bridges of the older literature. Good flux linkage is accomplished by an appropriate geometry of the windings, which are usually wound around a high-permeability toroidal magnetic core. Then the ratio of induced voltages or currents in the two windings can be to within a part in a million or so, simply the ratio of the number of turns of the windings. The small departures from the nominal turns ratio can be measured and allowed for in high accuracy work. Ratio devices should be subject to the same kind of precise electrical defining conditions as impedances (see section 5.3). The zero-voltage port should have no current flowing through a short, that connects the tap to the circuit of the outer conductors; that is, $I = 0$ and $U = 0$ at this port. Other ports that have defined voltages across them should have zero current flow through these ports. Ways of fulfilling these defining conditions will be discussed in section 8.1.

4.1.2 Impedance standards

We have seen in section 1.1.1 the need to construct standards that have no significant electric or magnetic fields outside their containers. They must be provided with terminal-pairs which are coaxial connectors; the inner connector of a coaxial connector is one terminal, and the outer connector the other terminal, of a pair. The value of the standard is strictly defined by voltage differences across one or more of the terminal pairs, and a current flow out of the inner terminal of another terminal pair, with an equal as possible current flowing back through the outer terminal. These specified defining conditions may be departed from, provided the departure can be shown to have no effect on the measured value. For work of the highest accuracy, the conditions at a single terminal pair may not be sufficiently reproducible, chiefly because of variations in contact resistance at the mating connectors. It is then necessary to establish *internal defining points* in the interior of the device and to provide two additional terminal-pairs which are coaxial connections. The resulting four terminal pairs are analogous to the four terminals of the familiar four-terminal DC resistor. Other definitions of standards are encountered, but they can be viewed as degenerate examples of the two cases above.

Just as with low-valued resistors measured with DC, where it is usually unsatisfactory for a standard to be provided with only two terminals because of uncertain contact resistance, and separate current and potential terminals are provided to make a four-terminal resistance standard, a standard having four terminal pairs or ports is required for AC work of the highest accuracy. The electrical conditions at each terminal pair (such as the voltage between inner and outer contacts when zero current flows through either) must be precisely defined. But just as a two-terminal construction may be adequate for DC high-valued resistors, a simpler two-terminal-pair standard may serve for some AC measurements.

The concept of a *direct* impedance or admittance is very important. In general, a component of admittance $1/Z$ will have admittances Y_1 and Y_2 from both its 'high' side and its 'low' side, respectively, to the surroundings or to a current return conductor, which, in the present context, is the outer shielding conductor of its container and of coaxial cables connected to it. These admittances and the currents flowing through them are distinct from the admittance of the component itself and the current flowing directly through it. If I_1 and I_2 flow to the same conductor, all three admittances form a Δ network, as shown in Figure 4.1 (see Appendix 1). We are often concerned only with the ratio of the current I flowing directly through the component to the voltage U across it, and this ratio is termed the 'direct admittance' of the component. Its reciprocal is the 'direct impedance' of the component. A network to measure the direct impedance Z must have a means of ignoring the shunt currents I_1 and I_2. For example, if the left side of Z is supplied from a voltage source of negligible output impedance, the extra current I_1 drawn from it will not significantly affect its output voltage. If the right side of Z is connected to a nulled detector, there will be no significant voltage across Y_2, and consequently, I_2 will be zero. Often, to obtain an equivalent situation in practice, auxiliary networks or balances are employed, which remove the effects of I_1 and I_2.

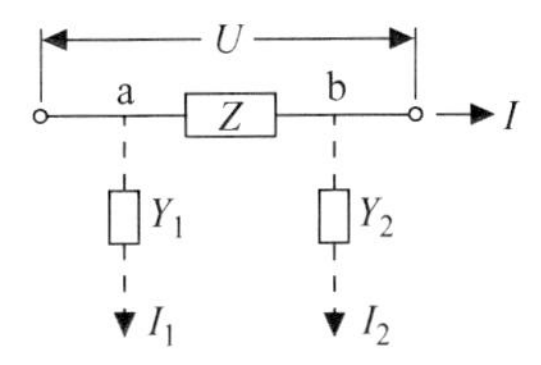

Figure 4.1 The direct admittance or impedance of a component

Section 5.3 is devoted to a careful treatment of the electrical definition of the values of standards and ratio transformers, and we must emphasise that this is most certainly essential. Once the defining conditions of standards are settled, the design of proper comparison networks will follow naturally. Neglecting the definition stage will, as in any other accurate measurement, inevitably lead to failure.

Standards can be represented by an equivalent circuit of lumped inductances, capacitances and resistances connected into a network. An equivalent network is such that its behaviour cannot be distinguished from the behaviour of the actual device by making any kind of measurement at the terminals of the device over the relevant frequency range.

Another minor advance in technique is that in most cases, the variable or decade switched impedances used in older bridge technology can be replaced with advantage by components of fixed value connected to a variable voltage generated by a multi-decade ratio transformer called an inductive voltage divider (IVD).

It is often useful to shunt an impedance with one of a much higher value so that the impedance of the combination has exactly the required value. This is straightforward if extra connections are made to the internal defining points, as shown in Figure 4.2, but unfortunately, standards are not usually constructed in this way.

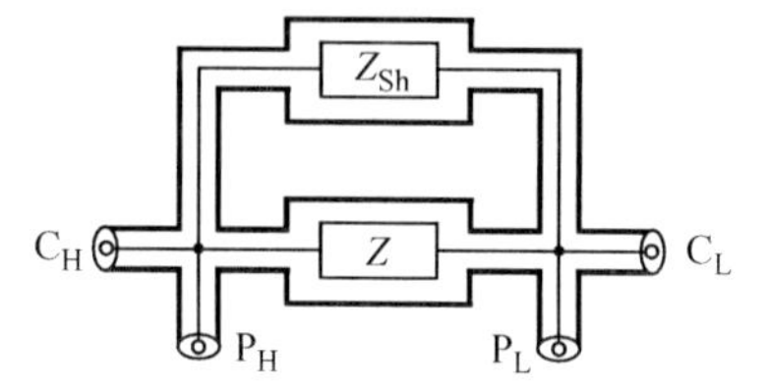

Figure 4.2 Shunting a four-terminal resistor via extra connections to its internal defining points

To shunt accurately in the absence of these extra connections is not quite straightforward, as we will illustrate by considering shunting the four-terminal DC resistor shown in Figure 4.3.

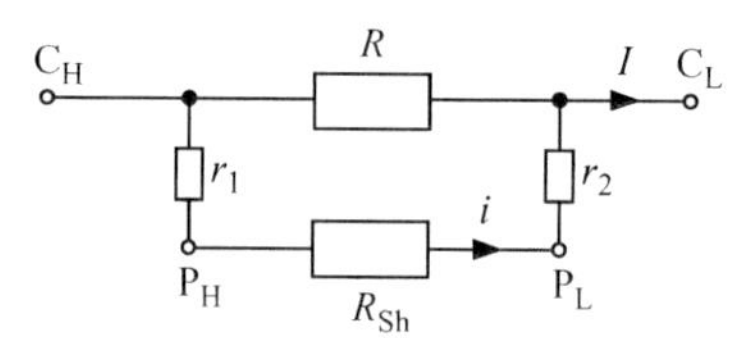

Figure 4.3 Shunting a four-terminal resistor ($R_{Sh} \gg R$)

Because a small current flows through the shunting resistor, the defining condition of zero current through the potential terminals P of the resistor being shunted is not met. The voltage between the potential terminals is decreased from the value

$$U = I\left(\frac{1}{R} + \frac{1}{R_{\text{Sh}} + r_1 + r_2}\right)^{-1} \tag{4.1}$$

(where I is the total current through the combination) which would be measured between the internal defining points to

$$U - i(r_1 + r_2) = U - \frac{U(r_1 + r_2)}{R_{\text{Sh}} + r_1 + r_2} \approx U\left(1 - \frac{r_1 + r_2}{R_{\text{Sh}}}\right) \tag{4.2}$$

between the potential terminals, where r_1 and r_2 are the resistances of the high- and low-potential leads, respectively, and i is the small current I flowing through the shunt. That is, the measured value of the shunted resistance is less than the value that would be found between the internal defining points, which is correctly calculated from (4.1), by approximately the fraction $(r_1 + r_2)/R_{\text{Sh}}$.

A similar decrease occurs if R_{Sh} is connected across the current terminals instead.

The same effect occurs if a four-terminal-pair impedance is shunted, and this is discussed in section 8.1.6.

The problem is avoided if the impedance has extra connections to the internal defining points, as shown in Figure 4.2.

Very high value impedances can be made from more convenient lower-valued ones by connecting them as a T-network as shown in Figure 4.4. Their direct impedance Z_D (section 4.1.2) can be calculated from a T-Δ transformation (Appendix 1).

$$Z_D = Z_1 + Z_3 + \frac{Z_1 Z_3}{Z_2} \tag{4.3}$$

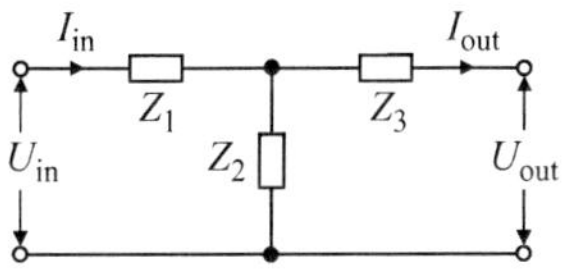

Figure 4.4 A direct very high impedance constructed from a T-network of lower-value impedances

Typical applications are the production of low-valued capacitances when Z_1, Z_2 and Z_3 are all moderate-valued capacitors, and high-valued resistors when they are all moderate-valued resistors (see section 6.4.10).

The operation of the network can be understood by viewing Z_1 and Z_2 as forming a potential divider to supply a lower voltage to Z_3 so that the current leaving Z_3 is attenuated. The direct impedance is the ratio of the voltage across $Z_1 + Z_2$ to this current and is consequently increased.

4.1.3 *Formal representation of circuit diagrams and components*

We, here, introduce the formal representation we use for circuit elements. The formal representations of some particular devices are introduced when these devices are described as they occur throughout the book.

A coaxial cable is represented as shown in Figure 4.5b. Thicker lines denote the outer conductors of cables and suggest their lower resistance and impedance. The cables connect at junctions with other cables and components by mating coaxial plugs and sockets. A junction, which is pictorially represented in Figure 4.5a, will be formally drawn as in Figure 4.5b. The circles emphasise that the mating contacts of *both* inner and outer conductors constitute terminals in the same sense as the single screw terminals of older bridge practice. In choosing a particular kind of coaxial connector, it is important that their mating surfaces make a reliable low-resistance contact.

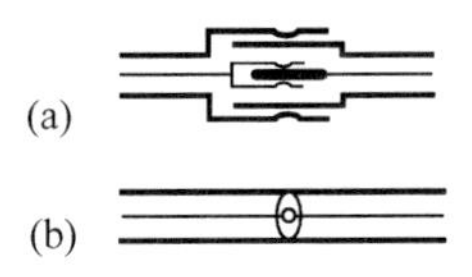

Figure 4.5 (a) A typical coaxial connector. (b) Its formal representation

In general, the symbol ◦ represents a single-conductor terminal or junction point to which additional external connections can be made, and the symbol • represents a junction point, often internal to a device, to which additional connections cannot be made.

Since the frequencies we are concerned with are less than radiofrequencies, impedance matching of cables and connectors is unimportant.

An immittance, represented diagrammatically as the first diagram in Figure 4.9d, may be either

(i) an impedance (Z) having reactive ($j\omega X$) and resistive (R) components or
(ii) an admittance (Y) having susceptive ($j\omega B$) and conductive (G) components.

Where the details of a general network are not of concern, the network will be represented as shown in Figure 4.6 as a cloud having one or more coaxial connections.

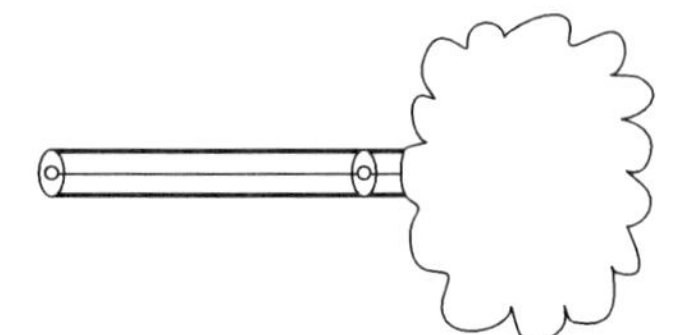

Figure 4.6 A network represented as a 'cloud'

The familiar symbols of electrical circuit theory represent resistors, capacitors and inductors, which comprise the lumped equivalent circuit representation of actual physical components; they will, in general, be surrounded by a screening and conducting case to which the outer connectors of coaxial cables are joined, and so a typical two-terminal-pair component can be represented as shown in Figure 5.12a.

There is perhaps an implication that the component and its enclosing shielding box also have coaxial geometry; this would be so in an ideal world, but the limitations of actual practical manufacture do not often permit it. The external fields generated by a non-coaxial geometry of a component can be eliminated if necessary by enclosing it in a conducting magnetic shield.

Conducting surfaces forming electrical screens are represented by broken lines.

Where single conductors are necessary, they will be represented as curved flowing lines with a loop to suggest their unpaired and unbalanced nature in that currents in them give rise to surrounding magnetic fields.

Transformers need special attention, both because they are principal components in most of the networks described and because their constructional topology is crucial to their accuracy. All of the transformers we shall be considering have a toroidal core made from high-permeability material. A toroidal form is preferred because of its circular symmetry so that if it is excited by a uniform winding, very little magnetic field exists external to the wound toroid. Some provision should be made for cancelling the effect of the advancing winding, which is equivalent to a single annular turn around the toroid and can be approximately cancelled by a returning turn as shown in Figure 7.17. See section 7.2 for actual constructional techniques.

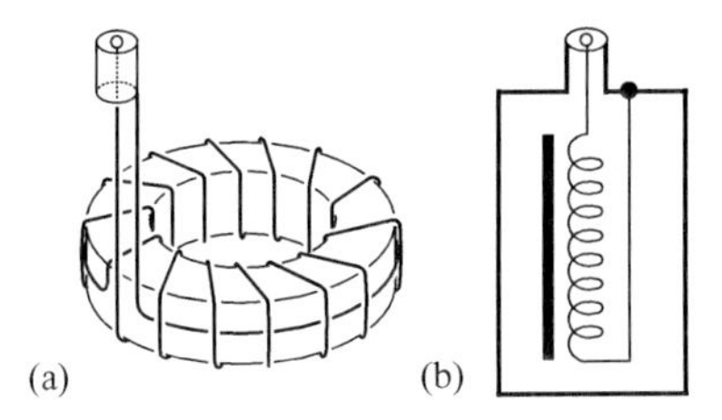

Figure 4.7 (a) A wound toroid. (b) Diagrammatic representation of a wound toroid

The windings of the transformer have to thread through the central space as each turn is made, as shown in Figure 4.7a. To show this diagrammatically, as shown in Figure 4.7b, the following convention will be used. First, on the left is the representation of the core as a heavy solid line; next to this is one winding shown as a spiral, and if this is completed by making a reverse annular turn next to the toroid as shown in Figure 4.7a, its return is shown as a straight line parallel to the core. The ends of the winding terminate on the inner and outer of a coaxial socket. In the more elaborate example drawn in Figure 4.8, there is an electrical screen over the first winding, shown as a broken line. This is followed by a second winding, and a final overall electrical screening enclosure is shown. It is to be understood that a given winding or screen threads all the windings or cores drawn to the left of it.

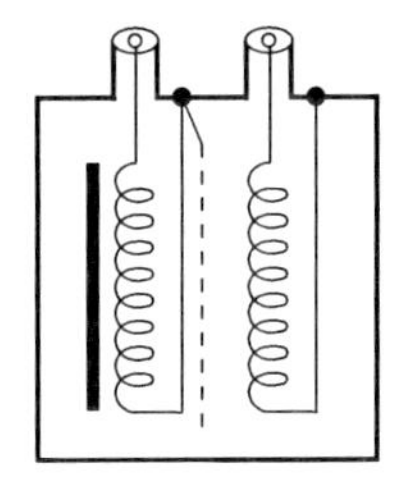

Figure 4.8 A toroid with two windings and a toroidal screen between them

Other specialised or more elaborate transformers will be depicted similarly – the order in which the component parts are shown in the diagram follows the order in which the transformer is assembled, starting with the core and working outwards through the windings and shields.

Windings applied simultaneously side by side or, for example, as twisted conductors are drawn above one another as shown in Figure 7.24.

Sometimes it is convenient to draw an assembly from right to left or from top to bottom or from bottom to top instead of from left to right, as given earlier, and to leave the context to clarify the intended construction.

We will always represent real bridge sources as voltage sources (i.e. sources of negligible series internal impedance) and add the lumped representation of the internal impedance of an actual source, if of any importance to the network in which it occurs. This is not usually the case, and so the internal impedance will then be omitted. Thus, an oscillator and its output coaxial terminations will be drawn as shown in Figure 4.9h. We will initially ignore the extra complication brought about

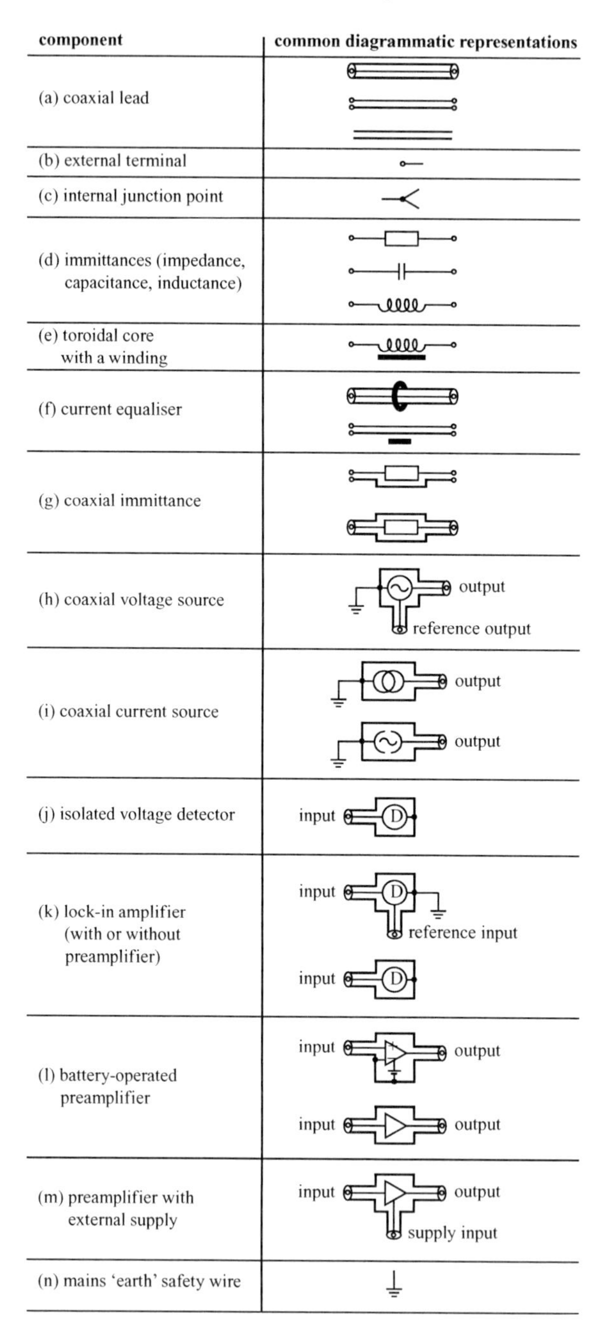

Figure 4.9 Diagrammatic representations of components

by connections to the mains supply, which are either direct or indirect via the inter-winding capacitance of supply transformers in sources or detectors. Isolation can actually be accomplished reasonably well by proper design of a screened transformer (see section 1.1.5). We discuss the attainment of isolation in sections 1.1.4–1.1.10, but for the present, it will be convenient to assume its accomplishment. For example, we assume that all active components (generators, detectors, amplifiers, etc.) are powered by batteries wholly contained within the screening case of the instrument.

In Figure 4.9, we tabulate the principal diagrammatic representations of the components occurring in other diagrams in this book. For want of a better symbol, we have represented a mains safety conductor as shown in Figure 4.9h, but this is the only meaning of this symbol. It is *not* to be interpreted as, for example, a connection to a chassis, an instrument case, a water tap or a 'ground plane'.

The source and detector for a bridge network will consist of an isolated box provided with a coaxial input connector. An isolated detector is represented as in Figure 4.9j. Information on using a non-isolated detector will be found in section 1.1.2.

We end with two particularly common components occurring in the construction of impedance-measuring networks ('bridges'). Figure 4.10a shows how the important isolation (see section 1.1.5) of a source represented diagrammatically as shown in Figure 4.10b can be accomplished in practice. Figure 4.11 illustrates the functions of in-line injection/detection transformers (see section 7.3.3), which can either inject a small voltage or detect the presence of a current in an inner conductor.

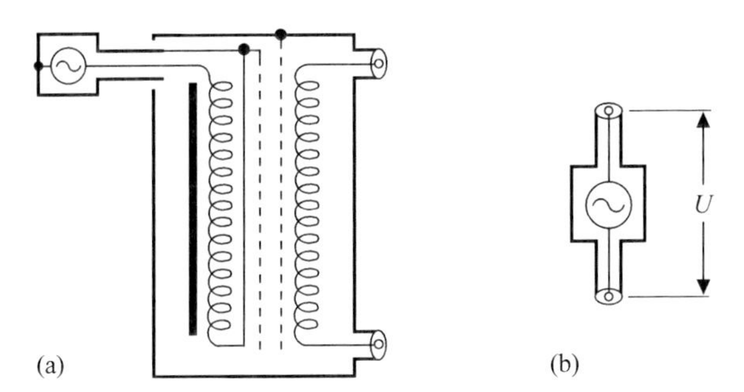

Figure 4.10 (a) An actual construction of an isolated source. (b) Its schematic representation

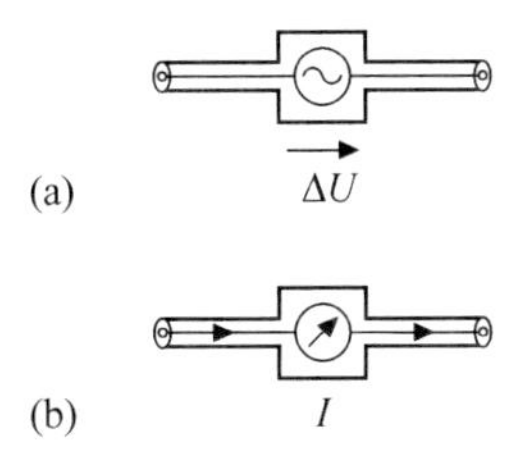

Figure 4.11 (a) A voltage ΔU injected into an inner conductor. (b) Detection of a current I in an inner conductor

Chapter 5
General principles of accurate impedance measurement

We now begin our discussion of balanced current coaxial techniques as applied to impedance measurement with AC bridges. As mentioned in the introduction, this provides a good example of how these techniques can be applied in general to sensitive electrical measurements of other parameters – voltage, current, power and so forth.

Before moving on, it is as well to note the advantage of a substitution measurement in comparing two nearly equal impedances, and indeed in metrology in general. We can make a formal statement of the philosophy of a substitution measurement as follows.

If two devices to be compared are so similar that the measuring system does not respond to minor differences in their construction, then the small difference in their values can be measured by substitution without incurring any systematic (type B) error arising from the measuring system.

For example, if two weights X and Y of similar value are compared by placing each in turn on the right-hand pan of a balance, which has a third, similar weight on the left-hand pan, by adding small makeweights to the right-hand pan to achieve the two separate balances for X and Y, the difference in the values of the makeweights will reflect the difference between X and Y irrespective of whether the balance arms are of equal length or whether the balance indicates zero with no weight on either pan.

In the present context, no significant type B (systematic) error will be incurred for a substitution measurement of two similar direct impedances Z_1 and Z_2 if the measuring system ('bridge') is insensitive to any difference between the shunt impedances at their terminal ports.

5.1 The evolution of a coaxial bridge

We begin with a discussion of the four-arm Wheatstone bridge and show how, after it has been converted to a coaxial circuit, the measurement can be made to respond only to the direct admittances or impedances (see section 5.3.2) of the four components in the bridge arms.

The reader will probably be familiar with the Wheatstone's bridge network generalised for AC comparison of two-terminal impedances, as drawn in Figure 5.1. We assume an isolated source (see section 1.1.5) and an isolated detector (e.g. one

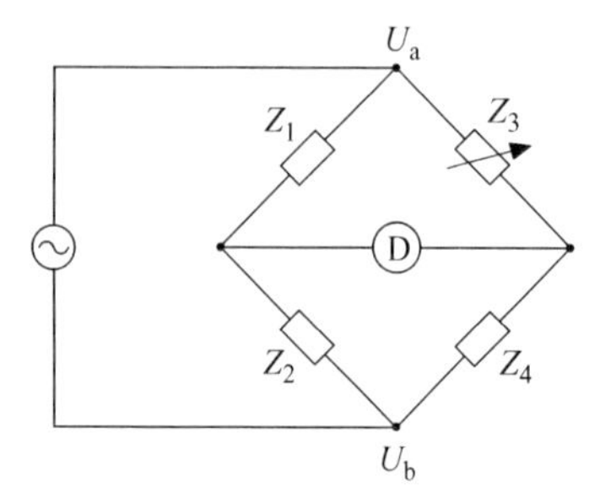

Figure 5.1 A single-conductor AC Wheatstone's bridge

powered by batteries contained within its complete screening case) in order to facilitate the following discussion.

The balance of this bridge made by adjusting Z_3 will be more or less affected by admittances (usually stray capacitances) between each corner of the bridge and its surroundings, and measurements made with it will be erroneous.

These admittances can at least be made much more definite by enclosing the components and the conductors in a complete screen, as shown in Figure 5.2.

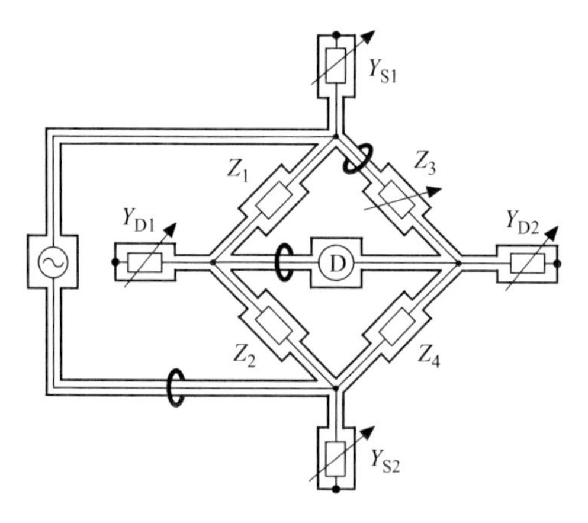

Figure 5.2 A completely screened single-conductor AC Wheatstone's bridge

If the network were to be physically arranged somewhat as in the diagram, the mutual inductive coupling (see section 1.1.1) between its three meshes could cause very significant error, particularly if the impedances involved are small. The circuit layout can be arranged so that conductors carrying equal and opposite currents are adjacent, to minimise inductive coupling, but at the expense of increased capacitive coupling.

Alternatively, current equalisers added to each mesh as shown will ensure equalised currents in the screen, so that the network is properly coaxial, and the physical layout should be immaterial.

The network will now be immune to external interfering electric and magnetic fields, but probably the admittances (in practice, lossy capacitances) Y_{S1}, Y_{S2}, Y_{D1} and Y_{D2} from the corners of the bridge will be increased. Y_{S1}, Y_{S2}, Y_{D1} and Y_{D2} and the cable capacitances are the adjustable circuit components rather than indefinite stray capacitances.

To facilitate discussion, the network has been redrawn as shown in Figure 5.3, where the extreme right-hand and lower lines represent the outer conductors of the cables and components. Y_{S1}, Y_{S2}, Y_{D1} and Y_{D2} will, in general, cause large errors, but their values can be adjusted by the following procedure so that they cause no significant error.

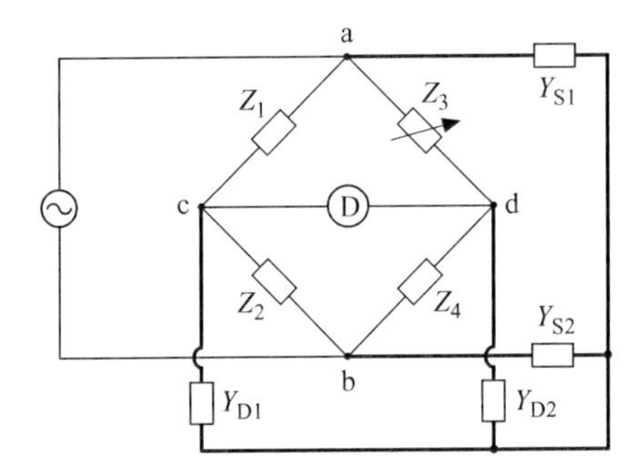

Figure 5.3 A simplified line drawing of a completely screened Wheatstone bridge

The main balance having been made by adjusting Z_3 to null the detector D, the bridge nodes c and d can be brought to the same potential as the outer conductors by adjusting the ratio Y_{S1}/Y_{S2} to null D with Y_{D1} temporarily shorted. It will probably be necessary to iterate the main balance with this auxiliary balance until both are simultaneously obtained. When this is true, Y_{D1} and Y_{D2} have no voltage across them and so do not affect the main balance whatever their value. The adjustment of the Y_{S1}/Y_{S2} ratio is called a Wagner balance in bridge literature, and there are many examples later in this book.

If now the Wagner balance is temporarily misadjusted by an amount that can later be reset, c and d will no longer be at the same potential as the screen, but they can, nevertheless, be set to be at the same potential as each other by adjusting the ratio Y_{D1}/Y_{D2} until D is again nulled. This adjustment can be termed a detector network auxiliary balance. The bridge balance is now immune to the values of Y_{S1} and Y_{S2}. If their temporary misadjustment is now restored, the bridge balance is immune to *both* auxiliary balances to an extent that is the *product* of the precision with which either auxiliary balance has been made. For example, if each has been made with a precision such that they individually do not affect the main balance by more than 1 in 10^3, their combination will not affect the main balance by more than 1 in 10^6. Also, Y_{D1}, Y_{D2}, Y_{S1} and Y_{S2} need only 1 in 10^3 stability in contrast to the 1 in 10^6 stability of the principle impedances Z_1, Z_2, Z_3 and Z_4. An example of the application of this philosophy in slightly different circumstances can be found in the quadrature bridge described in sections 9.2.1–9.2.3.

This simple four-arm bridge still has practical application for impedance measurement at higher frequencies at which it may be difficult to make satisfactory ratio transformers.

More often in the bridge designs described later in this book, the components Z_1 and Z_2 are replaced by the outputs of the windings of a voltage ratio transformer. Because the ratio of the output voltages is very little affected by loading the transformer outputs with shunt admittance, a detector network auxiliary balance is unnecessary. The node c is connected to the screen by a short-circuit, which can be temporarily removed, and Wagner components Y_{S1} and Y_{S2} can be adjusted so that the main bridge balance is unchanged whether the short is in place or not. When this condition is achieved, there will be no current through the short and no voltage across it, and this constitutes one of the defining conditions of the voltage ratio transformer. In the following section, we illustrate this principle by applying it in a simple bridge.

5.1.1 A simple coaxial bridge as an example of a coaxial network

Figure 5.4 is a partly pictorial section of a simple coaxial capacitance comparison bridge. It will enable the reader to appreciate the construction of a toroidal transformer having two screens between the windings. Figure 5.5 shows the schematic representation of the same bridge. The transformer construction (see section 1.1.5)

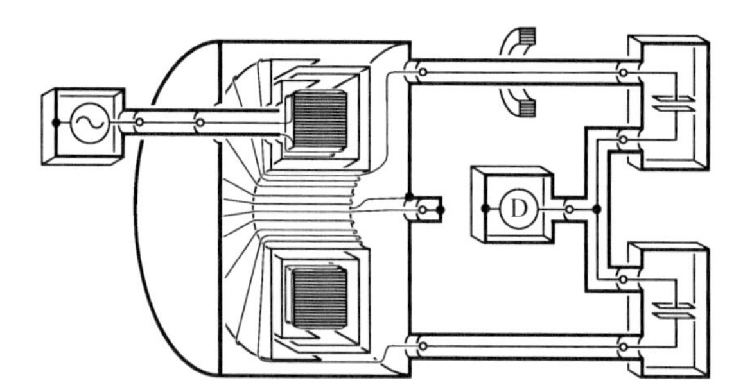

Figure 5.4 A pictorial representation of a simple capacitance ratio bridge based on a voltage ratio transformer

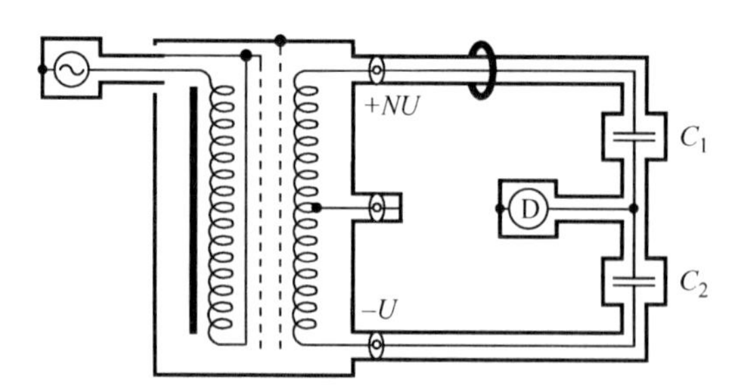

Figure 5.5 A diagrammatic representation of the bridge represented in Figure 5.4

ensures isolation between its primary and secondary windings, and transformers of this kind are used to isolate a commercial source from a measuring network.

Very accurate impedance comparison with a bridge does not need separate measurement of voltage and current; it is sufficient that the same current flows through both the devices to be compared. This is so in this simple coaxial bridge if the detector is nulled, that is, it draws zero current from the mesh. The ratio of the impedances is then the ratio of the voltages across them, and precise voltage ratios can be obtained from a properly constructed transformer. In the older bridge literature, the voltage ratio is provided by the potential drops across the known ratio of two other impedances in series, and Figure 5.6 shows how this bridge relates to the four-arm Wheatstone's bridge discussed in the preceding section.

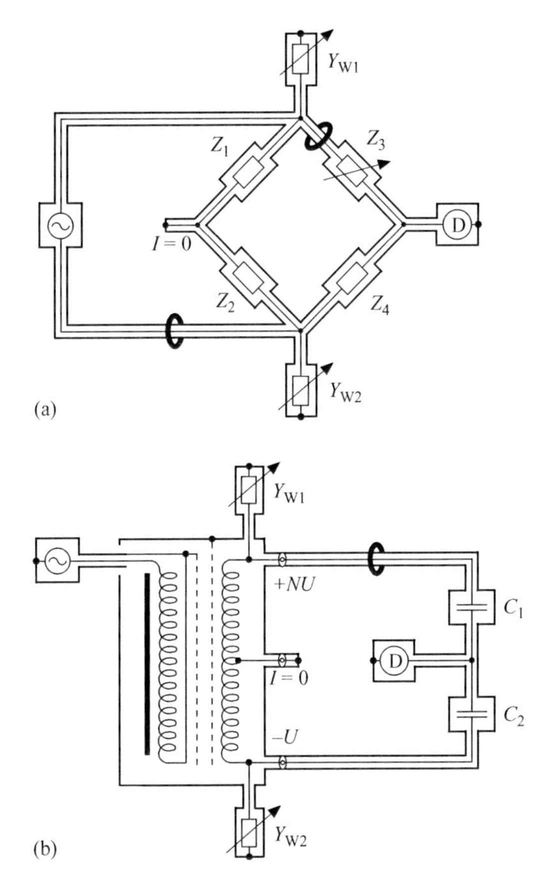

Figure 5.6 The relationship between (a) a four-arm bridge and (b) one based on a N:1 voltage ratio transformer

A properly designed voltage ratio transformer has a very small mutual output impedance of one winding with respect to the other. That is, if one output has its voltage altered, for example, by a shunt load, the voltage across the other winding will, to a very good approximation if it is constructed to have only a small leakage inductance (see section 7.1.3), also alters so that the ratio of its output voltages is unaltered. This ensures that the adjustment of the Wagner components Y_{W1} and Y_{W2}, which have been added in Figure 5.6b, is less critical.

As we noted earlier, it is one of the defining conditions of this kind of voltage ratio transformer that no significant current should flow through the short between the tap of the windings and the outer circuit. In the previous section, we saw how this can be achieved by temporarily removing this short and adjusting the Wagner components to restore the bridge balance. Alternatively, the detector can temporarily be connected in place of the short and nulled.

5.2 The validity of lumped component representations

The flow of electrical energy driven by an alternating voltage through an actual electrical component can be represented by considering the component to be a network of some or all of the idealised components of inductance L (in which the energy is associated only with a magnetic field), of capacitance C (in which electrical energy only exists in the electric field) and of resistance R (in which the energy is dissipated entirely as heat and there are no associated electric or magnetic fields).

In an actual component, all three processes take place together within the device and in the surrounding space. Consider a coil of insulated wire carrying a current. Each short section of the wire will be surrounded by a magnetic field so that it possesses inductance, will dissipate energy through its resistance and, as it is at a different potential from its neighbouring sections, there will be electric fields within and in the vicinity of the coil.

At sufficiently low frequencies, the magnetic fields from successive sections will have the same phase, and the total field can therefore be treated as a single entity as if it arose from an ideal inductance L. The effect of electric fields can be modelled in a similar way as an ideal capacitance C. The total resistive dissipation of the device can also be treated collectively as an ideal resistance in series with the inductance in the network of ideal components drawn in Figure 5.7. This equivalent circuit then behaves *in exactly the same way as the actual component* up to frequencies higher than the self-resonant frequency $\omega_r = (LC)^{-1/2}$ because the self-resonant frequencies of the individual sections of the coil are even higher, and the electric and magnetic fields from each add to give the total field of the device in a way that still produces a constant mean phase throughout the volume of the coil.

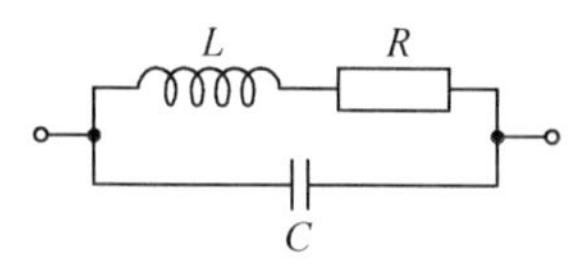

Figure 5.7 An equivalent circuit for a resistor

Thus, representing the actual behaviour of the coil by lumped component parameters is valid provided the wavelength of the oscillatory magnetic and electric fields associated with the coil is long compared with its dimensions. The values of the parameters may be frequency dependent.

If two devices are separated so that their electric and magnetic fields do not share a common region of space, their combined behaviour, when connected together, is represented by the connection of their respective ideal networks. This is not the case if their fields partially overlap. For example, two equal inductors of value L, when connected in series, have a combined inductance $2L$ when well separated, but if they are brought together, their combined inductance approaches one of the theoretical limits $4L$ or 0, depending on the sense of connection. In general, therefore, it is the spatial distribution and interaction of the electric and magnetic fields of a device with its conductors that should be considered when deciding whether a particular lumped parameter representation is appropriate.

5.3 General principles applying to all impedance standards

5.3.1 The physical definition of a standard

The value of capacitance C or inductance L is defined by the energy $\varepsilon = CU^2/2$ or $LI^2/2$ stored in its electric field E or magnetic field H created by the voltage U or current I, respectively. These energies can be expressed as integrals over a volume V in terms of the fields as $1/2\int \varepsilon\varepsilon_0 E^2 dV$ or $1/2\int \mu\mu_0 H^2 dV$. It, therefore, follows that any neighbouring object that modifies these electric or magnetic fields will alter the value of the capacitance or inductance. A good capacitance or inductance standard will be so constructed that no appreciable modification can occur. By the reciprocity principle explained in section 1.1.1, this will be so if the component is constructed in such a way that it itself has no appreciable external fields. The result of a satisfactory construction could be termed the *physical definition* of a standard.

Ideally, capacitors should be completely enclosed within a conducting screen, which is continuous with the outer conductors of coaxial terminals. The capacitor should be connected to the inner conductors so that it constitutes a two- or four-terminal-pair component (see sections 5.3.6 and 5.3.8). Its magnetic flux generated by these connections and the displacement current within the capacitor is usually negligible.

Inductors should be toroidally wound by one of the schemes of section 7.2.1, and, since electric fields emanating from inductors of millihenry values or greater are likely to be significant, they should be totally enclosed in a conducting screen with coaxial terminals in the same way as a capacitor.

A standard of resistance of value R should only dissipate energy at a rate $W = I^2R$; the geometry of its conductors should be such that when carrying an alternating current of frequency f, the stored energy per cycle of its electric or magnetic fields $(LI^2 + CU^2)/2$ is small by comparison with the energy I^2R/f dissipated in the resistance. Like standards of capacitance or inductance, it should also be completely enclosed in a conducting screen and have coaxial terminals.

Actual construction of commercially produced standards, so far, seldom even approach these ideals, and so the manner in which interactions with external fields are minimised during measurement needs to be clearly specified.

5.3.2 The electrical definition of a standard impedance

The considerations of this section apply equally to both AC and DC measurements. For DC measurements, replace 'impedance' and 'admittance' by 'resistance' and 'conductance', respectively.

The electrical conductors of a measurement network are connected to the terminals of standards. We will call 'the *electrical definition* of a standard impedance', the impedance defined as the ratio of the potential that exists between two designated terminals to the current that flows through one designated terminal. The admittance of a standard is the reciprocal of this ratio.

When an impedance is provided with separate potential- and current-measuring terminals, the junctions at which conductors from the impedance divide to go separately to the potential and current terminals are called 'internal defining points'. The junctions are not strictly points, being blobs of metal, but if current or voltage measurements are made remotely from them at terminals along conductors at a distance much greater than the cross-section of the conductors, the blobs behave as linear networks and the current and potential paths act as if they meet at an actual point within them. The reciprocity principle ensures that the location of these points remains the same if the roles of current and potential conductors are interchanged. The defined impedance is that which would be found by *direct measurement between these points* if they could be accessed. The purpose of the definitions below is to reproduce as closely as possible this defined impedance by measurements actually made at the remote current and potential terminals.

Internal defining points are explicitly labelled a and b in Figure 5.10, which represents a four-terminal component having separate current and potential terminals.

Additional potential or current conditions may be imposed at the same or other terminal points to increase the precision of a definition, and certain kinds of definition are advantageous. In an actual measurement, it may well be that the defining conditions are *not* fulfilled exactly, but instead it should be demonstrated that the result of the measurement is the same *as if* the defining conditions had been strictly fulfilled.

We emphasise again that the concept of a *direct* impedance or admittance (see section 4.1.2) is important. With reference to Figure 5.8, the direct impedance of the component is the ratio of the voltage U between its terminals a and b to the current I *flowing through one terminal*. The currents I_1 and I_2 internal to the component through various admittances Y_1 and Y_2 to its container or surroundings do not form part of its direct impedance. They should be so routed, or reduced to zero, or their effect removed by auxiliary balances, as to measure its direct impedance correctly.

The electrical definition of a standard, or defining conditions that can be shown to be equivalent, must be strictly realised in a balanced bridge network if accurate results are to be obtained. For example, consider the familiar case of a

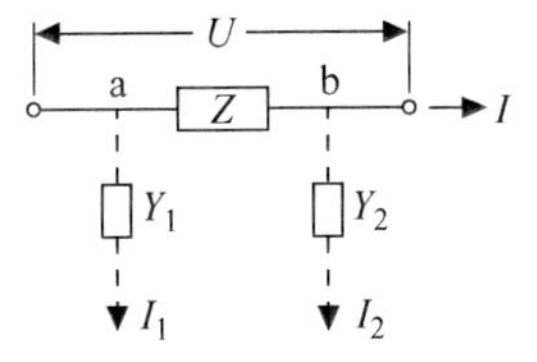

Figure 5.8 The direct impedance or admittance of a standard

four-terminal DC resistor where the electrical definition is realised in practice by the potentiometric method of comparison. In this method, the same current flows through two resistors in series, and the voltage differences between their respective potential terminals are compared under conditions of *zero current flow through these terminals*. The same defined value is obtained from Kelvin's double bridge described in section 5.6, even though in this network, appreciable currents flow through the adjacent potential terminals. The reason for the equivalence of these conditions to those of the exact electrical definition is discussed in that section.

5.3.3 Two-terminal definition

This definition applies when the component has only two terminals. Its impedance Z is defined as the ratio of the voltage U between these terminals to the current I leaving one of them, as shown in Figure 5.9. Since there are only two terminals, it is implicit in this definition that capacitance and conductance to the surroundings are negligible and that the current leaving the component is equal to the current flowing through it, otherwise the standard is ill defined.

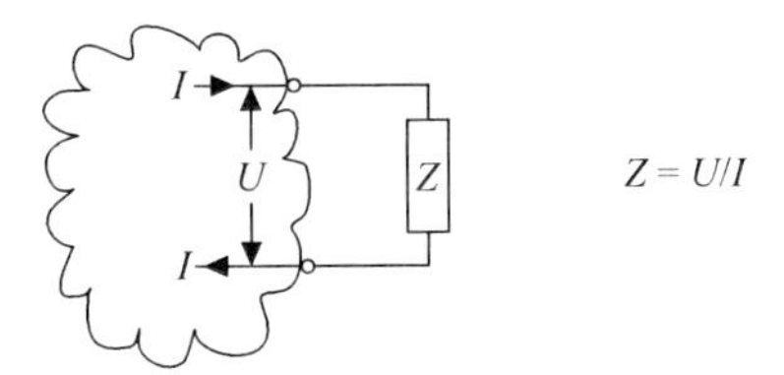

Figure 5.9 A two-terminal definition of a component

This simple definition is electrically unsatisfactory in that the terminals are only an approximation to internal defining points. Actual terminals are objects of finite physical size having different potentials between different points at which connections might be made to them. Moreover, uncertain potential differences arise through uncertain contact resistance, mutual inductance and capacitance between the measurement leads, and the terminals and conductors beyond them. Hence, more elaborate defining conditions, which overcome these problems, are needed for accurate standards.

5.3.4 Four-terminal definition

This definition, illustrated in Figure 5.10, may already be familiar as that used for accurate DC resistors because it eliminates any uncertainty caused by indeterminate terminal resistances. The standard is provided with two extra terminals connected to the internal defining points a and b between which the voltage difference is defined. The usual way of arranging the four terminals is only suitable for DC or for low-frequency measurements of medium-value impedances because considerable electric and magnetic fields emanate from them and the conductors going to them.

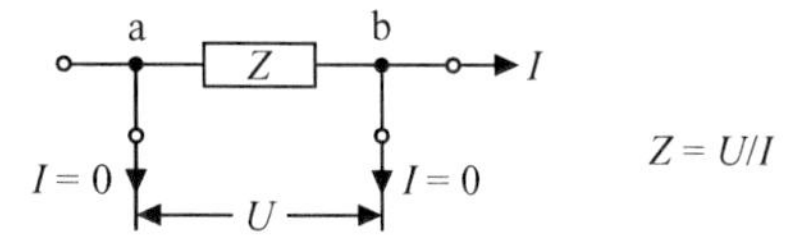

Figure 5.10 A four-terminal definition of a component

5.3.5 Four-terminal coaxial definition

This unfamiliar definition, illustrated in Figure 5.11, is formally identical to that of the previous section, but the physical arrangement of the terminals is altered to two terminal pairs. The connecting cables can be properly current equalised, and hence, cable corrections (see section 5.4.2) can be made to measurements. The comparatively large capacitance between inner and outer conductors of each coaxial cable shunts the resistor and gives it a phase angle. Consequently, this definition is only suitable for resistors measured at lower frequencies. It is particularly useful for resistance thermometers (see section 9.6).

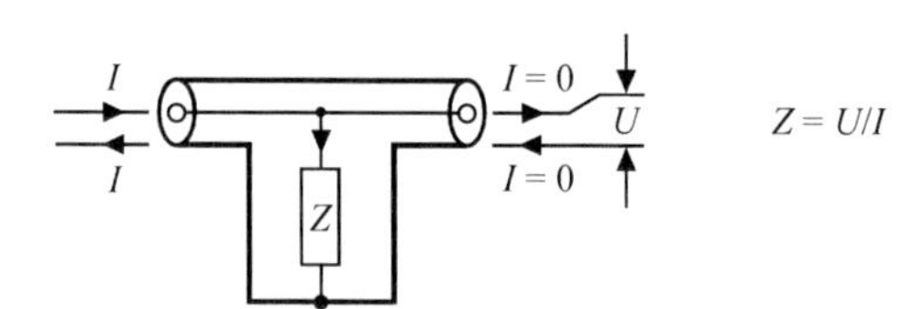

Figure 5.11 A coaxial four-terminal definition of a component

5.3.6 Two-terminal-pair definition

The more usual alternative arrangement of Figure 5.12 is more appropriate for higher-valued impedances. The admittances Y_{mn} and the impedance z internal to the impedance (Figure 5.13a) may be represented more simply as shown in Figure 5.13b as just three lumped components. The impedance is again defined as the ratio of the potential U_H between the two terminals of one terminal-pair to the current I_L out of and back into the terminals of the other terminal-pair, with the additional condition that there is zero potential between this terminal-pair. The currents into and out of the left-hand terminal-pair are also equal since the total current leaving the

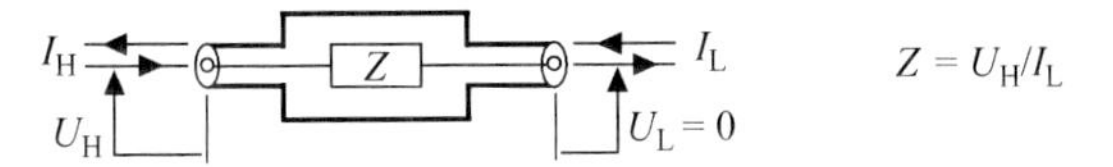

Figure 5.12 A two-terminal-pair definition of a component

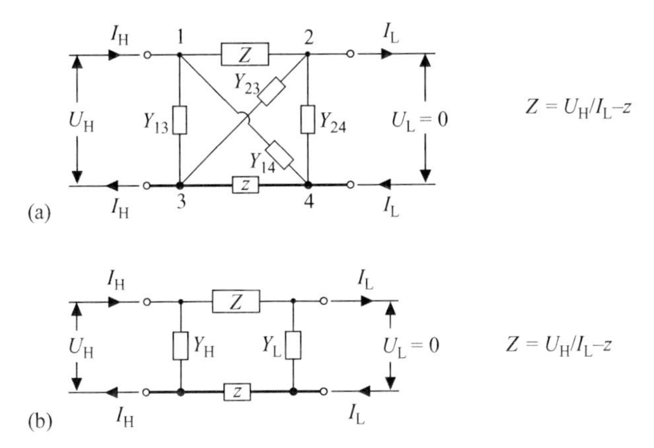

Figure 5.13 (a) The internal admittances of a two-terminal-pair component. (b) Simplified equivalent circuit

right-hand terminal-pair is zero, but in general, these currents will not equal I because of internal admittances Y_H and Y_L to the screen whose actual value plays no part in the defined direct impedance of Z (see section 4.1.2). This is because U_H is the potential difference measured across the left-hand terminal-pair, irrespective of how much current flows through Y_H, and since $U = 0$ between the right-hand terminal-pair, no part of the current I_L is shunted away through Y_L. This is in contrast to the previous section where Y_H and Y_L are cable capacitances added to the measured direct admittance of the device.

The terminals on both the right- and left-hand sides carry current, and the potentials measured between them will be subject to contact resistance variations. The total shunt admittance measured at the input terminal-pair with the output terminal-pair shorted is $Y_H + 1/Z$, and the similarly defined output shunt admittance is $Y_L + 1/Z$. The direct impedance $Z + z$, which is the defined admittance to be measured, is U_H/I_L when $U_L = 0$.

In section 5.1, we discussed how the effects of shunt admittances on measuring networks can be eliminated and conditions equivalent to this definition can be achieved.

5.3.7 Three-terminal definition

Wherever this phrase is encountered, the intended meaning needs to be examined with especial care. It is usually meant to be applied to a guarded (i.e. electrically screened) two-terminal device.

With reference to Figure 5.14a, the screen current I_G, which flows through the measurement network, should play no part in the definition of the component. Impedances defined in this way cannot be accurately measured in a current-equalised network. High-valued resistors are often provided with three single terminals, and their value is intended to be realised in accordance with the above definition. Some AC capacitors are also defined in this way, and despite being provided with two coaxial fittings, they cannot be used in a current-equalised network as the outer conductor of the high-potential port serves only as a termination of the screening of the high-potential lead and is not connected to the outer of the low-potential port. The comparison of impedances defined in this way is discussed in the older literature [1]; from our point of view, it is better to reconstruct and redefine them as a two terminal pair, as shown in Figure 5.12.

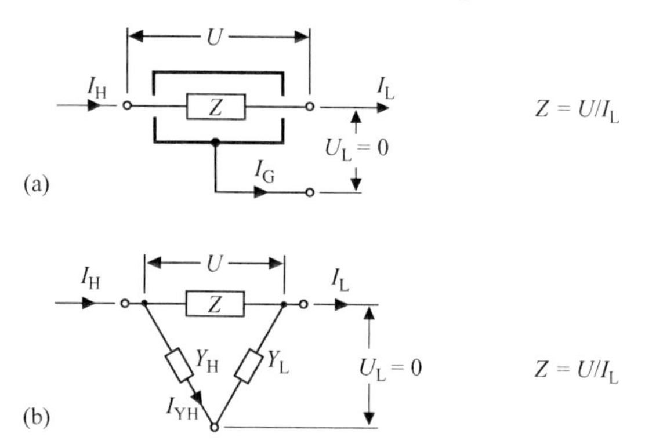

Figure 5.14 (a) The usual three-terminal definition of a component and (b) its equivalent circuit

5.3.8 *Four-terminal-pair definition*

To overcome the causes of uncertainty associated with a two-terminal-pair definition, it is necessary to combine the concepts of four- and two-terminal-pair definitions. The result, shown in Figure 5.15, is a four-terminal-pair definition, which is precise enough for all present practical applications and all values of impedances. It is shown in section 5.4.4 that the uncertainty in the product of the impedances and admittances associated with the coaxial cables and connectors used to connect the impedance into a network sets a limit to how precisely this definition can be realised, but this precision is more than adequate for all practical purposes.

The numbering of the terminal pairs or ports in Figure 5.15 follows that of the original paper by Cutkosky [2], and this has become the standard notation. Alternatively, and as we shall use in the rest of this book, they may be denoted as current (C) and potential (P) ports at the 'Hi' and 'Lo' ends of the component and labelled C_H, P_H, C_L and P_L as shown.

A generator may be connected across C_H to provide current through the component, which can be compared with another two- or four-terminal-pair impedance connected to it at C_L.

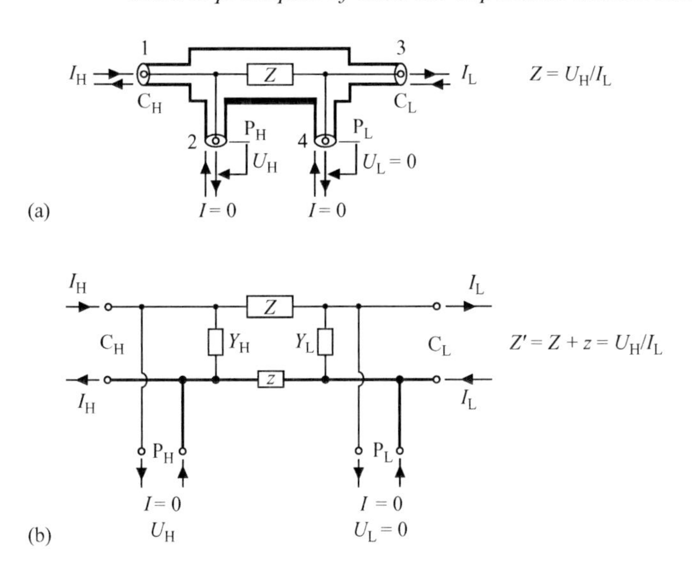

Figure 5.15 (a) The four-terminal-pair definition of a component. (b) Its equivalent circuit

The defining conditions for the device are not affected by the shunt impedances of cables or other components connected to C_H or P_L – all that matters is that the inward and outward currents at C_L are identical and that the defining conditions at P_H and P_L are fulfilled. We anticipate later discussion by noting that, for example, P_H could be connected to the known output potential across a port of a transformer and a generator connected to C_H could be adjusted until no current flows at P_H. The remaining defining conditions could be fulfilled by simultaneously bringing a detector connected to P_L to a null.

The four-terminal-pair definition appears, at first sight, to be both somewhat abstract and complicated, but it is easy to show that it has a close relationship to the familiar four-terminal impedance (see section 5.3.4), which has separate current and potential terminals.

Consider Figure 5.16 in which the impedance Z' and the inner conductors of the coaxial cables are drawn as a four-terminal impedance. This impedance can be seen as being in series with the small four-terminal impedance z formed by the outer conductors. Because of the current equalisation, the same current I flows through both, and because of the condition $U = 0$ between inner and outer at P_L, the potential difference U at P_H belongs to the *sum* of the inner and outer impedance. The actual defined impedance is therefore $U/I = Z = Z' + z$. The four-terminal-pair definition is

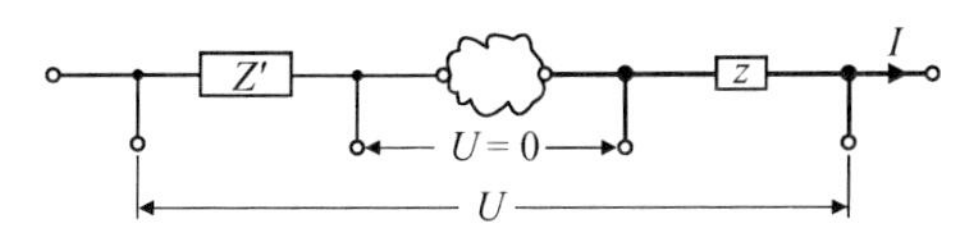

Figure 5.16 A four-terminal-pair component drawn as the sum of two four-terminal components in series

therefore a natural extension of the familiar four-terminal DC definition, which produces a carefully defined impedance suitable for measurement in an equalised current network.

With reference to Figure 5.15, the impedance $Z = U/I$ is fixed *completely* by the defining conditions at the internal defining points. The conditions at P_H and C_L are slightly different from those at the internal defining points because they include the effects of any internal cables between the internal defining points and these ports. The effects of these cables *add to those of external cables* connecting between P_H and C_L, and the measuring network and their combined parameters give rise to Z_2, Y_2, Z_3 and Y_4, as drawn in Figure 5.23. We will call the combination of internal and external cables 'defining cables'. The calculations of section 5.4.4 must take account of the internal cables, which are in series with the defining cables. *These conclusions are at variance with previous literature*, which assumed that the effect of any internal cables is negligible.

The exact point in the measuring network at which a defining cable ends is either the defining plane of a detection transformer (section 7.3.5) or a bridge T-junction at which connection to a detector is made or the defining plane of an injection transformer (section 7.3.4) which injects a voltage to balance a combining network or where an impedance is switched in to make this balance (e.g. in Figure 9.22).

Neither the internal impedance of the source at C_H nor any load impedance across P_L matters if the defining conditions are satisfied.

5.3.9 *Measuring four-terminal-pair admittances in a two-terminal-pair bridge by extrapolation*

If the defining conditions at a terminal pair (port) are not exactly fulfilled, we can, nevertheless, deduce the balance condition of a bridge, which corresponds to exact fulfilment by extrapolating the change produced on increasing the non-fulfilment by a known amount. This is an example of extrapolation techniques, which make use of the property of linearity and which can produce an accurate result as if the defining conditions were exactly fulfilled. We illustrate the principle by applying it to measuring a four-terminal-pair standard with a two-terminal-pair bridge, as shown in Figure 5.17.

We note the bridge balance readings in the original state and again when known admittances have been added to C_H and P_L, C'_H and P'_L. Then an extrapolation can be made to the balance readings to calculate those that would have been obtained under the correct defining conditions.

In the two-terminal-pair bridge shown in Figure 5.17, P_H is connected to an output tap of a transformer via a defining cable and a detector to C_L via another defining cable. As usual, the current returning via the outer conductors is made equal to that in the inner conductors by an equaliser, here shown in the defining cable going to P_H. The output taps of the transformer only produce their correct output voltages when zero current is drawn from them (i.e. their defining conditions are satisfied), and P_L and P'_L should have zero potential across them. Extrapolations

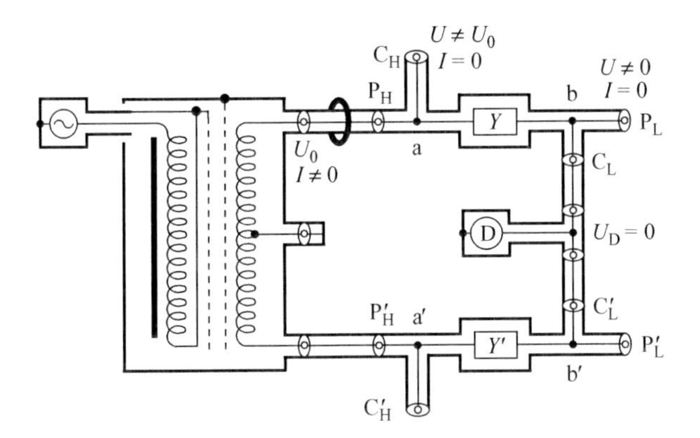

Figure 5.17 A two-terminal-pair bridge used to measure four-terminal-pair admittances

to these conditions are easily made by adding admittances to the open ports of each four-terminal-pair admittance Y and Y'. For Y, extrapolation to the defining condition of the output tap of the transformer can be made by connecting across C_H an extra admittance which, is for simplicity, equal to that measured between the inner and outer conductors at C_H with C_L shorted and the defining cable disconnected from the transformer output tap. Adding this extra admittance doubles the current taken from the transformer output port while leaving the four-terminal-pair admittance unaltered. Extrapolation back to zero admittance presented to the transformer output port is therefore equivalent to achieving its correct $I = 0$ defining condition. The correct defining condition of $I = 0$ at port P_H of Y is still violated because of current through the capacitance between its inner and outer conductors, which causes voltage drop down the defining cable. This can be accounted for by applying a cable correction as described in section 5.4.4, but note that the correction has the opposite sign to the case where $I = 0$ at P_H.

A correction for the error caused by violation of the defining condition at P_L can similarly be found. When the bridge is balanced, the voltage U_D between the inner and outer conductors at the T-junction where the detector is connected is zero, and no current flows there. Instead, a small portion of the current through the defining point b flows through the admittance from b to the detector T-junction between the inner and outer conductors and violates the defining conditions $U = 0$ and zero current between the inner and outer conductors at b. A second admittance Y_{CL}, equal in value to that measured at C_L with P_H shorted, when connected across P_L will double the potential there, and therefore, the current between inner and outer at b and extrapolation back to zero admittance will reproduce the proper defining conditions at b. Again, a cable correction with the opposite sign must be made for the defining cable at C_L.

The same procedure applied to Y' will yield its correct defining conditions.

It is not necessary to use the particular values of shunt admittances suggested, provided the values are known. Let the ratios of the shunt admittances added to C_H and P_L to those measured at P_H and C_L with C_L or P_H shorted be k_1 and k_2, respectively, and those added to C'_H and P'_L to those measured at P'_H and C'_L with C'_L

or P'_H shorted be k'_1 and k'_2, respectively. The individual changes of bridge balances brought about by adding such shunt admittances in turn or to the appropriate ports can be weighted by k_1, k_2, k'_1 and k'_2 and added algebraically to the bridge balance to give the total correction for the ill definition of Y.

Alternatively, if $k_1 = k_2 = k'_1 = k'_2$ the shunt admittances can be added simultaneously and the total correction obtained from just one extra bridge balance.

5.3.10 Adaptors to convert a two- or four-terminal definition to a four-terminal-pair definition

There are many situations where it is required to convert the terminal configuration of an existing device so that it can be measured as a four-terminal-pair component, with all the clarity of definition that it implies. It is instructive to examine the topology of an adaptor to do this because this same topology is a good way to mount an isolated component in a screening can and is employed in the construction of high-frequency standards as described in section 6.5.7–12. The two problems to be solved are, first, to bring the outer conductors of four cables together in such a way that their junction possesses zero four-terminal impedance and, second, to continue their inner conductors going to the component in such a way that there is acceptably small mutual inductance and capacitance between these leads.

The outer conductors can be joined at a four-terminal zero-impedance junction. A four-terminal zero impedance has the property that the potential between its potential terminals remains zero when a current flows between its current terminals. A zero DC resistance can have a conductor joining its current terminals and another joining its potential terminals and a link of small cross-section between a point on either conductor. Then, however much current flows between the current terminals, no potential difference appears across the potential terminals. A zero AC impedance must also have no significant mutual inductance between the current and potential links, and this can be achieved by an orthogonal geometrical layout.

The inner conductors need to be continued as the inner conductors of separately screened and insulated cables twisted together in current and potential pairs. A possible construction is illustrated in Figure 5.18a, where the outer conductors of P_L and P_H are connected together at the cross-shaped junction with the outer conductors of C_H and C_L. This orthogonal geometry minimises mutual inductance between the current and potential paths. The adaptor is shown with spade terminations suitable for connecting to the binding-post terminals of a four-terminal component, to convert its definition to four terminal pair, but they could equally well, for example, be connected to a single port, making a two-terminal connection by connecting C_H and P_H to one terminal and C_L and P_L to the other, or to other terminal arrangements [3]. If the component has a screening can, it should be connected to the single conductor from the centre of the cross.

Figure 5.18b, c and d are rigid adaptors devised ad hoc to make the conversion for particular designs of components. They might be sufficiently good for their purpose but exhibit small mutual inductive interactions. Figure 5.18b and d also have a small internal shunt capacitance.

Impedance standards for higher (microwave) frequencies are frequently single-port devices. They can be configured for four-terminal-pair measurement with an

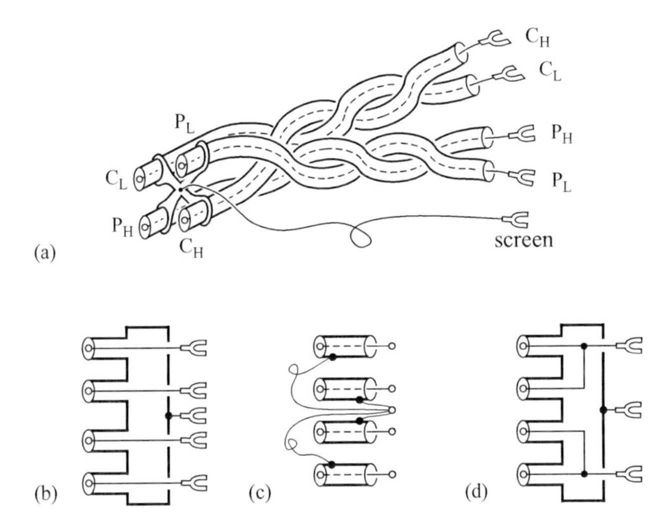

Figure 5.18 (a) An adaptor to convert two- and four-terminal components to a four-terminal-pair definition. (b, c and d) Rigid adaptors for particular applications

adaptor consisting of a mating connector mounted on, but insulated from, the top of a cylindrical or square conducting box of minimal size. The four-terminal-pair connectors are mounted 90 degrees apart at the same height on the sides of the box to provide orthogonal current and potential connections as in Figure 6.26. Radial connections go from the inner conductor of the mating connector to the 'low' and from the outer conductor to the 'high' inner conductors of the four-terminal-pair connectors. The outer conductors of the four-terminal-pair connectors are connected by the box. Shielding may need to be completed for interference elimination by a shield connected to the box which encompasses the standard. The residual four-terminal-pair capacitance of the adaptor up to the mating plane of the mating connector with no standard connected should be measured and subtracted.

5.4 The effect of cables connected to the ports of impedance standards

We have seen that impedance and admittance standards, that is, resistors, inductors and capacitors, possess inaccessible internal defining points where corresponding current- and voltage-measuring conductors from the accessible terminals of the device are connected together internally to the resistive, capacitive or inductive element. The value of the standard is defined in terms of the current and voltage differences at these internal defining points, and the effect of connecting cables between these points and the measuring bridge or instrument must be accounted for in accurate measurements. The following calculations build on the basic considerations of section 3.2.6.

We emphasise again that because cable corrections combine non-linearly, the correction for an individual cable added later cannot be directly applied. The properties of all cables connected in series up to the internal defining points of an

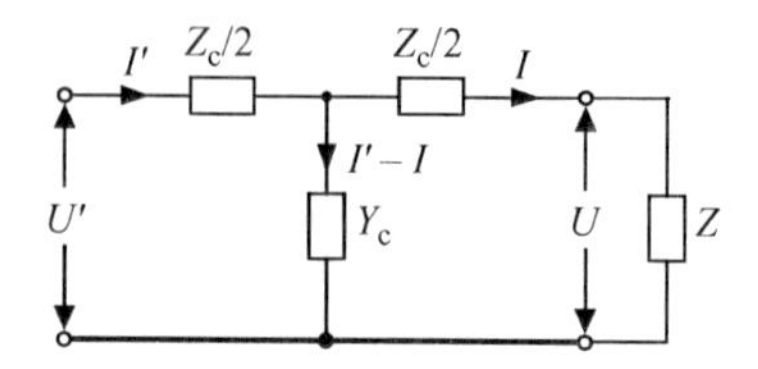

Figure 5.19 A cable added to a two-terminal component

impedance must be accounted for. This is because the effects of two cables in series, even if they have equal admittance and impedance per unit length, cannot be directly added because the effect of each cable is proportional to the *square* of its length. Therefore, the properties of both cables must be known, and the effect of the combination must be calculated (see section 3.2.6).

5.4.1 The effect of cables on a two-terminal component

With reference to Figure 5.19,

$$U' = \frac{I'Z_c}{2} + \frac{I' - I}{Y_c} \tag{5.1}$$

$$U = \frac{I' - I}{Y_c} - \frac{IZ_c}{2} \tag{5.2}$$

But $U/I = Z$, so (5.2) becomes

$$\frac{I}{I'} = \left(1 + \frac{Z_c Y_c}{2} + ZY_c\right)^{-1} \tag{5.3}$$

whence, from (5.1),

$$\frac{U'}{I'} = \frac{Z_c}{2} + \frac{(Z_c/2) + Z}{1 + (Z_c Y_c/2) + ZY_c} \tag{5.4}$$

Clearly, even if the YZ terms are negligible, there is a first-order addition of Z_c to Z. If we view a connector as being a 'cable' between Z and a network, even substitution measurement will be in error by the uncertain connector impedance Z_c, and if Z is a large impedance, the term $Y_c Z$ may be appreciable for even a short length of cable or a good-quality connector. Two-terminal standards having coaxial ports are only useful at the frequencies with which we are concerned in this book when moderate accuracy is required over a limited range of values.

Paradoxically, they are common at microwave frequencies where the defined conditions are those across a defined a plane of the conductor.

5.4.2 The effect of cables on a four-terminal coaxial component

A good definition of the arrangement of section 5.3.5 is obtained when connecting a component of this kind to a network with coaxial cables only if one cable carries the current to and fro from the component and the potential difference with zero current flow is maintained between inner and outer conductors of the other cable.

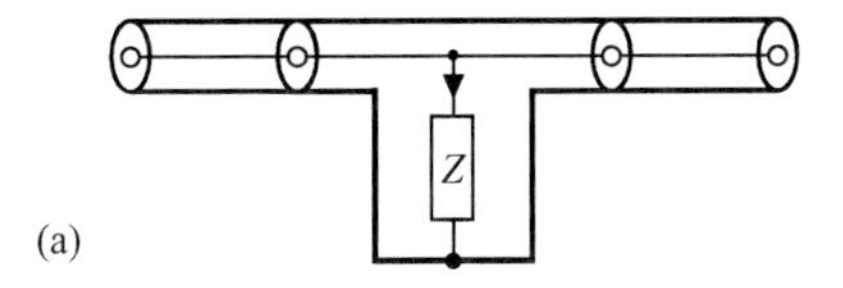

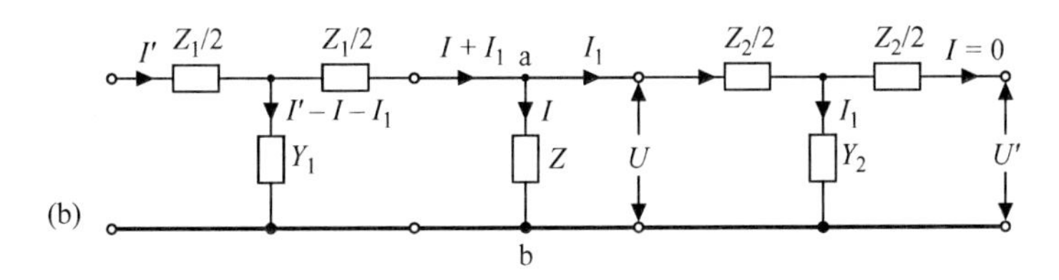

Figure 5.20 A (a) four-terminal component with cables attached (b) its equivalent circuit

Representing the impedance Z with cables attached as shown in Figure 5.20, from the left-hand mesh,

$$\frac{I + I_1 - I'}{Y_1} + \frac{(I + I_1)Z_1}{2} + IZ = 0 \tag{5.5}$$

And from the right-hand mesh,

$$I_1\left(\frac{Z_2}{2} + \frac{1}{Y_2}\right) - IZ = 0 \tag{5.6}$$

Eliminating I_1 between (5.5) and (5.6),

$$I' = I\left[1 + \frac{ZY_2 + ZZ_1Y_1Y_2}{1 + (Y_2Z_2/2)} + \frac{Y_1Z_1}{2} + Y_1Z\right]$$

From (3.8),

$$U' = \frac{U}{1 + (Y_2Z_2/2)}$$

Hence, the four-terminal admittance sensed at the end of the coaxial cables is

$$\frac{1}{Z'} = \frac{I'}{U'}$$
$$= \left(\frac{1}{Z}\right)\left(1 + \frac{Y_1Z_1}{2} + \frac{Y_2Z_2}{2} + Y_1Z + Y_2Z + \frac{Y_1Y_2ZZ_2}{2} + \frac{Y_1Y_2ZZ_1}{2} + \frac{Y_1Y_2Z_1Z_2}{4}\right) \tag{5.7}$$

since $1/Z = I/U$. The product correction terms can readily be evaluated from the properties of the cables and an approximate value of Z.

Typically, for a metre of coaxial cable of 50 Ω characteristic impedance, where Z is in ohms, Y is in ohms^{-1} and ω in rad/s.

$$Z = 10^{-2} + j\omega 2.5 \times 10^{-7}$$
$$Y = +j\omega 10^{-10}$$

($L_2 = 0.25\ \mu$H/m, $C_2 = 100$ pF/m, $R_2 = 0.01\ \Omega$/m)

The values of the term $Y_1Z_1/2$, $Y_2Z_2/2$ are then $-\omega^2 2.5 \times 10^{-17} + j\omega 2.5 \times 10^{-12}$.

At frequencies below $\omega = 10^5$ rad/s (≈ 16 kHz), these terms will not exceed 3×10^{-7}. The term $Y_1Y_2Z_1Z_2/4$ is usually quite negligible. These terms are independent of the value of the impedance Z being measured, but the remaining second-order terms Y_1Z and Y_2Z, and fourth-order terms are not. We consider three special cases with cables having the properties listed above.

In the following numerical examples ω is in rad/s, R is in ohms, C is in farads and L is in henries.

(i) Z is a pure capacitance, C.

$$Y_1Z = \frac{j\omega 10^{-10}}{j\omega C} = \frac{10^{-10}}{C}$$

as expected, because C is merely shunted by the cable capacitance. Clearly, the terminal configuration considered in this section is unsuitable for all but large values of capacitance.

(ii) Z is a pure inductance, L.

$$Y_1Z = j\omega 10^{-10} \cdot j\omega L = -\omega^2 10^{-10} L$$

This terminal configuration is not ruled out for moderate values of ω and L. For example, at $\omega = 10^4$ rad/s (≈ 1.6 kHz) and $L = 10^{-3}$ H, $Y_1Z = -10^{-5}$. This is almost a negligible correction to a measurement of an inductance because the accuracy is usually limited by both a lack of stability and, possibly, a large external magnetic field of the inductor, which can be affected by its surroundings. The term $Y_1Y_2ZZ_1/2 = -j\omega 10^{-20}(10^{-2} + j\omega 2.5 \times 10^{-7})L$, and for moderate values of ω, this is smaller than the term just evaluated.

(iii) Z is a pure resistance, R.

$$Y_1Z = j\omega 10^{-10} R$$

This is in quadrature with R, but at $\omega = 10^4$ rad/s (≈ 1.6 kHz), it becomes appreciable for values of R that exceed $10^3\ \Omega$ if the cable shunt admittance constituting Y_1 is a lossy capacitance. If this capacitance has a phase angle of 10^{-3}, this term will give an in-phase addition to R of $\omega 10^{-10} \times 10^{-3} R$, which, for $\omega = 10^4$ rad/s, is $10^{-9} R$. Therefore, this terminal arrangement is a poor way of measuring R accurately at this or similar frequencies if the value of R is greater than 1 kΩ.

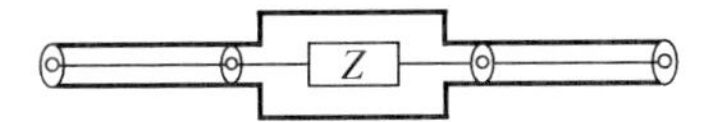

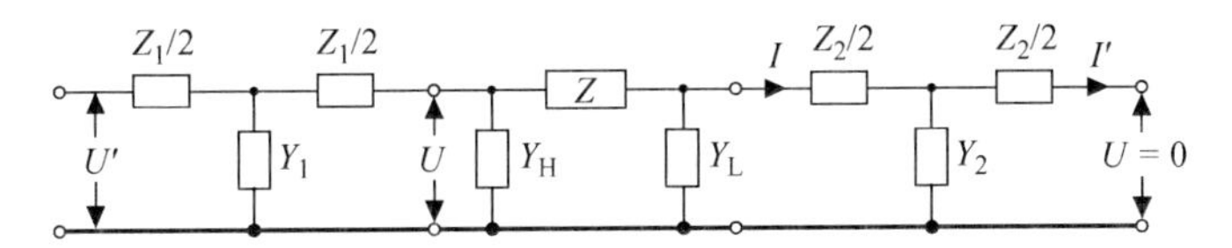

Figure 5.21 Cables added to a two-terminal-pair component

The fourth-order term $Y_1Y_2ZZ_1/2$ is again much smaller for moderate values of ω than the term just evaluated.

5.4.3 The effect of cables on a two-terminal-pair component

This case is shown in Figure 5.21, where the defining conditions are again to be fulfilled at the ends of added cables instead of at the internal defining points. From (3.7), we have $I = I'(1 + Y_2Z_2/2)$, if Y_1 is small compared to $1/Z_2$. Since current flows through the original terminals where the potential was defined, we cannot use (3.8) to deduce the relationship between U and U'. We can carry out an approximate analysis, however. If Y_1 and Y_2 are less than or of the order of $1/Z$, then the current through Y_H is UY_H and the current through Y_1 is approximately UY_1. Therefore,

$$U' - U = \frac{Z_1(UY_1 + UY_H + I)}{2} + \frac{Z_1(UY_H + I)}{2}$$

or

$$U' = U\left(1 + \frac{Y_1Z_1}{2} + Y_HZ_1 + \frac{Z_1}{Z}\right)$$

Since $Z = U/I$ and the apparent impedance of the component and added cables $Z' = U'/I'$,

$$Z' = Z\left(1 + \frac{Y_1Z_1}{2} + Y_HZ_1 + \frac{Z_1}{Z}\right)\left(1 + \frac{Y_2Z_2}{2}\right) \tag{5.8}$$

Hence, the correction to be made to the apparent impedance Z' to obtain Z depends to first order on the term Z_1/Z. This might have been anticipated since Z_1 is in series with Z. The correction is subject to the uncertainty in the connector impedance contained in Z_1. Either satisfactory corrections can be made to account for the other two terms, or the impedance can be regarded as having been redefined with the lengths of additional cable included (but take note of section 3.2.6 regarding the non-linearity of cable corrections if further lengths of cable are added to include the component in a measuring network). Because of these considerations, a two-terminal-pair definition is suitable only for high-impedances Z.

Since the way we define a three-terminal impedance is not different in principle from a two-terminal-pair definition, the above considerations apply to this definition also.

Many standard impedances made and sold for measurements of the highest accuracy have two-terminal-pair terminations, and no information is available for the length or impedance and shunt admittance of the internal cables between the impedance itself and these terminations. To improve their electrical definition, T-connectors can be added at the terminations. The impedance can then be regarded as being defined as a four terminal pair having its internal defining points at these T-connectors (see Figure 5.22).

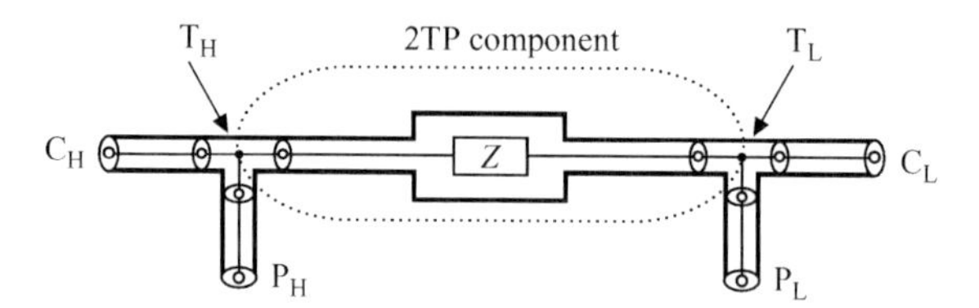

Figure 5.22 Converting a two-terminal-pair standard to four terminal pair by adding T-connectors at its output ports

The internal cables then become part of the defined impedance of the standard. They alter its value, frequency dependence and phase angle, but the redefined standard has four-terminal-pair completeness.

One design of commercial capacitance standard has only one side of the capacitor completely shielded. The other side is exposed to other circuitry through which noise currents can flow and make the measurement unnecessarily difficult. Worse, currents at the measurement frequency can alter the result by an unknown amount. Because the exposed side is, in most measurement networks and instruments, connected to a source of low-output impedance through which most of the interfering current is shunted away, the effect is often negligible. The exposed side is designated 'Hi', and the other 'Lo'. By taking care to connect the standard the correct way round, reasonably reproducible and accurate results can be obtained, but it is clearly better to modify these standards by completing the shielding. This involves dismantling the standard and adding a shield of a somewhat complicated shape.

5.4.4 *The effect of cables on a four-terminal-pair component*

With reference to Figure 5.23, the addition of a cable between the generator and port 1 has no significance as no quantities are defined at this port. A cable added to port 4 also has no effect since if $U = 0$ and $I = 0$ at one point on a cable, these conditions must hold everywhere and, apart from sensitivity considerations, its length is immaterial. We need to consider only the effect of adding cables to ports 2 and 3.

From (3.7), reversing the roles of I and I', we have

$$I' = I\left(1 + \frac{Y_3 Z_3}{2}\right)$$

and from (3.8),

$$U' = \frac{U}{1 + (Y_2 Z_2/2)}$$

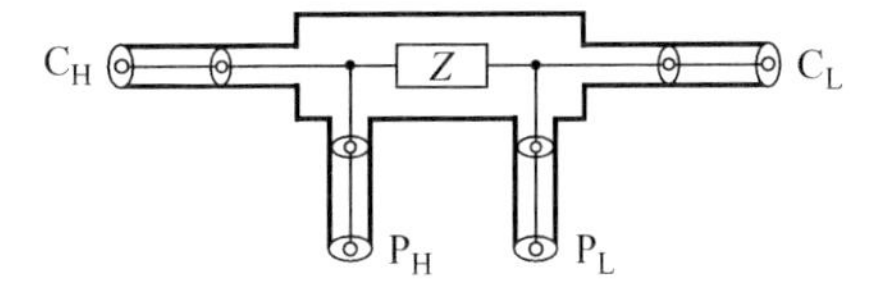

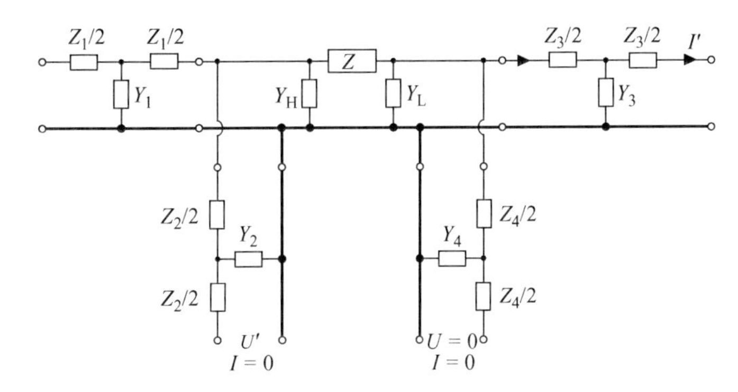

Figure 5.23 Cables added to a four-terminal-pair component

Hence, the apparent impedance Z', after the addition of cables, is

$$Z' = \frac{U'}{I'} = \frac{Z}{[1 + (Y_2 Z_2/2)][1 + (Y_3 Z_3/2)]} \tag{5.9}$$

Notice that the correction terms are now much simpler. It is especially important also to note that they do *not* involve the impedance Z. This means that a four-terminal-pair definition is suitable for *all* values of Z, that is, for all components at all frequencies within our restriction that the wavelength is negligible compared with the cable lengths. The effects of variable connector impedance, which we can view as being contained in Z_2 and Z_3, although still present, are the same for all values of Z. By making a sensible choice of connector types and restricting the values of Y_2 and Y_3, the uncertainty introduced into the value of Z can be kept less than 1 part in 10^9 at frequencies below 16 kHz.

For metre lengths of typical cable at 1.6 kHz, the correction terms are of the order of 1 in 10^9 and are easily calculated from the measured or quoted properties of the cables. The length is not very critical: at 1.6 kHz, a change of 100 mm in 1-m cables having the constant impedance and admittance per unit length assumed in section 5.4.2 changes the four-terminal-pair admittance by only 1 in 10^{10}.

Sometimes Z is taken as having been redefined at the new ports at the ends of the added cables, which then become part of the defined impedance Z'. Due regard must again be paid to the non-linear addition of the effects of cables external and internal to the device, as pointed out in section 3.2.6.

There are three ways of configuring the internal cables of a four-terminal-pair impedance.

1. Two coaxial leads go from the impedance to an insulating panel on which four coaxial connectors are mounted, and appropriate connections are made there

between inner conductors and between outer conductors to produce a four-terminal-pair standard. The four-terminal-pair definition can be strictly obeyed at these connections and any corrections must be made for defining cables, between them and a measuring network. The problem with this method of cabling is that the two cables between the impedance and the panel modify the defined impedance and, in particular, its frequency dependence and phase angle.

2. Four coaxial leads go from internal defining points of the impedance to four ports mounted on an insulating panel. Making correct cable connections for the defining cables, which connect this impedance to a measuring network, now requires some care. The properties of the relevant cables between the impedance and the panel have to be known and must be combined with those of the defining leads outside the device as described in section 3.2.6, and then the effect of the internal leads subtracted in order to produce a measurement as if the four-terminal-pair defining conditions had been strictly fulfilled at the ports mounted on the panel. Alternatively, the defining conditions can be regarded as holding at the internal defining points rather than at the panel connectors.
3. With the configuration of 2 above, the defining conditions can be regarded as holding at the potential ports on the panel. They can be strictly fulfilled by detecting zero current through the high-potential defining port and connecting a combining network (see section 5.6) to the low-current defining port. This approach has the disadvantage that the high-potential lead between the panel and the ratio device (usually a transformer) loads this ratio device, and this loading has to be accounted for. This can be done either by T-connecting an admittance having a known ratio to that of the loading cables and extrapolating back to the condition of zero loading.

Although the complexity of networks for measuring four-terminal-pair impedances is greater than that for the other definitions, the use of impedances defined in this way is to be encouraged, and in work of the highest accuracy, it is essential. The planes where a voltage difference or a current is defined can be moved along a cable by employing tri-axial cable with the intermediate screen driven by an auxiliary source to a potential equal to that of the inner conductor, as explained in section 3.2.7.

5.5 An analysis of conductor-pair bridges to show how the effect of shunt admittances can be eliminated

In this section, we examine how a bridge network can compare the direct admittances of devices whilst ignoring their input and output shunt admittances. (These quantities have been defined in section 5.3.2.) It is possible to so arrange the components that the bridge balance condition can be made to depend mostly on the *direct* admittances but with a correction term that involves the shunt admittances.

This term can be factorised into the product of two other terms, both of which depend on shunt admittances but both of which can be made small by auxiliary adjustments. The product is then usually negligible. We will discuss two-terminal-pair components for simplicity, as the shunt admittance aspect of bridge design is not significantly different for four-terminal-pair components. We will consider just the three most commonly occurring cases, namely bridges based on a voltage transformer ratio (see section 9.1.5), the classical generalised four-arm bridge discussed in section 5.1 (which is still useful, particularly at higher frequencies) and the network known as a quadrature bridge (see sections 9.2.1–9.2.3) because it compares voltages across resistances with voltages in quadrature across capacitances. Some elements of the discussion can be recognised as the Wagner and detector auxiliary balances of the older bridge techniques (see section 5.1) now applied to the better-defined conductor-pair measurements.

Thompson [4] has also given an elegant treatment.

5.5.1 Comparing direct admittances using voltage sources

Consider the arrangement shown in Figure 5.24 where two-terminal-pair components are connected to separate sources. Their direct admittances Y_1 and Y_2 are to be compared by adjusting either U_1 or U_2 or the ratio U_1/U_2 to null the detector.

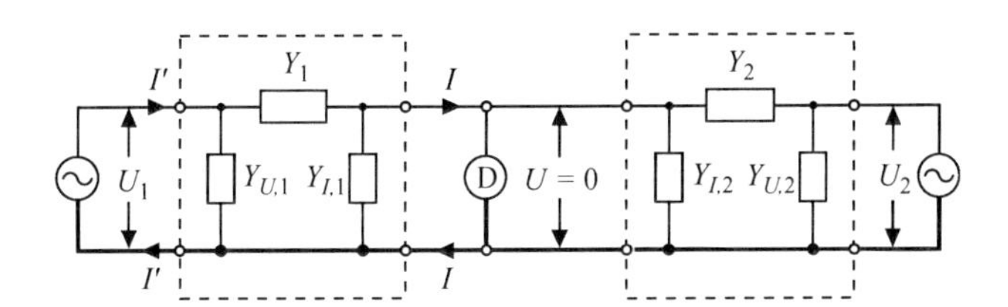

Figure 5.24 Comparing two-terminal-pair components

If the sources are of negligible output impedance (e.g. if they are the outputs of two separate windings of a well-constructed voltage ratio transformer), their output voltages U_1 and U_2 are not greatly affected by the presence of the shunt admittances $Y_{U,1}$ and $Y_{U,2}$. That is, to a good approximation, $Y_{U,1}$ and $Y_{U,2}$ do not affect the bridge balance. Also, when the detector is nulled, $Y_{I,1}$ and $Y_{I,2}$ have no voltage across them. Hence, no current flows through these admittances, and their value also does not affect the bridge balance condition, which depends only on the well-defined direct admittances Y_1 and Y_2. The effect of $Y_{I,1}$ and $Y_{I,2}$, which shunt the detector terminals, is to degrade the detector sensitivity and noise performance.

The error arising from the impedances and shunt admittances of the cables connecting Y_1 and Y_2 to the detector and to the voltage sources has already been calculated in section 5.4.3.

5.6 Combining networks to eliminate the effect of unwanted potential differences

5.6.1 The concept of a combining network

Historically, the concept of a combining network was first employed in Kelvin's double bridge. This network was invented to solve the problem of comparing accurately the values of low-valued four-terminal DC resistors. The four-terminal defining conditions of these resistors are that the defined resistance is the ratio of the voltage between the two designated potential terminals (under conditions that no current flows through either) to the current leaving one designated current terminal. The network is drawn in Figure 5.25.

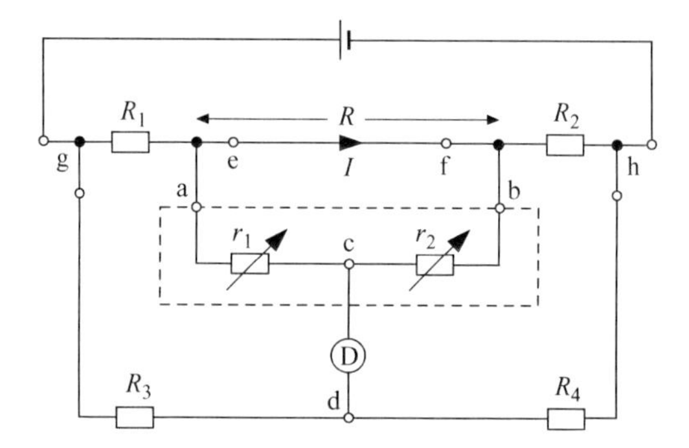

Figure 5.25 A DC Kelvin double bridge. The components within the dotted box are a combining network

The network is basically a Wheatstone bridge. When the detector is nulled by adjusting R_4, the ratio of the four-terminal resistances R_1/R_2 approximately equals the known ratio of the much higher-valued resistances R_3/R_4. The potential terminals g and h of R_1 and R_2 have to supply the small but finite current that traverses the higher-valued resistances R_3 and R_4. Consequently, the defining conditions at these terminals are only approximately fulfilled, but we are not concerned here with this imperfection.

The current I traversing R_1 and R_2 causes a potential difference IR across the resistance R, which is the total resistance between the low internal defining points of R_1 and R_2. This potential difference appears across the potential terminals a and b between which the resistors r_1 and r_2 are connected, and the bridge detector is connected to their junction point. A value of r_1/r_2 can be found such that the detector does not respond to changes in the potential between the low internal defining points of R_1 and R_2 brought about, for example, by temporarily opening the link between e and f. Then we can appeal to linear network theory to remark that since the bridge balance no longer depends on the value of the potential difference between the low internal defining points of R_1 and R_2, the situation is the same as if this potential difference were truly zero, the low internal defining points of R_1 and R_2 were actually connected together and the defining condition of zero current through their low-potential terminals a and b had actually been achieved. The resistance of these potential leads between the internal defining points and a and b forms part of the resistances r_1 and r_2.

Evidently, r_1 and r_2, together with the rest of the network, form an auxiliary Wheatstone's bridge balanced so that the detector does not respond to the 'source' IR. If the main balance condition $R_1/R_2 = R_3/R_4$ has also been achieved, by iteration if necessary, their values are such that

$$\frac{r_1}{r_2} = \frac{R_1}{R_2} = \frac{R_3}{R_4} \tag{5.10}$$

That is, the voltage developed across the resistance between the low internal defining points of R_1 and R_2 is divided by the resistances r_1 and r_2 at the point c in the same ratio as the voltages between the internal defining points of R_1 and R_2 and the bridge balance is independent of R and the network behaves *as if* the correct defining condition of zero current through the low-potential terminals of R_1 and R_2 has actually been achieved.

r_1 and r_2 form a *combining network*, which combines the two potentials at the two internal defining points to which they are connected with the intention of eliminating the effect of the difference in these potentials from the bridge balance condition.

Warshawsky [5] has devised a network, which uses four combining networks at the corners of a Wheatstone bridge, which allows fully correct four-terminal definitions of all of R_1, R_2, R_3 and R_4.

The technique of incorporating a combining network to eliminate the potentials between internal defining points of components has found widespread application to AC networks ('bridges'), which compare impedances accurately. There are many examples later in this book.

5.6.2 A general purpose AC combining network and current source

If we consider the T-network of Figure 5.26a, where nodes a and b are connected by an ideal transformer of infinite input impedance and zero output impedance, the current I flowing into a third node c is

$$I = [U_b + \rho(U_a - U_b) - U_c]Y$$

Consider the network of Figure 5.26b.

$$I' = \rho(U_a - U_c)Y + (1 - \rho)(U_b - U_c)Y = I$$

for all values of ρ. Therefore, the two networks are equivalent.

Note that if a or b are taps on the transformer, values of $\rho < 0$ and $\rho > 1$ can be obtained. Therefore, if $\rho < 0$, the admittance ρY between a and c is negative, and could be either a negative conductance or a negative susceptance. The admittance between b and c would then be greater than Y.

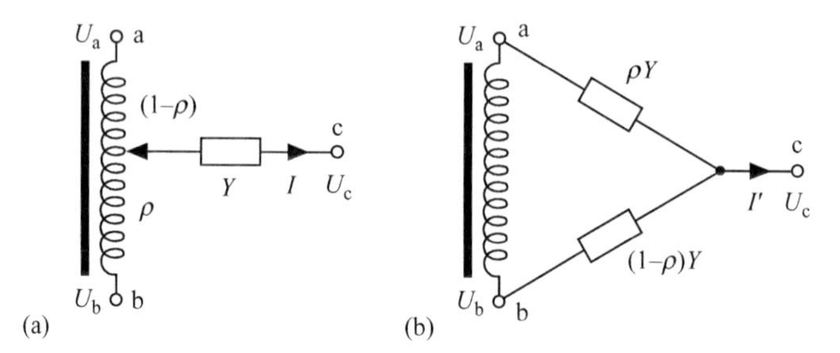

Figure 5.26 (a) An AC combining network and (b) its equivalent circuit

If we take two such networks and connect them in parallel, as shown in Figure 5.27a, the resulting network will be equivalent to that of Figure 5.27b, and since ρ_1 and ρ_2 are fully adjustable, the network is a general combining network having ratios of admittances of both in phase and in quadrature. It can therefore eliminate the effect on the rest of the network of any potential difference between a and b.

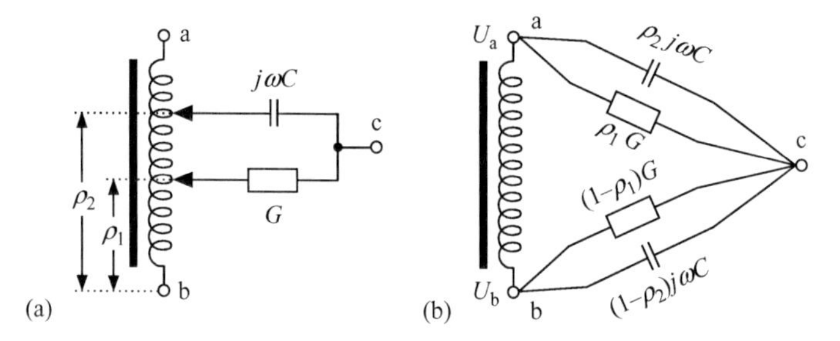

Figure 5.27 (a) Connecting two AC combining networks in parallel to obtain an in-phase and quadrature adjustment. (b) The equivalent circuit

A sufficiently close approach to an ideal transformer can easily be made in practice, and sufficiently large admittances G and $j\omega C$ can be used so that there is no significant effect on the noise (see section 2.1.3) and sensitivity of the total network containing the combining network. We give several examples of its use in the networks to be described in chapter 9. It is interesting that this AC combining network is one of a dual pair of components, of which the other is an adjustable quasi-current source used extensively in the bridge networks described in chapter 9.

Where two four-terminal-pair impedances are connected in series via their C_L terminals as shown in Figure 5.28 and the effect of the potential difference between their internal defining points is eliminated with a combining network, the effect is as if a and b were at the potential of the outer conductor when the detector D is nulled.

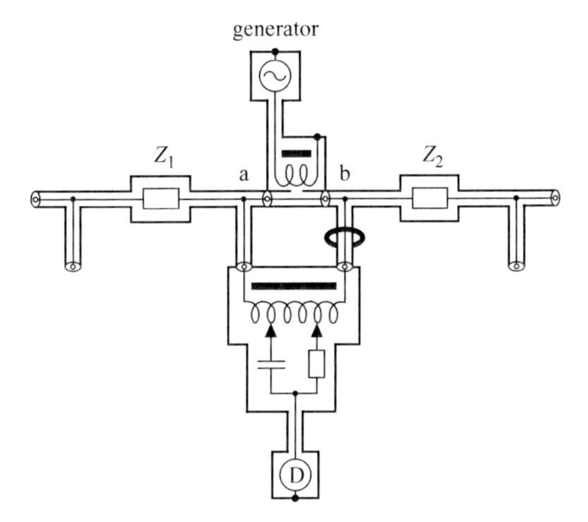

Figure 5.28 Connecting four-terminal-pair impedances in series using a combining network

5.7 Connecting two-terminal-pair impedances in parallel

While it is not in general easy to connect impedances in series and compute their combined impedance because of the influence of the somewhat uncertain shunt capacitances of the joining cables, connection of two-terminal-pair impedances in parallel is possible. For example, connection of two or more two-terminal-pair capacitances in parallel is a practicable way of producing a larger capacitance, calculable from the individual two-terminal-pair values as measured at the ends of their cables. The junction points of these cables, when the capacitances are connected in parallel, then become the internal defining points of the parallel assembly, which can therefore be defined as a four-terminal-pair impedance if required. The meshes formed by the parallel connection, in principle, need current equalisers.

References

1. Hague B. (revised by Foord T.R.). *Alternating Current Bridge Methods*. 6th edn. London: Pitman; 1971
2. Cutkosky R.D. 'Four-terminal-pair networks as precision admittance and impedance standards'. *Trans. IEEE Commun. Electron*. 1964;**83**:19–22
3. Kibble B.P. Four terminal-pair to anything else! IEE Coll. On Interconnections from DC to Microwaves (No. 1999/019) pp.6/1-6/6, 1999
4. Thompson A.M. 'AC bridge methods for the measurement of three-terminal admittances'. *IEEE Trans. Instrum. Meas*. 1964;**IM-13**:189–97
5. Warshawsky I. 'Multiple bridge circuits for measurement of small changes in resistance'. *Rev. Sci. Instrum*. 1955;**26**:711–15

Chapter 6

Impedance standards

This chapter details the key SI primary and secondary standards of capacitance, resistance and inductance for applications covering the frequency range DC to 100 MHz. These standards include the Thompson–Lampard calculable cross-capacitor, Gibbings quadrifilar resistor, Campbell mutual inductance standard as well as the latest standard of impedance derived from the quantum Hall effect (QHE). Impedance measurements span one of the widest ranges of any physical quantity, from nano-ohms to tera-ohms (i.e. a ratio of 10^{21}). This chapter also discusses the limitations of the standards and their evolution to meet the contemporary scientific and industrial needs.

Oliver Heaviside was the first to introduce the concept of electrical impedance in the 1880s. Impedance is a more general concept than resistance because the latter may be regarded as the value of a particular physical quantity (resistance) at a specific point (i.e. at DC or $f = 0$ Hz) on the frequency axis. In addition, impedance takes phase differences into account for $f > 0$ Hz and is a fundamental concept in electrical engineering. That is, it must be quantified by two parameters rather than the one parameter of DC resistance. Thus, impedance may generically be expressed in terms of complex numbers (or displayed on a polar or Argand diagram) such that $Z(\omega) = R(\omega) + jX(\omega)$, where $\omega = 2\pi f$ is the angular frequency, $j = \sqrt{-1}$, $R(\omega)$ is the frequency-dependent resistance and $X(\omega)$ is the frequency-dependent reactance. The impedance $Z(\omega)$, in general, can be any combination of resistance, inductance or capacitance and can be measured by either energising a device with a known voltage $U(t)$ and measuring the current $I(t)$ through it or vice versa, such that $Z(\omega) = U(t)/I(t) = |Z|e^{j\omega t}$. The real and imaginary components of the impedance are then given by

$$\mathrm{Re}(Z) = |Z|\cos(\phi) \quad \text{and} \quad \mathrm{Im}(Z) = |Z|\sin(\phi) \tag{6.1}$$

where

$$|Z| = \sqrt{R^2(\omega) + X^2(\omega)} \quad \text{and} \quad \phi = \tan^{-1}\left(\frac{X(\omega)}{R(\omega)}\right) \tag{6.2}$$

There are four fundamental properties of impedance, as defined earlier and discussed throughout this book; *causality*, the response of the system is due only to the applied perturbation; *linearity*, the impedance is independent of the magnitude of the perturbation provided the perturbation is not too great; *stability*, the system returns to its original state after the perturbation is removed; *finite-valued*, the impedance must be finite-valued as $\omega \to 0$ and $\omega \to \infty$ and must also be a continuous and finite-valued function at all intermediate frequencies. As discussed in section 6.4.2, these four conditions are implicit in the Kramers–Krönig integral relations (for an ideal system), which relate the real and imaginary parts of the impedance spectra over a wide range of frequencies. In this chapter, we also briefly discuss the empirical Debye and Cole–Cole relaxation models for gas/solid dielectrics (or insulators with negligible conductivity), because any impedance standard will inevitably be surrounded by, or will principally involve, a dielectric in its physical construction. These complex theoretical developments are complemented by equally complex and elaborate array of measurement techniques, which are discussed in more detail in the other chapters.

6.1 The history of impedance standards

Historically, the primary impedance standard was entirely arbitrary. In a laudable attempt to include it in the number of standards based on naturally occurring phenomena, the ohm was defined as the DC resistance of a column of pure mercury 1 m long and 1 mm^2 in cross section. Wire resistance standards copied and disseminated this for practical everyday use. The accuracy was of the order of 1 in 10^3.

Later, in the early 1900s, when AC impedance comparison bridges became commonplace, the ohm was redefined in terms of a Campbell mutual inductor – a device whose impedance depends almost entirely on its geometry, which had to be determined by making many dimensional measurements on the coils of wire it was made from. The accuracy was much better – of the order of 1 in 10^6.

Later still, in 1956, the Australians Thompson and Lampard [1] devised the calculable cross-capacitor that realised the SI ohm from the capacitance between four long, parallel cylindrical electrodes. The improvement this standard was achieved chiefly because the capacitance depends on the length of the capacitor so that only a single length measurement is required, and this can be made very accurately with a laser interferometer. But a long chain of bridge measurements is needed to relate the 2×10^8 Ω impedance of the capacitor at 1.592 kHz ($\omega = 10^4$ rad/s) in decade steps to the 1–1000 Ω values of maintained resistance standards. Even so, the accuracy attained with different versions of this device in various national laboratories is of the order of 5 parts in 10^8. Another reason for preferring the calculable capacitor to the older mutual inductor is that the inductor must be in the middle of a considerable space free of conducting or magnetically permeable materials, whereas the electric fields of a capacitor can be confined and completely defined by the conducting walls of its enclosure.

Recently, attention has focussed on the quantised impedance of a QHE device. At first, measurement chains begin with its DC resistance of about 12 906.4035 Ω. This resistance is known in SI units because the von Klitzing constant, h/e^2, which governs this value, is closely related to the fine structure constant of atomic physics [2]. The fine structure constant has been measured in SI units with an accuracy of

0.68 parts in 10^9 by frequency observations made on the spectrum of atomic hydrogen and by an independent method based on photon recoil from atoms. A chain of measurements is needed to relate the 12 906.4035 Ω to the value of practical working DC resistances, and another chain of AC measurements to relate this value to that of working capacitance standards. It is important for the coherence of the SI to discover whether the impedance unit set up in this way agrees with that derived from a calculable capacitor because this would imply that the QHE is indeed governed by h/e^2 to the accuracy of agreement obtained, with no unexpected corrections to the theory of the QHE. At the moment, the best evidence for this is that quantum Hall devices of various kinds including one based on the different physical system of graphene have been shown to yield the same DC resistance within a few parts in 10^{10}. Figure 6.34 shows a quantum Hall device mounted for DC or AC measurement.

6.2 The Thompson–Lampard theorem

Dimensionally, capacitance is the product of a length, a shape factor and the permittivity of the space exposed to its electric field. Thompson and Lampard's achievement was to show how to avoid having to make any measurements to determine the shape factor by finding a geometrical arrangement of the electrodes for which the capacitance is not dependent on the shape or size of the cross-sectional dimensions perpendicular to the principal axis, provided that these cross-sectional dimensions remain the same as the principal axis is traversed over the measured length.

With reference to Figure 6.1, if the intersections of four electrode surfaces with a plane perpendicular to their generators lie on the perimeter of a two-dimensional region of arbitrary and the shape of the region does not change, and if the gaps

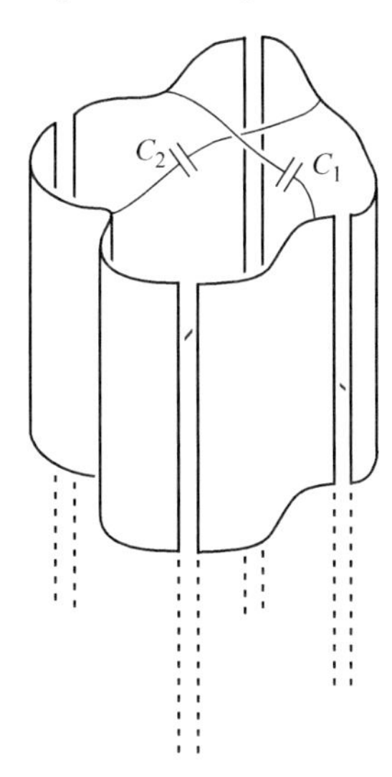

Figure 6.1 The Thompson–Lampard capacitance theorem

between the electrodes are negligibly small, in principle, over an infinite distance in the perpendicular direction, there is a relationship between the two capacitances per unit length of an electrode and the non-adjacent one that is independent of the shape of the region.

If the 'cross-capacitances' are C_1 and C_2 between the two opposite pairs of electrodes per unit length of the infinite system in a vacuum, this relationship is

$$\exp\left(\frac{-\pi C_1}{\varepsilon_0}\right) + \exp\left(\frac{-\pi C_2}{\varepsilon_0}\right) = 1 \tag{6.3}$$

where ε_0 is the permittivity of vacuum. Practical Thompson–Lampard capacitors are invariably evacuated to avoid having to know the relative permittivity of air between the electrodes. The relationship is not altered by the presence of thin, longitudinally uniform layers of dielectric material on the electrode surfaces – an advantage of considerable practical importance since oxide layers and contaminating coatings on polished metal surfaces cannot be entirely avoided [3]. We note later that this property enables the device to constitute a phase angle standard.

Usually, the electrodes are made as symmetrical as possible so that $C_1 - C_2 = \Delta C$, where ΔC is small. Then (6.3) can be expanded as a series for the mean value $C = (C_1 + C_2)/2$.

$$C = C_0\left[1 + \frac{\ln 2(\Delta C/C_0)^2}{8} - \cdots\right] \approx C_0\left[1 + 0.0866\left(\frac{\Delta C}{C_0}\right)^2\right] \tag{6.4}$$

where C_0 is the value obtained when $\Delta C = 0$, that is,

$$C_0 = \frac{\varepsilon_0 \ln 2}{\pi} = 1.953\,549\,043\,\mathrm{pF/m}$$

assuming the defined value of $\mu_0 = 4\pi \times 10^{-7}$ and the defined value of the velocity of light $c = (\varepsilon_0/\mu_0)^{-1/2}$ is 299 792 458 m/s. In practice, it is not difficult to adjust the cross-capacitances C_1 and C_2 to be equal to better than 1 in 10^4 (i.e. $\Delta C/C_0 < 10^{-4}$) so that (6.4) can be used to an accuracy of 1 in 10^9.

Because circular cylinders are the easiest mechanical objects to make accurately, they are the electrodes in all existing Thompson–Lampard capacitors. The operative surfaces are the inward-facing arcs shown in Figure 6.2b.

The need for electrodes of infinite length is overcome by inserting centrally up the axis between the electrode surfaces, another conducting cylinder, as drawn in Figure 6.2a. This movable electrode has a diameter that is slightly smaller than the distance between opposite electrodes, and thus effectively blocks the cross-capacitances. The electrostatic field does not change along the axis of the electrode system except in the region that lies just beyond the end of this inserted cylinder. Here, great distortions occur, but the distortion becomes negligible after a further distance down the axis of two or three times the diameter of the inserted cylinder. Therefore, since these end effects are the same with respect to the inserted cylinder no matter

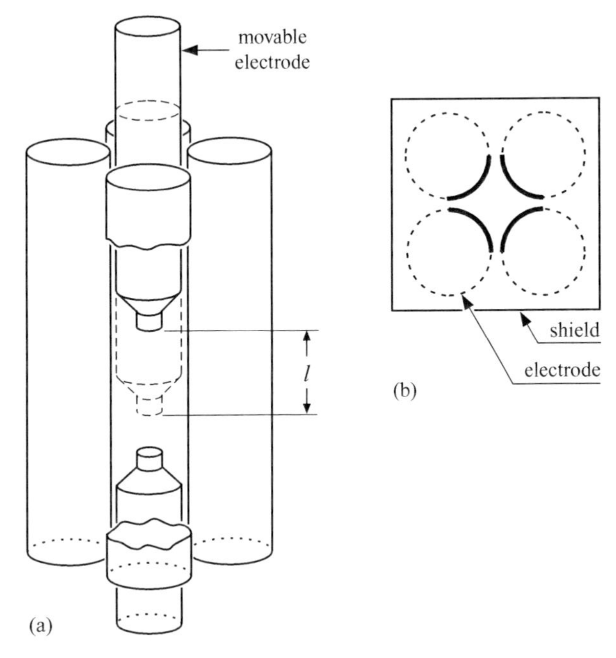

Figure 6.2 (a) A practical Thompson–Lampard capacitor. (b) The inward-facing surfaces of four cylinders

what its position with respect to the fixed electrodes, they can be eliminated by inserting the cylinder a further distance l and observing the change in the cross-capacitances. l is of the order of 0.25 m, corresponding to a capacitance change of about 0.5 pF, and is usually measured with a laser interferometer. The remote end of the electrode system can be terminated by a similar fixed inserted cylinder as shown in Figure 6.2a.

The main difficulty in the practical implementation of this beautiful concept arises from the initial assumption that the electrode surfaces have straight generators all parallel to an axis so that the cross section is always the same whatever the distance along the axis. Imperfections in the manufacture of the electrodes will inevitably mean that the electrode surfaces are not everywhere parallel to the axis, and consequently the non-uniform electric field at the end of the inserted cylinder will vary slightly with the position of this electrode, and the 'mechanical' distance l may no longer be quite identical with the 'electrical' distance. It is usual to improve this situation by providing a short extension to the cylinder of lesser diameter. For an empirically determined ratio of length to diameter of this extension, the effect of a slight linear divergence of the electrodes (e.g. if the electrodes are slightly inclined to one another) can be compensated so that the mechanical and electrical distances can again be made identical. The use of a properly adjusted extension is essential if uncertainties of less than a part in a million are to be achieved with electrodes whose departure from perfect straightness is of the order of a micrometre.

The operation of this extension can be understood by appreciating that displacing the four electrodes outwards increases the amount of electric field that passes around the extension, and this compensates for the small extra amount of the fringing field intercepted by the end of the extension. It is advantageous to provide the fixed cylinder at the remote end with an identical extension because then the cross-capacitances of the assembly will be immune to sideways displacements of the electrodes from whatever cause (including thermal expansion).

The electrodes are not exactly equipotential surfaces because they possess finite resistance and inductance. In particular, their inductance will cause errors of the order of parts in 10^7 if it is not minimised, or a numerically calculated correction made.

The type of bridge for relating the cross-capacitances to a fixed stable capacitor is discussed in sections 9.1.4 and 9.1.5.

6.3 Primary standards of phase angle

Calibrations of electrical power measuring equipment, of power loss in reactive components and of the phase angle of resistive components, need to be traceable to a primary standard of phase angle. Fortunately the phase angle of a sealed two-terminal-pair gas capacitor with clean electrode surfaces can be relied on to be less than 10^{-5}. This is sufficiently accurate for all ordinary purposes, provided that the electric field of the observed direct capacitance does not penetrate any lossy solid dielectric used in the mechanical construction. There is nevertheless a need for a primary standard of phase angle so that this statement can be made with confidence. There have been three main approaches to this problem.

In the past, mutual inductors at low or moderate frequencies have provided a standard of phase angle. Departure from the ideal phase angle of the mutual reactance arises from eddy current losses in the conducting materials of the windings and of any adjacent metallic material used for mechanical support or terminations and from the capacitance of the windings combined with their resistance. The energy associated with the electric field surrounding mutual inductors is small, but it also causes loss in the dielectrics of insulation or the supporting structure. Nevertheless, loss effects can be kept small and a phase angle standard with a phase angle approaching 10^{-6} radians can be obtained. A mutual inductance phase angle standard is particularly useful in calibrating wattmeters for the condition when the current and voltage waveforms are in quadrature. The instrument should register zero power under this condition.

A gas-dielectric capacitor in which the spacing of the electrodes can be varied is another approach [4]. A guard ring surrounds the low-potential electrode to ensure that no part of the electric field of the measured direct capacitance encounters solid dielectric material. A correction can be made for the loss in the resistance of the conductors connecting the electrodes to the terminals of the device. The only significant source of loss, which cannot be directly accounted for, arises from dielectric films of water, metallic oxide, etc. on the surfaces of the electrodes. The phase defect caused by this loss is inversely proportional to the electrode spacing and can be evaluated by varying it. Polished stainless steel is a suitable material for the electrodes, and the construction should minimise acoustic movement resulting from the forces exerted on the electrodes by the electrostatic field. As these forces are proportional to the square of the applied voltage, this subtle cause of loss can be investigated by observing the effect of varying the voltage applied to the capacitor.

The most precise primary phase angle standard makes use of a property of the symmetrical Thompson–Lampard cross-capacitor. If one of the electrodes is coated uniformly with a lossy dielectric film, its contribution to the phase angle per unit length of the two cross-capacitances is equal and opposite [3]. Neglecting small end effects, it follows that the phase angle of the mean capacitance is independent of axially uniform, but not necessarily identical, lossy films on the electrode surfaces. The end effects can be eliminated by measuring the difference between two inserted lengths of the fifth guard electrode, as in its usual role as a capacitance standard. Shields [5] has adopted a slightly different approach by constructing a toroidal cross-capacitor that has no ends. The cancellation of phase angles is not theoretically perfect as with the linear cross-capacitor, but is smaller than other causes of uncertainty such as acoustic loss, non-uniformity of any electrode surface film. A toroidal device is also useful as a secondary maintained capacitance standard.

Rayner [6] has shown that the self-inductance and capacitance of an ordinary electronic component resistor whose construction does not involve a cylindrical spiral path, whose value R is of the order of 100 Ω and which is mounted in a very loosely fitting coaxial tube, approximately cancel. The time constant τ $(=L/R$, where L is the effective inductance) of this two-terminal-pair device is less than 10^{-9}. Therefore, although it does not constitute a primary standard, this device makes a very simple, practical standard whose phase angle is less than 10^{-5} at frequencies below a few kHz.

Once a primary standard of phase angle has been established, the phase angle of other standards can be calibrated from it by means of the bridges described in sections 9.1.5 and 9.2.1.

A quantum Hall device in a proper measurement configuration (see section 6.7.5) probably constitutes an excellent phase standard, but this aspect has not yet been fully investigated.

6.4 Impedance components in general

In this section, we describe the desirable attributes of resistance, capacitance and inductance components, whether these are to be used as secondary standards in impedance standardising laboratories or as circuit components in critical parts of sensitive sensing and measurement circuits in general.

6.4.1 Capacitors

Capacitors have been exploited in a whole plethora of practical applications (such as dielectric measurements, resonators, filters, energy storage, power conditioning and sensing) ever since their inception over 250 years ago. Capacitance standards for impedance metrology are selected for their stability, small phase angle, small frequency dependence and small change of value with applied voltage. The main types of capacitance standards will be considered from the standpoint of their basic physical and theoretical properties, which dictate their particular application. These types are parallel-plate capacitance standards (section 6.4.2), compact high-frequency transfer standards between high and low frequencies (section 6.5.8) and a new type of capacitance standard that is coaxial and has a *calculable* frequency dependence (section 6.5.9).

Electrical energy E is stored in a capacitor in the space between and immediately adjacent to its two conducting electrodes. If the potentials of these electrodes, as shown in Figure 6.3a, where the electric field is indicated by light broken lines, differ by U, $E = U^2/2C$, where C is the capacitance. For example, for parallel plates of area A separated by a distance L, which is small compared to the size of the plates, $C = \varepsilon\varepsilon_0 A/L$ farads, where ε_0 is the permittivity of free space (vacuum) and ε is the permittivity relative to this of the dielectric between the plates.

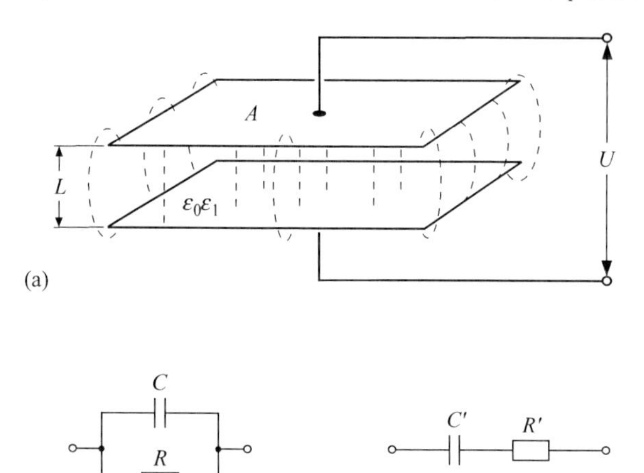

Figure 6.3 (a) A parallel-plate capacitor. (b) Parallel equivalent circuit. (c) Series equivalent circuit

For a stable standard, it is clear that (i) a stable geometry of the conductors and (ii) a constant dielectric relative permittivity are required.

Inevitably, some part of the electric field will spill out from the capacitor, in the fringing field at the plate edges for example, or from unscreened terminals. Any object in the vicinity of the capacitor that intercepts or modifies this field will alter the measured capacitance. Hence, it is best practice to enclose the whole capacitor and its terminals in a conducting screen. The effect of the screen on the capacitance is large but *definite* and becomes part of the *defined* capacitance. Objects outside the screen no longer affect the capacitance.

Two alternative lumped circuit representations of a lossy capacitor are drawn in Figure 6.3b and 6.3c. The capacitive and resistive parameters C, C' and R, R' of these equivalent representations are connected by the relationships

$$C' = C(1 + D^2) \quad \text{and} \quad R' = \frac{RD^2}{1 + D^2} \tag{6.5}$$

where

$$D = \frac{1}{\omega CR} = \omega C'R' = \tan\delta = \frac{1}{Q}$$

6.4.2 Parallel-plate capacitance standard

Parallel-plate capacitors have been employed to study the response of a variety of materials (solid, liquid or gas) to both quasi-static and alternating electric fields in order to ascertain their dielectric properties. The dimensionless real and imaginary parts of the complex relative permittivity $\varepsilon = \varepsilon_0\varepsilon_r = \varepsilon' - j\varepsilon''$ of a dielectric, where $\varepsilon_0 = 8.85 \times 10^{-12}$ F/m is the permittivity of free space, contain a great deal of information about the micro- and macroscopic nature of the material under investigation. The bulk of the work in this field, to date, has been mostly concerned with dipolar, polar and ionic relaxation effects (where the response of the induced polarisation in the dielectric is linear with the magnitude of the applied stimulus – voltage or current) up to around 10 GHz. For this purpose, the capacitor plates can be moderately large and have a variety of shapes relevant to the particular application, such as rectangular, circular, coaxial or interdigitated. Although the effect of the metallic electrodes on the measured dielectric properties of the material is not usually a major concern, they can play a significant role in electrolyte measurements, where interfacial polarisation effects can dominate the measured response. Only in the last few years efforts have been directed, at various research institutes, towards developing micro- and nanoscale capacitive, resistive and inductive devices and structures [7]. These make possible investigations at frequencies as high as 1000 THz (i.e. optical frequencies) where the dielectric properties can be dominated by electronic and atomic relaxation effects. Clearly the subject of capacitive devices and their applications in material measurements as well as in sensors and transducers is a very broad one. Therefore, the focus here will be confined to the use of parallel-plate capacitors as impedance standards and on their applications at frequencies ranging from power frequencies (50–100 Hz) to about 100 MHz, that is, the radiofrequency part of the electromagnetic spectrum.

A capacitor can be a sensor responding to a number of physical conditions, such as temperature, pressure, force, displacement, acceleration and humidity, but in designing or selecting a secondary standard of capacitance these responses must be minimised and, where necessary, specified in its calibration defining conditions. Stability is important if the calibrated value of a capacitance standard needs to be maintained as long as possible, in order to minimise the expense and time spent in recalibration.

The original unscreened two-terminal connection configuration of a parallel-plate capacitor is still widespread, but, for accurate and reproducible measurements it is inadequate because of fringing field effects. This limitation became apparent when demands on electrical metrology were increased after improved realisation of the SI units. Moreover, commercial electronic instrumentation required more accurate calibration from national metrology institutes. Thus, a three-terminal definition emerged (sometimes also referred to as 'guarded'), which improved the reproducibility of the capacitance standards by introducing a shield conductor that surrounds the main two-terminal capacitive element (see Figure 5.14). The *direct* capacitance (see section 5.3.2) between the two active electrodes could now be measured highly reproducibly in the presence of the shield conductor, with negligible influence from external interference. Reproducibilty is gained if the extra shunt capacitance between the high electrode and the shield merely loads the generator and that at the low terminal has no potential across it. Therefore, neither shunt capacitance affects the measured direct capacitance. It was not long before the limitations of the three-terminal definition and realisation also became apparent. The contact resistance between the terminals of the standard and the connecting cables to the measurement system (such as a bridge) lead to irreproducible changes in the measured dissipation factor (or loss tangent) of the capacitance standard. Both the real and imaginary parts of the dielectric constant must be determined for a complete measurement of a given capacitor. To do this accurately, Cutkosky in 1964 [8] devised the four-terminal-pair definition and realisation of a capacitance standard. The beauty of this definition lies in the universality of its application. The definition can be applied to capacitance, resistance and inductance standards of all values. For formal two-, three- and four-terminal-pair definitions of impedance standards, see sections 5.3.6, 5.3.7 and 5.3.8, respectively.

As noted earlier, the electrostatic energy of a parallel-plate capacitor is $\varepsilon = CU^2/2$, where $C = \varepsilon\varepsilon_0 A/d$, the area of the electrodes being A and their separation d. The permittivity ε of the dielectric in the capacitor has real ε' and imaginary ε'' components, and the latter represents an energy dissipation or real power loss P, given by $P = \omega C_0 U^2 \varepsilon''/2$, or equivalently $P = \omega\varepsilon_0\varepsilon'' AdE^2/2$, since $E = U/d$ and $C_0 = \varepsilon_0 A/d$ is the capacitance of a lossless capacitor of the same dimensions having no dielectric. The impedance Z of a dielectric-filled capacitor may simply be represented by a lumped electrical parameter model of a capacitor C in parallel with a shunt resistor R

$$Z = \frac{1}{G + j\omega C} = \frac{d}{j\omega\varepsilon_0 A(\varepsilon' - j\varepsilon'')} \tag{6.6}$$

where $G = 1/R$. Equation (6.6) is overly simplistic because in practice a real capacitor will also have lead resistance, inductance, shunt capacitance and even eddy current effects in the electrodes. These additional effects need only be included in a more sophisticated model if greater accuracy is required for meaningful comparison with measurements. From (6.6), it is clear that the capacitance is increased by a factor of ε' over that of a vacuum gap (or air/gas-filled) capacitor. The real and imaginary parts of Z are plotted in Figure 6.4 for a 1-pF standard

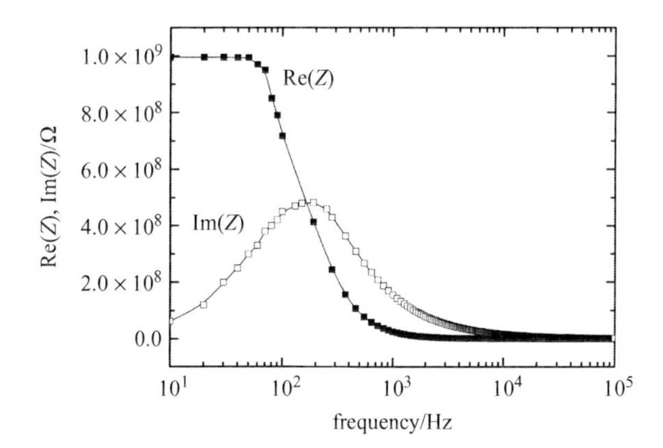

Figure 6.4 Frequency dependence of the magnitude of real and imaginary components of a 1 pF parallel-plate capacitor at frequencies up to 100 kHz

having an artificially low shunt resistance (for the purposes of this example only) of 1 GΩ at frequencies up to 100 kHz. The corresponding dissipation factor is then given by $D = \varepsilon''/\varepsilon' = 1/\omega CR = 0.159$ rad at 1 kHz.

The graphs in Figure 6.4 show that the real part of the impedance is constant at $f < 100$ Hz, whereas at higher frequencies it decreases quadratically with frequency. In contrast, the imaginary part of the impedance initially increases linearly with frequency and peaks around 200 Hz before eventually decreasing linearly with frequency.

The real and imaginary parts of the permittivity can be calculated from the parameters of a capacitor (measured with an impedance meter at, for example, 1 kHz) so that

$$\varepsilon' = \frac{Cd}{\varepsilon_0 A} \text{ and } \varepsilon'' = \frac{Gd}{\omega \varepsilon_0 A} \tag{6.7}$$

As shown in Figure 6.4, the real and imaginary parts of (6.6) are both frequency dependent, therefore the expressions in (6.7) may be regarded as low-frequency limits of (6.6). For example, for a 1-pF capacitor with a more realistic dissipation factor of 10^{-6} rad and a ratio of $A/d = 0.029$ (typical for a fused-silica standard), it can readily be shown using (6.7) that $\varepsilon' = 3.8$ and $\varepsilon'' = 3.8 \times 10^{-6}$. This compares well with data for fused silica from the literature of $\varepsilon' = 3.8$ and $\varepsilon'' = 20 \times 10^{-6}$ at 100 kHz and 20 °C.

The frequency dependence of the real and imaginary parts (or equivalently $\varepsilon'(\omega)$ and $\varepsilon''(\omega)$) of the impedance given in (6.6) can, with care, also be established from the Debye empirical model if three constants are known; the static or low-frequency (ε_S) and high-frequency (ε_∞) permitivities and the characteristic time constant (τ) of the dielectric, such that

$$\varepsilon_r = \varepsilon_\infty + \frac{\varepsilon_S - \varepsilon_\infty}{1 + j\omega\tau} \tag{6.8}$$

For a fused-silica dielectric:

$$\varepsilon' = \frac{\varepsilon_S + \omega^2\tau^2\varepsilon_\infty}{1 + \omega^2\tau^2} \tag{6.9}$$

and

$$\varepsilon'' = \frac{\omega\tau(\varepsilon_S - \varepsilon_\infty)}{1 + \omega^2\tau^2} \tag{6.10}$$

whereas in the Debye model for an *electrolyte* capacitor (i.e. materials having a small but finite conductivity, $\sigma = \varepsilon''\omega$, as well as permittivity), the numerator in (6.9) is replaced by ($\varepsilon_S - \varepsilon_\infty$). The use of a single time constant can be justified since the average response of the dipoles in the dielectric shows a resonant response at a particular frequency (or relaxation time $\tau = CR$). For completeness, (a) the relative permittivity ε_r is related to the index of refraction n and the absorption coefficient k (used at optical frequencies) through $\varepsilon_r = (n - jk)$, such that $\varepsilon' = n^2 - k^2$ and $\varepsilon'' = 2nk$, and (b) recently *metamaterials* have been developed that can have negative relative permittivity as well as negative permeability – topics that are beyond the scope of this book. A good review can be found in Reference 9. In practice, the Debye dispersion model is found not to be particularly accurate (since it is based on empirical analysis and each dielectric material can have substantially different features at the microscopic scale). As a result, Cole–Cole suggested a slightly modified empirical model:

$$\varepsilon_r = \varepsilon_\infty + \frac{\varepsilon_S - \varepsilon_\infty}{1 + (j\omega\tau)^{1-\alpha}} \tag{6.11}$$

where $0 \le \alpha \le 1$ is a dimensionless fitting parameter that accounts for a distribution of relaxation times, as found in real dielectric materials.

A fundamental theoretical model, which may be regarded as underpinning the Debye and Cole–Cole empirical models for dielectrics, is the Kramers–Krönig transforms (which correspond mathematically to the Hilbert integral transformation). The Kramers–Krönig transformations are very general because they are derived by assuming only the principle of causality (an effect cannot precede its cause). They enable the imaginary part $Z''(\omega)$ of an impedance to be calculated from a knowledge for all values of the real part $Z'(\omega)$ of the impedance for ω between 0 and infinity, or vice versa. It should be emphasised that none of these models actually predict what the frequency-dependent response of a given dielectric will be. The former models are normally fitted to a set of determined measurement data and relaxation constant, whereas the Kramers–Krönig transformations enable calculation, usually numeric rather than analytic, of either the real or the imaginary parts of the impedance, provided one or the other is known over a given frequency range. Thus, the real part $Z'(\omega)$ of the impedance is given by (6.12) in terms of the imaginary part $Z''(\omega)$ of the impedance:

$$Z'(\omega) = Z'(\infty) + \frac{1}{\pi}\int_{-\infty}^{\infty} \frac{\omega_0 Z''(\omega_0) - \omega Z''(\omega)}{(\omega_0)^2 - \omega^2} d\omega_0 \tag{6.12}$$

where $Z'(\infty)$ is the high-frequency value of the real part of the impedance. Similarly, the imaginary part of the impedance is given by (6.13) in terms of the real part of the impedance.

$$Z''(\omega) = -\left(\frac{\omega}{\pi}\right)\int_{-\infty}^{\infty} \frac{Z'(\omega_0) - Z'(\omega)}{(\omega_0)^2 - \omega^2} d\omega_0 \tag{6.13}$$

A difficulty that arises in evaluating these integrals is that a singularity (or a pole) occurs when $\omega = \omega_0$. However, there are mathematical methods available that circumvent this difficulty using the Cauchy principal integral. The key point to note about (6.12) and (6.13) is that they enable an important cross-check as to their accuracy, validity and self-consistency of any impedance spectra ranging from DC to optical frequencies, or whether there are any systematic errors in the measured data.

A further difficulty with Kramers–Krönig transforms is the infinite frequency range of measurements needed for accurate prediction of the conjugate impedance spectra. Some mathematical methods combined with assumptions about the impedance spectra being finite-valued as $f \to \infty$ or $f \to 0$ are usually used to circumvent this problem. It can be shown that the Debye model can be derived using (6.12) and (6.13). To date, it seems very little use of the above three models has actually been made in the audio- to radiofrequency range to complement impedance metrology. This may be because these models are more applicable to electrochemical and optical spectroscopy, but we see no reason why this should hinder their exploitation in radiofrequency impedance metrology, albeit with care. To illustrate the application of the Kramers–Krönig transforms, we apply them to a simple air-dielectric capacitor of $C = 1$ nF, which has an overall series inductance of $L = 0.05\ \mu$H and a dissipation factor of 10^{-4} rad at $\omega = 10^4$ rad/s (or equivalently a series resistance of $R = 10\ \Omega$). This circuit will produce a series resonance at approximately 22.5 MHz, given by

$$f_0 = \frac{1}{2\pi\sqrt{LC}} \tag{6.14}$$

The calculated real and imaginary parts of the impedance of the capacitor at frequencies up to 1000 MHz are shown in Figure 6.5a as well as the corresponding numeric evaluation of the Kramers–Krönig (6.12) and (6.13). Figure 6.5b shows the difference between the calculated and the Kramers–Krönig relations (6.12)

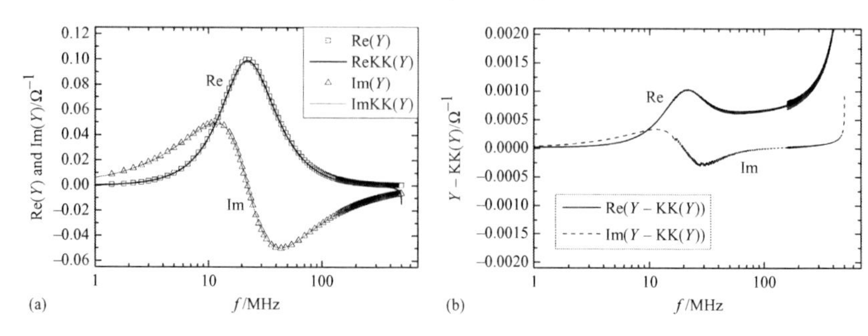

Figure 6.5 (a) Calculated and estimated (from (6.12) and (6.13)) frequency dependence of the real and imaginary components of a 1-nF parallel-plate capacitor having 0.05 μH series inductance and $D = 10^{-4}$ rad at frequencies up to 1000 MHz. (b) The corresponding difference between the calculated and estimated values

and (6.13). The agreement between the two methods is better than 0.5 mS (or 0.5%) near the resonant frequency f_0, but falls to 1.5 mS (or 1.5%) when the frequency of interest is well below or above f_0. Despite the differences at the higher and lower frequencies, these results illustrate that comparisons with actual measured data will give greater insight into the validity and accuracy of the assumed ideal-component inductance–capacitance–resistance (LCR) model and whether, for a given application, it requires modification.

6.4.3 Two-terminal capacitors

Capacitors in electronic circuits are usually unscreened two-terminal components. Their conducting surfaces are often two foils separated by a thin solid dielectric film, the sandwich then being rolled up tightly into a cylinder. Wires connected to each foil are brought out at the ends. Another common construction is a stack of alternating thin conducting and solid dielectric plates. Odd numbered conducting plates are connected together to one terminal and the even to the other. Because the plates or films are close together and the film between them has a high relative permittivity, the fringing field outside the assembly is weak compared to that inside; consequently, the capacitance associated with it is small, and neighbouring conducting or dielectric objects do not have any great influence on the apparent capacitance. Nevertheless, these effects are sufficiently large to exclude these components as accurate standards. For this purpose, they ought to be enclosed in a screen that is usually connected to the high terminal to which a source is connected. Then currents can flow from the screen to surrounding objects and return to the source without forming part of the measured current flowing out of the low terminal.

6.4.4 Three-terminal capacitors

A much better reproducibility is achieved by connecting a two-terminal capacitor to the inner conductors of ‘high’ and ‘low’ coaxial sockets and enclosing the capacitor in an enveloping guard screen that is connected to the outer of the 'low' coaxial port. In the usual way of defining its measuring conditions, the guard is to be maintained at the same potential as the low terminal, but not by being connected to it. It should be connected elsewhere in the measuring network in such a way that any current from it is diverted and does not form part of the measured current.

A configuration commonly employed in some gas-dielectric parallel-plate standards is illustrated in Figure 6.6. The screen is connected to the outer of the 'high' coaxial connector, and a short insulated overlap limits current to the outer of the 'low' coaxial connector. This break in the screen is not readily apparent, and the standard must not be mistaken for a two-terminal-pair component.

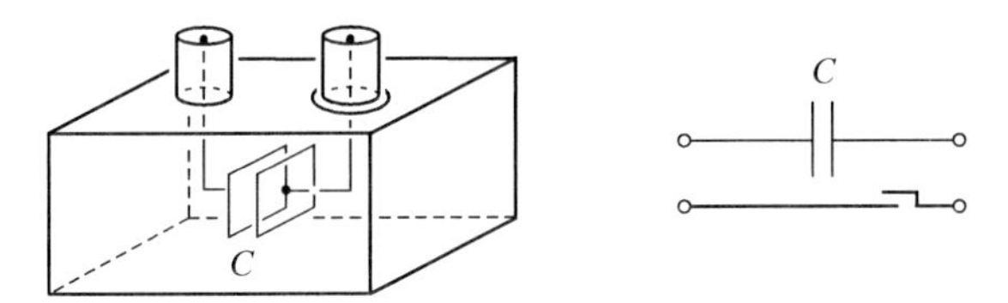

Figure 6.6 A three-terminal capacitor and its diagrammatic representation

6.4.5 Two- and four-terminal-pair capacitors

The screen is connected to the outers of all the coaxial connectors. The added complexity has the advantage, already explained in section 4.1.2, that the bridge networks into which such a standard is connected can, by current equalisation, also be made immune to external changing magnetic fields. The electrical definition of these standards (see sections 5.3.6 and 5.3.8) is sufficiently precise for even the most accurate measurements.

6.4.6 The mechanical construction and properties of various types of capacitors

The stability and phase of capacitors under environmental influences such as ambient temperature and humidity depend on their mechanical construction and whether they have a solid dielectric or whether the plates are supported so that they are separated and immersed in a dry gas or a vacuum in a sealed container. The latter construction yields low-loss capacitors of 10 nF or less whose vacuum or gas-dielectric has a relative permeability that is independent of frequency.

For values greater than this, the large plate areas and small separations required make construction impossible and so the electrodes must be separated by a solid dielectric. Some capacitors of value less than 1 nF are made that have pure fused silica as the dielectric, which have a phase angle of only 1×10^{-6} and which are exceedingly stable, but unfortunately the fragile nature of this dielectric makes it impracticable for constructing larger valued capacitors.

The principal types of high-quality capacitors that can be used for standards are listed in Table 6.1, which lists as a guide only the properties of various types of

Table 6.1 Types of capacitors and their properties

Type and dielectric	Range of values	Temperature coefficient per K	10% change in relative humidity at 20 °C	Barometric pressure effects per kPa	Stability per year	Phase angle at 1 kHz, μrad
Sealed; dry gas	0–1 nF	1–25	–	–	1–5	1–5
Unsealed; air	0–10 nF	1–25	20	6	1–5	1–25
Fused silica	0–100 pF	10	–	–	1	1–5
Mica	1 nF–1 μF	20–50	5–25 (unsealed)	–	5–25	50–200
NPO or COG ceramic	1 pF–0.1 μF	10	<1	0.05	1	100
Polystyrene	1 pF–1 μF	−200	–	–	50	60

capacitors that might be used as standards or in other critical applications. The units are parts-per-million for the changes indicated in the table headings. The properties of switched decade capacitance boxes are those of the capacitors contained in them, with some increase in phase angle at the higher settings of capacitance because of the resistance of the switch contacts and internal wiring.

The variation of the apparent value of a capacitor with frequency principally arises from two causes. An example of the first cause would be the finite inductance of the connections to the electrodes of large-valued capacitors. The finite resistance of these leads increases the apparent phase angle of the capacitor. A lead geometry having orthogonal current and potential connections, which minimises these effects, is described in section 6.5.7. The second cause is the frequency dependence of the relative permittivity of any dielectric between the electrodes. This is related to the loss in the dielectric, and a rule that is often applicable is that the order of magnitude of the proportional decrease in capacitance on increasing the applied frequency ten times is the same as the phase angle at the higher frequency.

The electrode structure of vacuum or gas-filled capacitors can deform slightly under their own weight. This distortion will depend on their orientation so that the capacitance will change as the capacitor is tilted. In extreme cases for calibrations of the highest accuracy, it may be necessary to specify the orientation at which the capacitor is calibrated.

Electrode assemblies may also be prone to disturbance by vibration or accidental jarring through handling, and changes in capacitance may result. This problem can be minimised either by making the electrode structure more rigid or by ensuring that the capacitance as a function of electrode displacement in particular directions is a maximum or a minimum and is independent of small electrode displacements in others. A satisfactory design can result from a combination of these techniques, and will have the additional advantage that the capacitance change resulting from differential thermal expansion between parts of the electrode structure will also be minimised. The temperature coefficient of capacitance then arises from thermal expansion of the electrode structure as a whole. Thermal expansion can be reduced by manufacturing from low thermal expansion materials such as Invar or fused silica. Another approach is to design the geometry of the electrodes so that thermal expansion of one part produces an equal and opposite effect on the capacitance to the expansion of another part. These thermally compensated capacitors are unfortunately sensitive to transitory temperature changes that produce temperature gradients within the electrode structure, and a period of several hours at a constant temperature may be required for the whole capacitor to be at the same temperature and the capacitance to become constant.

The best standards for work of the highest accuracy have a built-in resistance thermometer; it is then possible to use as their temperature coefficient the ratio of their change of capacitance to the change in resistance of the thermometer and to refer their defined capacitance to a certain value of this resistance. This simplification means that it is never necessary to consider their temperature per se, but does have the slight disadvantage that any drift in the resistance of the thermometer appears as if it were a drift in the capacitance itself.

The Andeen-Hagerling company makes a range of instruments for measuring two-terminal-pair capacitances in the range 0–1 μF with an accuracy of 3 ppm, a resolution of up to 0.5 aF, and accompanying loss tangents down to 10^{-8}. This remarkable performance is achieved by basing these instruments on an internal temperature-controlled parallel-plate capacitance standards having fused quartz as their dielectric and comparing this with the capacitor to be measured via a voltage ratio transformer. Models are available that can measure over a frequency range from 20 Hz to 20 kHz.

They also supply temperature-controlled capacitors in values of 1, 10 and 100 pF, housed in separate enclosures to serve as laboratory standards for general calibration and inter-comparison purposes. Other values can be made to special order. Their remarkable properties include a relative capacitance stability of the order of 10^{-7}/year, an exceedingly small hysteresis of capacitance when temperature-cycled, for example, between their 55 °C controlled temperature and ambient temperature if the temperature control is switched off. They also transport well between laboratories when relative changes of as little as 2 in 10^8 are found.

There is one characteristic to be aware of. As supplied, the internal wiring between the inner conductor of the 'Hi' coaxial port and the internal capacitor has no screen between it and the temperature-control circuitry. That is, the capacitors have a three-terminal configuration with exposed 'Hi' connections. This can lead to measurement error and serious interference problems unless the 'Hi' port is driven from a voltage source of small internal impedance such as a ratio winding of a voltage ratio transformer or the 'Hi' connection of an Andeen-Hagerling bridge. Even if this is done, there can still be a small measurement error.

Some laboratories have therefore dismantled their Andeen-Hagerling capacitance standards and added shielding to the 'Hi' side to convert the capacitors to a two-terminal-pair configuration. For national and international comparisons of the greatest possible accuracy, the standards can easily be further modified to a four-terminal-pair configuration by adding T-connectors at their output ports (see section 5.3.8). This modification has the advantage that no account need be taken of the unknown cable properties within its enclosure when calculating cable corrections – these properties simply become part of the defined value of the standard.

Two- or four-terminal-pair capacitors of very small value, known as Zickner capacitors, can be made by interposing between two plates a screen having a small aperture in it. This screen is part of the surrounding conducting case of the capacitor as shown in Figure 6.7.

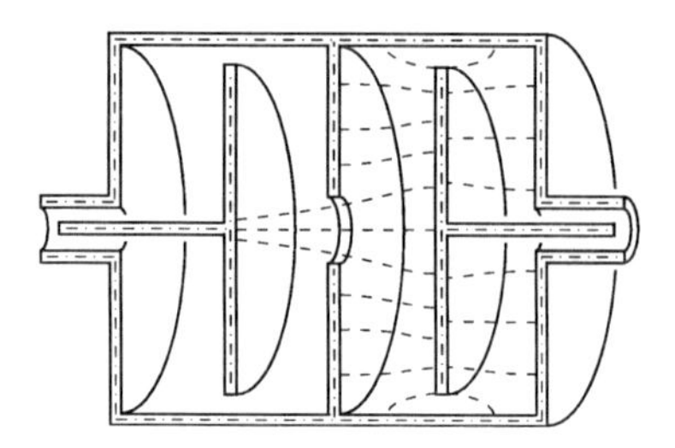

Figure 6.7 A Zickner capacitor

By interposing interchangeable screens with apertures of different sizes any capacitance from 1 pF to 0 (a screen with no aperture) can be obtained.

6.4.7 Capacitance standards of greater than 1 μF

The electronics industry uses large numbers of electrolytic capacitors and similar types of large storage capacity. Values are typically from 1 μF to 1 F at frequencies from DC to a few hundred hertz. Quality control and standardisation require that the capacitance of these components be measured to a modest accuracy of the order of 1%. To verify the accuracy of the commercial bridges that make these measurements, stable standards of at least the decade values within this range are needed. Electrolytic capacitors themselves are not stable enough as a function of time, applied voltage (AC or DC) or temperature, to be useful as standards. Non-electrolytic capacitors such as polystyrene film – metal foil–rolled capacitors – are good enough but are only available as single components in values up to 1 μF. Other plastic film materials such as polycarbonates enable rolled capacitors to be made in values up to 10 μF, and they can be connected in parallel to provide standards of up to, say, 100 μF in value. Although their stability and phase angle are not quite as good as polystyrene-rolled capacitors, they are good enough for the present purpose.

Standards of apparently greater capacitance can be made as a transformer-aided component. An elegant arrangement with two autotransformers tapped at fractions α and β is shown in Figure 6.8a.

The direct admittance Y_T of this device is the ratio of the current I flowing into and out of one pair of terminals to the open-circuit voltage U that appears as a result at the other pair. The device is therefore a four-terminal component having separate

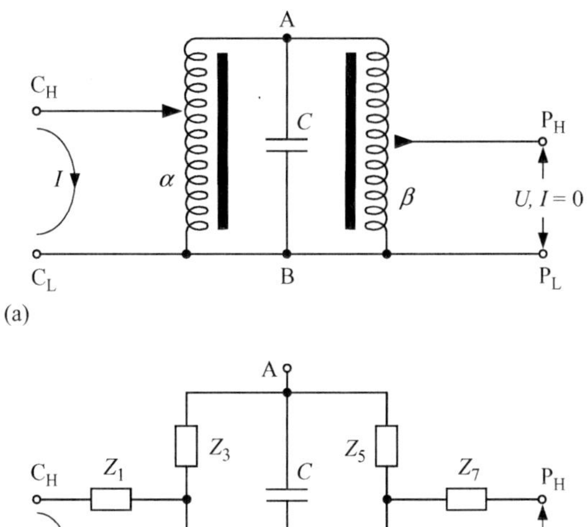

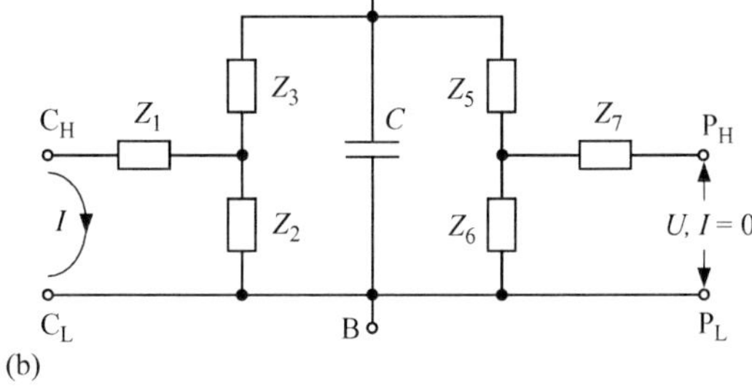

Figure 6.8 (a) Diagrammatic representation of a transformer-enhanced capacitance. (b) Its equivalent circuit

pairs of current and potential terminals. Y_T is principally the admittance of a large equivalent capacitance C_T.

The transformers can be represented as equivalent T-networks, as shown in Figure 6.8b, where Z_1 and Z_7 are small impedances arising mainly from the resistance of the transformer windings with a small contribution from the transformer leakage inductances. $Z_{2,\ 3,\ 5\ \mathrm{and}\ 6}$ are primarily composed of the transformer winding inductances that are designed to be much larger than $1/\omega^2 C$.

$$Y_T = \frac{j\omega C[1 - (1/\omega^2 LC)]}{\alpha\beta}$$

where $j\omega L$ is approximately the paralleled impedance of the two transformers:

$$(j\omega L)^{-1} \approx (Z_2 + Z_3)^{-1} + (Z_5 + Z_6)^{-1}$$

If we define $C' = C[1 - (1/\omega^2 LC)]$,

$$Y_T = \frac{j\omega C'}{\alpha\beta}$$

and we have the very simple result that

$$C_T = \frac{C'}{\alpha\beta} \tag{6.15}$$

C' is the apparent two-terminal capacitance between A and B with the transformers connected and the current and potential terminals open-circuited. C' varies slightly with the voltage U because L depends on U, which affects the magnetic flux density in the transformer cores. Provided that the voltage between A and B is measured and a small predetermined correction made, this is not a serious limitation at the accuracies envisaged.

The principal disadvantage of this otherwise elegant idea is the impedance present in the upper current and potential leads. These impedances can be expressed as

$$Z_C = \frac{Z_{Co} + \alpha(\alpha - \beta)}{j\omega C'} \tag{6.16}$$

and

$$Z_P = Z_{Po} + \frac{\beta(\beta - \alpha)}{j\omega C'} \tag{6.17}$$

where Z_{Co} and Z_{Po} are the impedances measured in the C_H and P_H leads with A shorted to B. The first term of each expression is the resistance and leakage inductance of the transformer, and the second is a capacitance that is positive or negative, or zero if $\alpha = \beta$.

To obtain a ratio between C and C_T that is an odd power of 10 with this restriction requires $1/\alpha = 1/\beta = 10^n\sqrt{10}$. Since $\sqrt{10}$ is irrational, this cannot be accomplished with a finite number of turns on the transformers, but a close approximation can readily be obtained.

The General Radio model number 1417 standard is based on this principle. To provide all the decade values from 1 μF to 1 F in one unit, the transformers have switch-selected tapping points. The transformers are wound with a large number of turns of fine wire to ensure that the difference between C and C' is not too great because the impedance of the inductance L of the transformers in parallel is made much greater than that of the capacitance C so that variations in L do not cause unacceptably large variations in C_T. Unfortunately, as a consequence, the windings have large resistances that result in somewhat large, mostly resistive, impedances in the C_H and P_H leads for values other than 1 μF. These impedances are of the order of the direct impedance for the 10 μF range at 100 Hz and rise to about 50 times the direct impedance for the 1 F range at 100 Hz. Note that the impedances in the C_L and P_L leads are much lower as they arise from only a short length of connecting wire, and this enables a standard of this kind to be compared with another using a comparatively simple bridge such as that in section 9.1.3 and the four-terminal to four-terminal-pair adaptor in section 5.3.10.

The actual construction of the device incorporates networks that are associated with the switching of the transformer taps and that provide a constant small phase angle for the complete device that is independent of the value selected. By other switched networks, the same phase angle is achieved at the three frequencies of 100 Hz, 120 Hz and 1 kHz.

As an alternative design, if C is a 100-μF capacitance made from polycarbonate components, the number of turns needed on the transformers can be greatly reduced. For example, the transformer ratios α and β could both be 127/400 turns for a 1-mF value and 40/400 turns for a 10-mF value. By having α and β equal the second terms in (6.16) and (6.17) above are zero, and since C is 100 times the value of that in the General Radio design, 1/10th of the turns are needed on the transformers to give L a suitably high value. Therefore, they can be wound with a heavier gauge of wire with a resulting lower value for the C_H and P_H lead resistances. The approximation of 400/127 to $\sqrt{10}$ results in a capacitance that is slightly different from 1 mF, but it can be trimmed if required by shunting C with a small shimming capacitance.

6.4.8 Voltage dependence of capacitors

Even well-designed and constructed standard capacitors are not quite perfectly linear circuit elements; their apparent value depends to some small extent on the voltage applied to them. Mechanical movement of the electrodes under the force exerted by the electric field between them and dielectric effects arising from the strong field gradients at the edges of thin-film electrodes coated onto solid dielectric separators are among the causes of this phenomenon. For the best standard capacitors whose dielectric is vacuum, dry gas or solid pure fused silica and whose values are in the range from 0 to 1000 pF, changes are much less than a part in a million for the applied voltages in the range from 1 to 100 volts, which are commonly used in high-accuracy applications. Nevertheless, in the most accurate work, the change with voltage must be accounted for, in particularly, when calibrating voltage transformer ratios by the permuting capacitors method (see section 7.4.6). The value of a standard should be defined as that measured at a specified applied voltage. This value might be the value obtained by extrapolation to a negligibly small voltage. It is not possible to make a reference capacitance standard having a guaranteed known small change of value with applied voltage, but the change in value of a capacitor can be determined in terms of the increments produced by factors of two increases in applied voltage, provided that three capacitors of the same nominal value are available, together with 2:$-$1 and 1:$-$1 ratio transformer bridges. The method of Shields [10], which is fully described in Kibble and Rayner [11], assumes that the ratio of a properly constructed 2:$-$1 voltage transformer (Figure 7.46) can be measured to a sufficient accuracy, say 1 in 10^9, either by the method of section 7.4.3 or by exchange of nominally equal capacitors in to 1:$-$1 voltage bridges centred on the 0 and +1 taps.

Recent improvements in making AC measurements of capacitors directly in terms of the QHE [12] offer an alternative. The voltage coefficient of quantum Hall devices can be adjusted to be very small by comparing them with resistors whose voltage coefficient is negligible at the very low power applied to resistor and device. Then the variation in the values of capacitors with applied voltage can be obtained by comparison with quantum Hall devices at even lower values of power dissipation.

6.4.9 Resistors

The primary requirement for a resistor is that its resistance should be as constant as possible with time and for the range of currents and frequency and with ambient changes of temperature, pressure humidity, etc. within its design limits. For high-accuracy impedance measurements, the electric and magnetic fields that inevitably accompany current flow in the device should be minimised so that the reactive part of its impedance is small. Any energy loss associated with these fields should also be small so that as much as possible of the energy dissipation takes place in the resistance element. Undesirable losses include dielectric loss or loss arising from eddy currents induced in either the surrounding metal such as other parts of the resistance element (the proximity effect) and each individual part of the resistance element (the skin effect). These effects are frequency dependent and therefore lead to a frequency dependence of the resistance.

The principle cause of change with time is change in the state of strain of the wire. All resistors are to a certain extent also strain gauges. The resistance element should be mounted with as little mechanical restraint as possible, consistent with a sound mechanical construction, and joints or connections between the resistive material and other conductors must be made by a stable technique. Any jointing technique such as soft-soldering, crimping, pressure-welding and brazing will produce a surrounding region of strain. The extent of this region should be minimised.

Only chemically inert substances should be used in the construction. The resistance element is often immersed or embedded in oil or a flexible potting medium to exclude the corroding effects of the atmosphere and aid heat transfer.

The most successful material for the resistance element is a four-component or quaternary alloy of 75% Ni, 20% Cr, 2.5% Al, 2.5% Cu made under the brand name of Evanohm. Its change of resistivity with temperature is of the order of ppm/°C and is parabolic. The parameters of the parabola can be modified to alter the resistance and its temperature coefficient to a limited but useful extent by the heat treatment given to the wire as part of the strain-relieving process during manufacture of the resistor. By positioning the maximum of the parabola near the desired working temperature, the temperature coefficient at that temperature can be made very small. Evanohm has a small thermoelectric Peltier coefficient with respect to copper, which is important for DC and very low frequency applications where the combination of Peltier or Thompson effects can alter the apparent resistance. Peltier heating or cooling at alloy–copper junctions can combine with the thermal capacity of the junction to produce a thermoelectric emf that can persist into the next cycle of current reversal.

The main constructional techniques that attempt to put these principles into practice are described in the following paragraphs.

The most stable resistors having values in the range from 0.1 to 10^4 Ω are constructed with minimally supported bare wires. They are the resistance standards of national and other calibration laboratories. The resistance wire is stout enough (0.1–3 mm is usual) to be self-supporting without touching neighbouring turns between the supporting points on the framework. The aim is to ensure that the wire is not under tension so that no change in strain, and therefore in resistance value, occurs. The winding on its supporting framework should enclose only a small open-loop area to minimise its inductance; less attention need be paid to stray capacitance as it is less important for these lower resistances. These 'S-class' resistors are now designed to be suitable for AC impedance measurements, although DC methods are still the principal means of intercomparison. Binding-post terminals provide a four-terminal definition that ought to be modified to a four-terminal-pair definition by connections to added coaxial sockets or by the adaptor described in section 5.3.10 if accurate AC measurements are to be made.

The resistance element is sealed in an oil-filled metal container and the terminals are fixed in an insulating top, or through insulating bushes. A well for a thermometer lies alongside the resistance element. The entire resistor can be immersed in a temperature-controlled oil-bath or metal block.

Figure 6.9 shows the construction of a premier class standard resistor of this type. The resistance element is supported on a skeletal insulating framework. Somewhat loose winding and annealing at about 180 °C for several days ensures the vital strain-free condition.

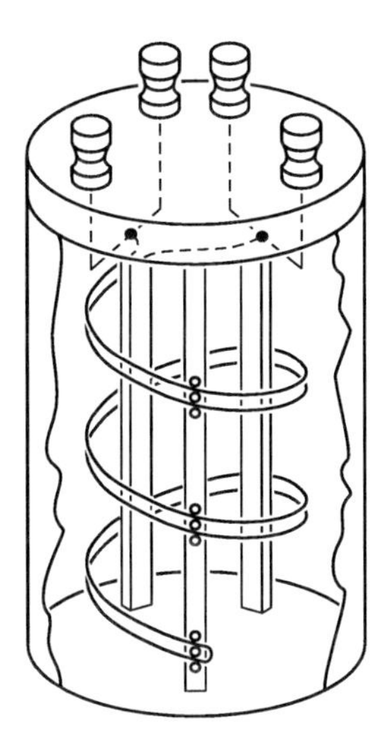

Figure 6.9 An S-class standard resistor

Standard resistors in the range from 10^3 to 10^6 Ω are made by winding enamel-insulated wire on to small thin cards, usually of mica. The reactance of these resistors, particularly at the higher end of the range, will arise principally from inter-turn capacitance. The dielectric loss associated with capacitances should be small enough to ensure that the resistance is not altered significantly. The inductance of the wound resistor should also be kept small, but the fine wire, the large number of turns and their close proximity set a limit to what can be achieved. Careful annealing helps ensure the stability of the resistance.

A range of resistors from 1 Ω to nearly 2 MΩ manufactured by Vishay are relatively inexpensive and have properties that make them suitable as secondary standards or components in networks where very good stability is required. Their meandering ribbon thin-film construction on a few millimetres square substrate of a material that is closely matched to the temperature coefficient of expansion of the resistive material confers low inductance and self-capacitance so that they are useful for both AC and DC applications. Their temperature coefficient is typically 2 ppm/°C, and there is a more limited range of decade values having a lower temperature coefficient of 0.3–0.5 ppm/°C. When protected by a suitable varnish and potted in a flexible medium, these resistors can be remarkably stable and are capable of a similar performance to S-class resistors. But as a cautionary note, this method of construction is similar to that of electrical strain gauges, and this should be borne in mind when mounting and using them.

The very inexpensive laser-trimmed miniature chip versions made by the million that are designed for automated soldering into printed-circuit boards are useful if a stability of some tens of ppm is adequate. They can exhibit even smaller changes of value with increasing frequency even up to hundreds of MHz if terminated as in section 6.5.7.

Another way of constructing stable resistors in the range from below 1 to 10^7 Ω or greater is to wind insulated resistance wire on a bobbin. In the best construction, the winding sense is reversed one or more times to minimise the self-inductance, which usually dominates the self-capacitance of this type of resistor.

6.4.10 *T-networks*

If, in designing a bridge network, inconveniently large resistances or small capacitances are needed, these components can be generated by T-networks as shown in Figure 6.10a and 6.10b.

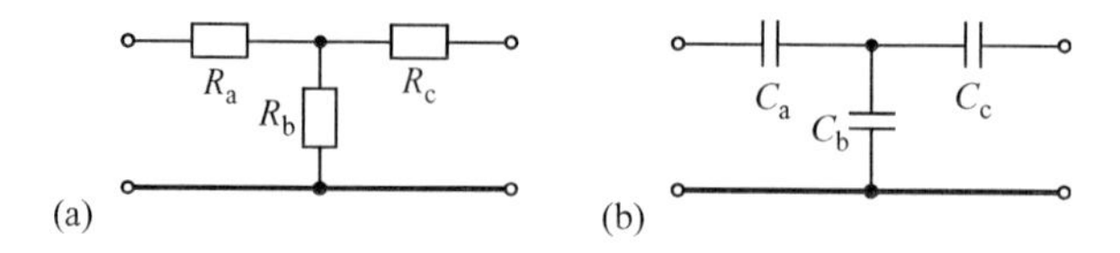

Figure 6.10 T-networks to generate (a) high-valued direct resistances or (b) low-valued direct capacitances

R_a and R_b can be thought of as forming a potential divider ($R_a \gg R_b$) to supply a reduced voltage to R_c that does not greatly load R_b if $R_c \gg R_b$. An exact value for the direct resistance is obtained from a T–Δ transformation (see Appendix 1).

Similar considerations apply to the network b.

6.4.11 Adding auxiliary components to resistors to reduce their reactive component

The energy stored in the electric and magnetic fields associated with a resistor is a small fraction of the energy dissipated per cycle by the resistance. In this book, we are principally concerned with well-defined resistors in which these fields lie within an outer conducting enclosure. The stored energy may be represented as a lumped reactive component either in parallel or in series with the resistance.

If the energy is stored principally as a magnetic field, it is more appropriate to represent its effect as a lumped inductive component L in series with R as shown in Figure 6.11. An alternative representation is a capacitance C which has a negative value in parallel with a resistance R' as shown in Figure 6.12. If the energy is stored principally as an electric field C has a positive value.

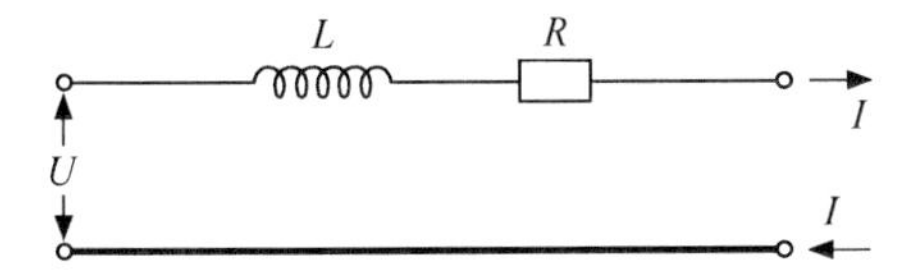

Figure 6.11 Equivalent circuit of an inductive resistor

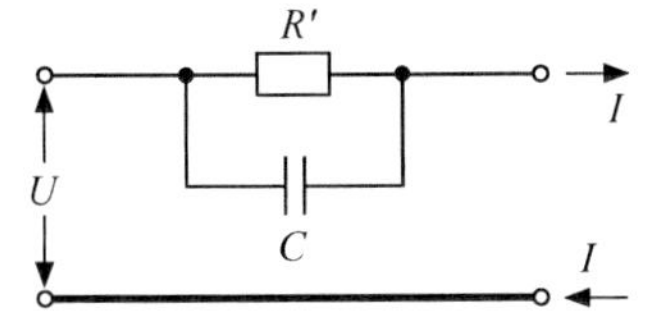

Figure 6.12 Equivalent circuit of a capacitive resistor

These two representations are equivalent if

$$\frac{R'}{1 + j\omega CR'} = R + j\omega L$$

That is, if

$$R = \frac{R}{1 + \omega^2 C^2 R'^2} \tag{6.18}$$

and

$$\begin{aligned} L &= \frac{-CR'^2}{1 + \omega^2 C^2 R'^2} \\ &= -CR'^2 \quad \text{if } \omega CR' \ll 1 \end{aligned} \tag{6.19}$$

The phase angle of the resistor is the dimensionless quantity $\omega L/R$ or $-\omega CR'$. The quantity $\tau = L/R = -C/R'$ has the dimensions of time and is termed the time constant of the resistor. Its value should be 10^{-7} seconds or less for a well-designed AC standard resistor.

Whether a series inductance or a parallel capacitance is appropriate to represent the phase angle of a resistor modified by both magnetic and electric fields usually depends on whether the phase angle is positive or negative. The capacitance across one of the types of resistor described above is likely to be of the order of a picofarad or less; hence from (6.19) above, for a 10^4-Ω resistor,

$$L = -10^{-12}(10^4)^2 = -10^{-4}\,\text{H}$$

The actual residual inductance is unlikely to exceed 10^{-5} H so that the shunt capacitance representation is appropriate. On the other hand, for a 1-Ω resistor, the equivalent inductance for a shunt capacitance of 1 pF is

$$L = -10^{-12}(1)^2 = -10^{-12}\,\text{H}$$

and the actual residual inductance will certainly exceed this value; hence, the series inductance representation is now likely to be appropriate.

The capacitive and inductive contributions can cancel, resulting in a very small time constant. This is likely for a resistance value of the order of 100 Ω, the exact value depending on the method of construction. This property is made use of in section 6.3.

Auxiliary components can be added to resistors to reduce their phase angle. The two cases are treated separately.

(i) If the resistor has a predominately inductive residual reactance, it can be shunted by a capacitance C, which, if $\omega L \ll R$, can be adjusted to have the value $L/(R^2 + \omega^2 L^2) \approx L/R^2$. Then the total impedance is $Z \approx R$ (see Figure 6.13).

A more elegant solution that compensates over a wide range of frequencies is shown in Figure 6.14. An auxiliary resistor whose value is approximately equal to R is connected in series with C. If C is adjusted to equal L/R^2, $Z = R$ for *all* frequencies for which the lumped parameter representation is valid.

(ii) The resistor has a predominantly capacitive residual reactance.

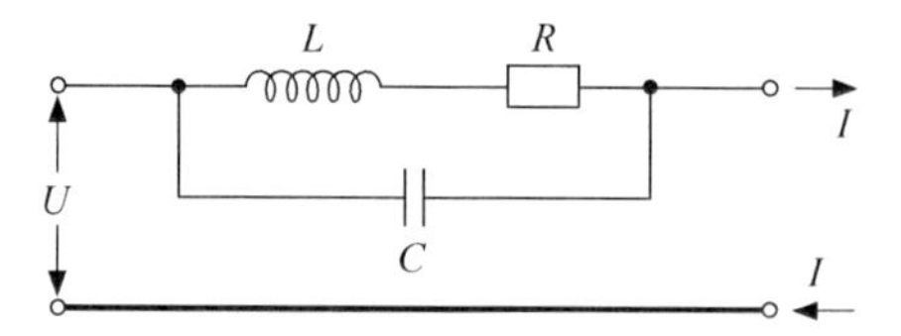

Figure 6.13 Compensating an inductive resistor

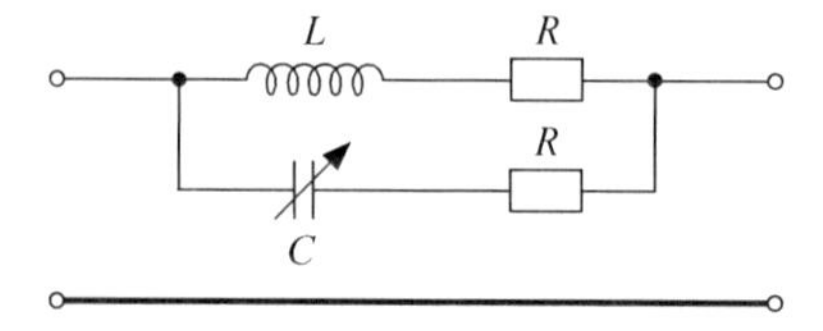

Figure 6.14 Compensating an inductive resistor over a wide frequency range

The roles of inductor and capacitor in case (i) cannot readily be reversed for this case. Whereas an adjustable capacitor of high impedance and small phase angle can readily be obtained, this is not the case for an inductor.

A practicable solution is to design the resistor so that it has two nominally identical sections with a tapping point between them, each section having a capacitive phase defect. Phase angle compensation can then be carried out by connecting an adjustable capacitance C' from the tapping point to the outer conductor, as shown in Figure 6.15. A T–Δ transformation of the network then enables it to be represented as in Figure 6.16 and the direct admittance to be calculated as

$$Z = \frac{U}{I} = \frac{2R}{2 + j\omega CR} + \frac{j\omega C' R^2}{(2 + j\omega CR)^2}$$

For resistors of small phase angle, that is, if $\omega CR \ll 1$ and $C' = 2C$, Z will be real and equal to R.

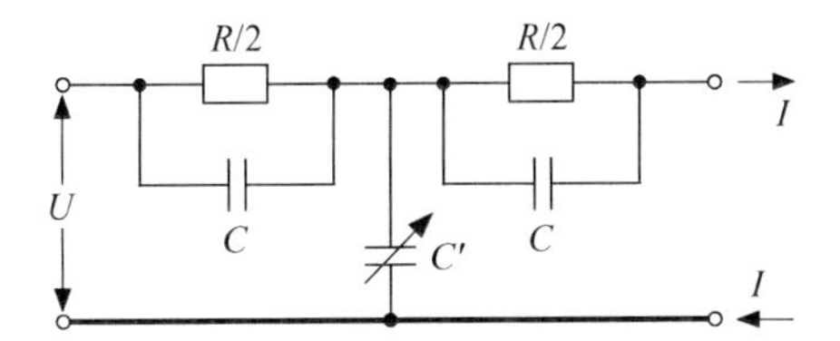

Figure 6.15 Compensating a capacitive resistor

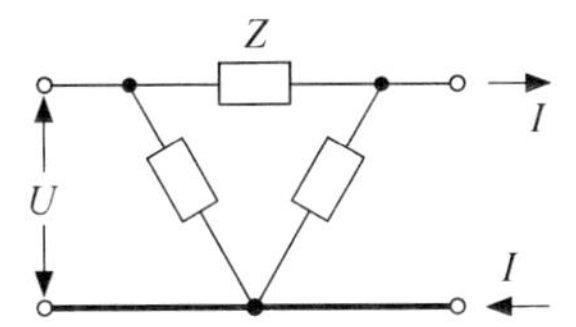

Figure 6.16 The equivalent direct impedance

In practice, the two sections will not quite be identical but C' can nevertheless be adjusted to make the phase angle of the direct impedance of the device zero.

6.4.12 Mutual inductors: Campbell's calculable mutual inductance standard

Before the advent of the Thompson–Lampard calculable capacitor in 1958 [1], various NMIs employed either self- or mutual inductance standards in order to realise the unit of resistance and reactance. Self-inductance standards were relatively straightforward to construct and use in practice, because the value of resistance standards could be derived from their calculated values using bridge comparison techniques. The accuracy of the calculated values for such standards was severely limited due to the number of dimensional measurements that were needed. In addition, the mechanical instability of the standards also limited their accuracy. Consequently, Campbell in 1907 [13] proposed a calculable mutual inductance standard based on a coaxial set of coils.

A current I in a primary coil produces a proportional magnetic flux ϕ that threads an adjacent secondary coil and induces a voltage U. That is, $\phi = MI$, where the quantity M is called the mutual inductance between the primary and secondary coils, and is in quadrature with I, since if $I = I_0 \sin \omega t$,

$$U = \frac{-d\phi}{dt} = -I_0\omega M \cos \omega t = -I_0\omega M \sin(\omega t + 90^\circ)$$

Mutual inductance has attributes that, in the past, led to mutual inductors being widely used in bridges for reactance measurements. It is an additive quantity in that if a primary coil induces magnetic fluxes in two or more secondary coils, their individual induced voltages may be added. Therefore, neglecting capacitive effects, a precise linear scale of mutual inductance can be obtained from a primary coil and tapped secondary coils.

The primary winding of a Campbell mutual inductance standard consisted of two single-layer coils connected in series, wound on the same former, but separated along the former by a distance equal to the radius of the coils, as shown in Figure 6.17. The secondary of the mutual inductance standard was wound centrally in between the two primary coils and consisted of several layers. This design conferred several advantages. In particular, the calculated mutual inductance only required five accurate dimensional measurements (such as the mean primary coil diameter and pitch length, and secondary coil diameter, width and depth). In addition, at a certain value of the radius of the secondary winding, the mutual inductance almost becomes independent of small changes of this radius or axial displacement (i.e. $dM/da = 0$, where M is the mutual inductance and a the radius of the secondary winding). In this configuration, the secondary coil is positioned such that all round its mean circumference of the field due to the primary coils cancel. Campbell built and

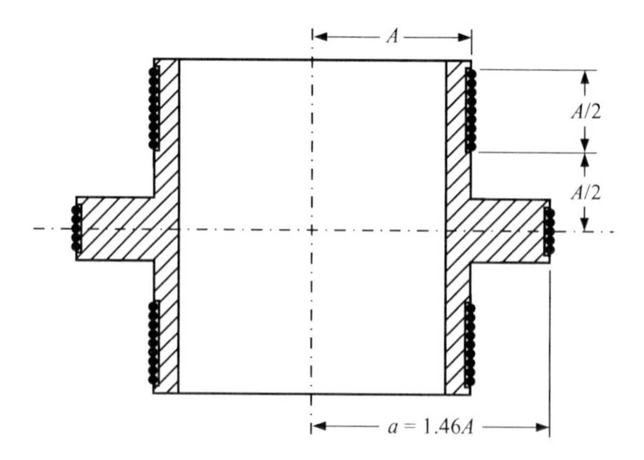

Figure 6.17 A Campbell mutual inductor

tested a nominal 10 mH mutual inductance standard that had 75 turns each for the two primary windings wound on a marble cylinder of 150-mm radius and a secondary winding with 488 turns wound in a channel of 100 mm^2 with a mean radius of 219 mm (at a centre-to-centre distance between the primary and secondary windings of $b = 150$ mm).

The mutual inductance of the standard can be expressed analytically using elliptic integrals. Thus, if A is the radius of the primary winding with n_1 turns and a is the radius of the secondary winding with n_2 turns, then the mutual inductance is, in SI units, given by

$$M = \mu_0 n_1 n_2 (A + a)\left\{\frac{c}{k}[F(k) - E(k)] + \frac{A - a}{b}\psi\right\} \tag{6.20}$$

where $\mu_0 = 4\pi \times 10^{-7}$ H/m is the free-space permeability, $E(k)$ and $F(k)$ are the complete elliptic integrals of the first and second kind to modulus k, respectively, b is the distance between the primary and secondary coil centres with

$$k^2 = \frac{4aA}{(A + a)^2 + b^2} = \sin^2(\gamma) \quad c^2 = \frac{4aA}{(A + a)^2} \tag{6.21}$$

and

$$\psi = \frac{-\pi}{2} - F(k) \cdot E(k', \beta) - [F(k) - E(k)]F(k', \beta) \tag{6.22}$$

where $E(k', \beta)$ and $F(k', \beta)$ are the incomplete elliptic integrals of the first and second kind, respectively, $k' = \cos(\gamma)$, $c' = \sqrt{1 - c^2}$ and $\beta = \sin^{-1}(c'/k')$. Thus (6.20) shows that the mutual inductance is calculable using only μ_0 and linear dimensions (hence, the emergence of the Thompson–Lampard calculable capacitor that similarly requires ε_0 and, crucially, measurement of only one linear dimension, see section 6.2). In practice, small corrections are made to the ideal expression for mutual inductance given by (6.20) for the finite relative permeability of the former and the non-uniform distribution of the current in the wires. Consequently, the initial accuracy of the standards was limited to around a few parts in 10^5, but was later improved in the 1960s by, for example, G.H. Rayner at NPL, to around 1 part in 10^6.

Measurements of practical mutual inductance standards, for precision work, were typically performed at low frequencies between 10 and 125 Hz using a bridge comparison method because (6.20) is accurate only as the frequency f, at which the inductor is operated, tends to zero. Therefore, operation at a higher frequency than this also requires a small but finite allowance for the frequency dependence of the mutual inductance. In a mutual inductance bridge measurement system, for the ideal inductor the potential of the primary and secondary windings is made definite by joining one end of both windings together as shown in Reference 11 p.82.

Consequently, the equivalent circuit of such a mutual inductance standard may be represented as shown in Figure 6.18.

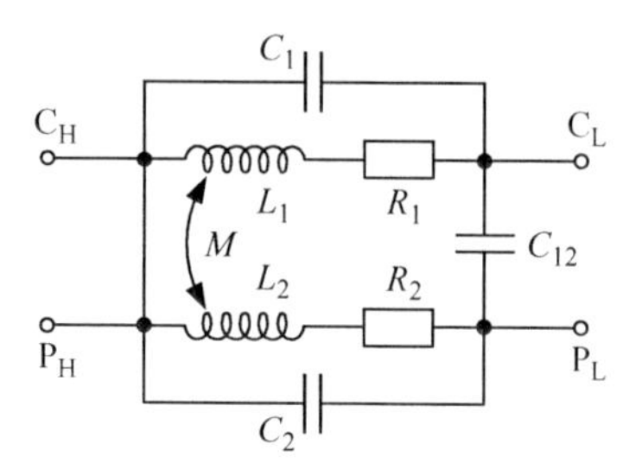

Figure 6.18 The equivalent circuit of a mutual inductor

The equivalent mutual inductance (M_E) is then given by

$$M_E = \pm M - C_{12}R_1R_2 \pm M\omega^2\left[C_1L_1 + C_2L_2 + C_{12}\frac{(L_1 \pm M)(L_2 \pm M)}{\pm M}\right] \quad (6.23)$$

where M is given by (6.20) and the $\pm$ signs refer to the series-aiding or series-opposing connections, respectively. A typical relative correction to M at 10 Hz can be of the order of 0.5 ppm. Similarly, the resistance of the mutual inductance standard also has a frequency dependence and is given by

$$R_E = \omega^2\{MC_1R_1 + MC_2R_2 + C_{12}[R_1(L_2 + M) + R_2(L_1 + M)]\} \quad (6.24)$$

which may be regarded as the leakage resistance of the mutual inductance and is distinct from the much larger resistances R_1 and R_2 of the primary and secondary windings, respectively. That is, for a given current in the primary winding, the induced voltage in the secondary winding will depart, slightly, from the ideal 90° (i.e. voltage in quadrature with the current). This slight departure from ideal quadrature voltage is equivalent to a voltage component that is in-phase with the current. Therefore, the ratio of the in-phase voltage to the current is equivalent to a resistance R_E in (6.24), which is real and hence dissipative. This can be as small as 1 $\mu\Omega$ for a typical 10 mH mutual inductance standard at a frequency of 10 Hz.

An actual mutual inductor can easily be made so that it complies closely, at low frequencies of the order of 50 Hz, with the conditions for an ideal mutual inductor, that is, additivity and negligible departure from 90° of the phase between the primary current and the secondary voltage. The device is also inherently four terminal in that it has separate current and potential terminals, and for the definition of M to be realised in a measuring circuit, no current must be drawn from the terminals of the secondary coil. More information on these increasingly historic devices can be found in Kibble and Rayner [11].

6.4.13 Self-inductors

Self-inductors are not usually carefully defined; they are still used as two-terminal components and therefore need conversion to two- or four-terminal-pair components for measuring in coaxial networks. The adaptor described in section 5.3.10 will do this. Networks of twisted, individually screened wires are often used instead. Inductance is a quantity that is defined only for a complete circuit, and the quoted values of inductors are in fact the change of inductance they produce when inserted into a complete circuit. For a measurement to be valid, therefore, there must be negligible magnetic interaction between the inductor and the rest of the circuit.

The property of inductance is not, in general, additive because the inductance of two inductors connected in series is modified by the common magnetic flux linking them both. More information on self-inductors can be found in Kibble and Rayner [11].

The self-inductance of a solenoidal inductance standard with tightly wound turns in a solenoidal geometry with radius r, length l and number of turns N is given by

$$L = \frac{\mu_0 \mu_r \pi r^2 N^2}{l} \tag{6.25}$$

where μ_0 is the free-space permeability and μ_r is the relative permeability that is unity for air but for magnetic iron, for example, it can be as high as 200. Equation (6.25) is only accurate under the long solenoid assumption that $l \gg r$. The frequency dependence of such a standard will be affected by numerous factors, such as skin and proximity effects in the winding, inter-winding capacitances and losses in the magnetic core (which may be regarded as negligible for an air-core inductor). Furthermore, as for a capacitance and resistance standard, the inductance standard, unless hermetically sealed, will also display a finite pressure, temperature and humidity dependence.

Sullivan solenoidal inductance standards have a Q-factor between 10 and 50 and a resonant frequency between 1 MHz and 5 kHz, depending on their value that is in the range from 1 μH to 1 H. They have an external magnetic flux and nearby conductors or magnetically permeable objects can alter their value.

The self-inductance of a toroidal inductance standard having an average radius r equal to the mean of the inner and outer radii, a and b, respectively, is

$$L = \frac{\mu_0 \mu_r r N^2}{2} = \frac{\mu_0 \mu_r N^2 (b-a)^2}{8r} \tag{6.26}$$

This expression is accurate provided $r \gg (b-a)/2$.

The General Radio inductance standards, with nominal values of 100 μH to 10 H are wound toroidaly on a non-magnetic core. They have lower Q-factors and self-resonant frequencies (so the frequency-dependence of their inductance is greater), but negligible external magnetic flux.

6.5 Resistors, capacitors and inductors of calculable frequency dependence

The primary concern in this section is the definition and construction of impedance standards with calculable frequency dependence. With sufficient care to ensure strain-free construction and sound joints, they will usually be found satisfactory. How the various standards are affected by temperature, humidity, pressure, mechanical resilience, etc., though important for a standard, are aspects that we have not addressed here. The performance of impedance standards over a greater frequency range than the historical 20 Hz to 10 kHz is important because many applications now demand this (see sections 9.4.8 and 9.5).

6.5.1 Resistance standards

Although the DC resistance of four-terminal standards can be measured with an uncertainty of a few parts in 10^9, it is far more difficult to achieve this level of accuracy at the 1 or 1.592 kHz frequencies normally employed in standards laboratories. This is principally due to the limitations of the bridge measurement systems at these frequencies. Nevertheless, the reduction in accuracy is only moderate – a few parts in 10^8 is achievable. At higher frequencies, particularly in the 1–100 MHz range, the measurement accuracy reduces significantly to about 1% at 100 MHz, although this is still adequate for most applications. Again, the limitations arise from the measurement system, whereas the standards are probably defined with an order of magnitude better uncertainty. Moreover, so far as high-frequency measurements are concerned, the DC or few kHz measurements of a resistance standard provide an excellent reference data point on the frequency axis. This is because, in some applications, measurements that can only be performed at the higher frequencies can then be extrapolated to DC or lower frequencies to provide some qualitative indication of the variation of resistance with frequency at intermediate frequencies.

Because of the very large range of applications of resistors and resistance standards in a variety of fields, many diverse types of standards have been developed, particularly in the last 50 years. In this section, we detail some of the main reference standards that are either already being extensively used in standards laboratories throughout the world or are beginning to be deployed for new and emerging applications.

Recently, AC measurements of QHE resistance standards have attained an accuracy comparable to DC measurements (see section 9.3). The double-shielding technique applied to a quantum Hall device results in a linear variation of its resistance that is less than a few parts in 10^9/kHz, at least up to a few kHz. This performance equals or even exceeds that of the best resistors of calculable frequency dependence.

6.5.2 Haddad coaxial resistance standard

Haddad [14] and Cutkosky, at the National Bureau of Standards (now NIST), USA, developed a coaxial resistance standard during the late 1960s as part of the effort, based on the Thompson–Lampard calculable capacitor, to realise the ohm from the farad (see also section 6.2). This required a resistor with a known AC/DC difference having only a few parts in 10^9 uncertainty from DC to a frequency of 1.592 kHz. Their design had straight Evanohm either 100-Ω or 1-kΩ resistance wire elements housed in a split-cylinder copper coaxial return conductor. The resistance wire was welded to thicker copper wires at either end. One end was fed through, and supported by, a copper-backed poly-tetraflouride-ethylene (PTFE) disc mounted in the split cylinders. Welding the resistance wire to the flat ends of copper wires was considered to offer two main benefits; the temperature of most of the length of the resistance wire would be unaffected and changing the length of the copper wire provided a means of adjusting resistance of the standard by up to 100 ppm towards its nominal value. The four-terminal-pair connections were made to the copper support wires with the high-potential and high-current terminals located at one end of the standard and the low-potential and low-current terminals at the other end. To complete the standard, the whole standard was immersed in mineral oil for thermal stability. Figure 6.19 shows a Haddad resistance standard with the resistance wire mounted within one of the two split cylinders. It also shows the four-terminal-pair connections and the copper-backed PTFE disc that defines one end of the coaxial standard. In addition, at one end of the standard, there is a terminal screw thread. Adjusting this terminal screw not only fine-tuned the resistance of the standard close to the required nominal values (100 Ω or 1 kΩ) but also aligned the wire with the centre of the coaxial return conductor, without applying any significant stress.

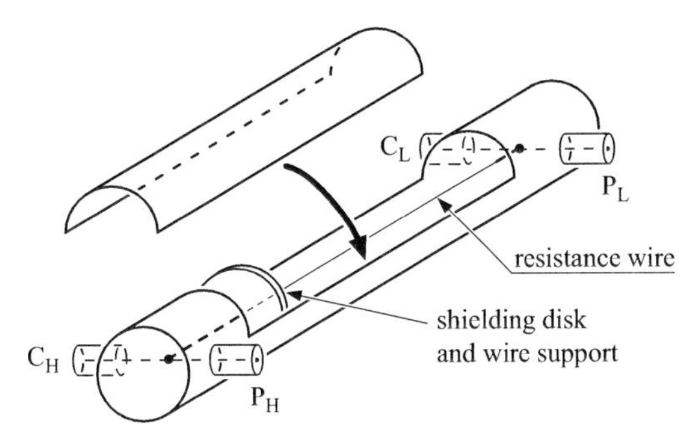

Figure 6.19 A Haddad coaxial resistance standard

Haddad and Cutkosky calculated the frequency dependence of the standard taking into account most of the significant contributing factors, such as eddy current losses in the resistance wire and the coaxial return conductor, shunt capacitance, series inductance, dielectric losses in the oil and any small but finite contributions due to the resistance element being off-centre in the coaxial shield.

They calculated the change in the real part of the resistance R for the standard as a result of the distributed capacitance as

$$\frac{\Delta R}{R} = 8\omega^2\varepsilon_0^2 abR^2 \sum_{k=1}^{\infty} \frac{(-1)^k}{k^2} \phi_{10}(kb)\phi_{10}(ka) \tag{6.27}$$

where $\varepsilon_0 = 8.85 \times 10^{-12}$ F/m is free-space permittivity, a and b are the radii of the resistance wire and the inner of the coaxial return conductor, R is the nominal value of the standard at DC, and

$$\phi_{10}(kt) = \frac{[I_1(k\pi t/c)/I_0(k\pi b/c)] + [K_1(k\pi t/c)/K_0(k\pi a/c)]}{[I_0(k\pi a/c)/I_0(k\pi b/c)] - [K_0(k\pi a/c)/K_0(k\pi b/c)]} \tag{6.28}$$

where $I_0(x)$, $K_0(x)$, $I_1(x)$, $K_1(x)$ are the modified Bessel functions of order zero and one, respectively, and c is the length of the resistance wire. In general, expression (6.27) can be equivalently rewritten as $\Delta R/R \sim \omega^2 C^2 R^2$, where C is the total distributed capacitance of the coaxial standard. End-plate calculations were not given – but a rough estimate was made and for this standard at frequencies between 10 and 10 kHz the contribution of the end plates was not considered to be a major factor (whereas in the HF calculable resistance standard to be described in section 6.5.3, it is an important factor that should be accounted for). Haddad estimated the change in resistance due to the distributed capacitance of the standards at 1.592 kHz (or 10^4 rad/s) to be 6.9×10^{-14} and 5.2×10^{-12} for the 100 Ω and 1 kΩ resistance standards, respectively. Furthermore, these values increase *quadratically* with frequency, as Haddad's calculations at 15.92 kHz show. Similarly, the change in resistance due to inductance of the resistance wire and the return coaxial conductor is given by

$$L_0 = \frac{\mu_0}{2\pi}\left[\ln\left(\frac{b_1}{a}\right) + \frac{2(b_1/b)^2}{1-(b_1/b)^2}\cdot\ln\left(\frac{b}{b_1}\right) - \frac{3}{4} + \ln(\zeta)\right] \tag{6.29}$$

where b_1 is the outer radius of the coaxial return conductor and the term $\ln(\zeta)$ is a function of the ratio (b_1/b) tabulated in Grover's text book on inductance calculations [15, Table 4]. The inductance given by (6.29) varies with frequency, such that L_0 at DC approaches the high-frequency limit L_∞, given by

$$L_\infty = \frac{\mu_0}{2\pi}\left[\ln\left(\frac{b_1}{a}\right)\right] \tag{6.30}$$

This inductance L (the limit of whose value is given either by (6.29) or (6.30)) leads to a change in the resistance R of the standards in two ways; first L combines with R, such that $\Delta R/R \sim \omega^2 L^2/R^2$, and second it combines with R and the distributed capacitance C of the standards mentioned above, such that $\Delta R/R \sim \omega^2 C^2 L^2$. The first factor leads to a resistance change of 2.1×10^{-11} and 2.2×10^{-11} for the 100 Ω and 1 kΩ standards, respectively, at 1.592 kHz. The second factor leads to a resistance change of 1.1×10^{-9} and 1.6×10^{-11} for these standards at the same frequency.

Haddad also considered the change in resistance due to eddy current losses in the Evanohm (an Ni:Cr:Al alloy) resistance wire, copper support wires, copper return conductor and the end plate. Only in the 100 Ω standard at 1.592 kHz did the eddy current losses in the copper support wires and the copper return conductor result in significant resistance changes, of 8.9×10^{-10} and 5×10^{-10}, respectively. For the copper support wires, the resistance changes quadratically with frequency, whereas for the copper return conductor, it changes as $\omega^{3/2}$. Furthermore, mineral oil also makes a relatively small contribution to the change in resistance of the standards because of two factors; first it increases the distributed capacitance by a factor of its relative permittivity, $\varepsilon_r = 2.16$, and second its loss angle, $\tan\delta =$ 3.5 μrad produces a distributed resistance between the Evanohm wire and the copper return conductor, thus producing a relative change in resistance of $\Delta R/R \sim \omega RC \tan\delta = 2.2 \times 10^{-12}$ and 1.9×10^{-11} for the 100 Ω and 1 kΩ standards at 1.592 kHz, respectively. Haddad also considered other factors that could influence the resistance of the standards at 1.592 kHz relative to their DC values, such as misalignment of the resistance wire from the centre of the coaxial return conductor and unequal current distribution between the two split cylinders. These were estimated to contribute about 1×10^{-9} each to the resistance change. However, Haddad regarded the contribution from mutual inductance between the voltage and current leads to be negligible. In contrast, at high frequencies, the mutual inductance is expected to be significant, since its contribution depends quadratically on frequency. Similarly, the end-plate effects are also expected not to be negligible at high frequencies. Both of these aspects are discussed in more detail in section 6.5.3 for the HF calculable coaxial resistance standard. Furthermore, Haddad also estimated the phase angle of the resistance standards as

$$\varphi = \omega\left(CR + \frac{L}{R}\right) = \omega\tau \tag{6.31}$$

where $L = (L_0 + L_\infty)/2$ and τ is the equivalent time constant of the standards. For the 100 Ω and 1 kΩ standards, the calculated phase angles were estimated to be 33.6 ± 0.8 and 9.4 ± 0.6 μrad, respectively, at 1.592 kHz. As (6.31) indicates, both vary linearly with frequency.

Finally, Haddad also compared these calculations with measurements of the 100 Ω and 1 kΩ standards, at 1.592 and 15.92 kHz, on a four-terminal-pair bridge. For the in-phase change in resistance of the ratio of the 100 Ω and 1 kΩ standards, he found a measured difference of $(28.7 \pm 10) \times 10^{-8}$ between the two frequencies. The calculated difference was $(20.3 \pm 3.7) \times 10^{-8}$. Although these two results agree within their stated uncertainties, Haddad noted that the relatively large uncertainty in the measurements was due to two current equalisers not being as effective as they should have been in their measurement set-up. In contrast, Haddad found a very good agreement between the calculated, 217.8 μrad, and measured, 225.6 μrad, phase angle change between the ratio of the 100 Ω and 1 kΩ standards at the two frequencies.

Recently, Delahaye [16] at the International Bureau of Weights and Measures (BIPM) used a 1290.6 Ω (equal to $R_K/20$, where R_K is the von Klitzing constant) Haddad resistance standard to measure the frequency dependence of two 51.6-kΩ resistance standards at frequencies between 500 Hz and 6 kHz. He then measured with a quadrature bridge the frequency dependence of two sets of Andeen-Hagerling 10 and 100 pF fused-silica capacitance standards in terms of the frequency dependence of the 51.6-kΩ resistance standards. The relative change in the four-terminal-pair capacitance value from 500 Hz to 6 kHz was a decrease of between 70 and 120 parts in 10^9 for all four capacitance standards, with a standard uncertainty ranging from 13 parts in 10^9 at 500 Hz to 34 parts in 10^9 at 6 kHz. The measured data was fitted to a simple dielectric relaxation model that confirmed the origin of the frequency dependence in the capacitance standards. The internal inductances of the standards, which would normally lead to an increase in the apparent capacitance of a standard at higher frequencies, were found to be no more than 0.05 μH as confirmed by resonance frequency measurements. The corresponding frequency dependence of the dissipation factors was not reported.

Despite the elegance, simplicity and usefulness of the Haddad coaxial resistance standard for measurements at low frequencies, as it will be shown in section 6.5.3, this type of standard has several limitations at higher frequencies.

6.5.3 A nearly ideal HF calculable coaxial resistance standard

This standard was devised [17] because of the limitations of the Haddad (section 6.5.2) [14] and Gibbings (section 6.5.5) [18] resistance standards at frequencies above 16 kHz. These limitations include the difficulty of accurately calculating the frequency dependance of these standards above 16 kHz and the relatively large frequency dependance (proportional to ω^2) resulting from the mutual inductance between the current and potential leads. By identifying suitable solutions and improvements, a nearly ideal calculable standard of resistance has been realised for the DC to 100 MHz frequency range (see Figure 6.20). In contrast to the Haddad and Gibbings standard, the measurement plane and the calculation plane are at exactly the same spatial

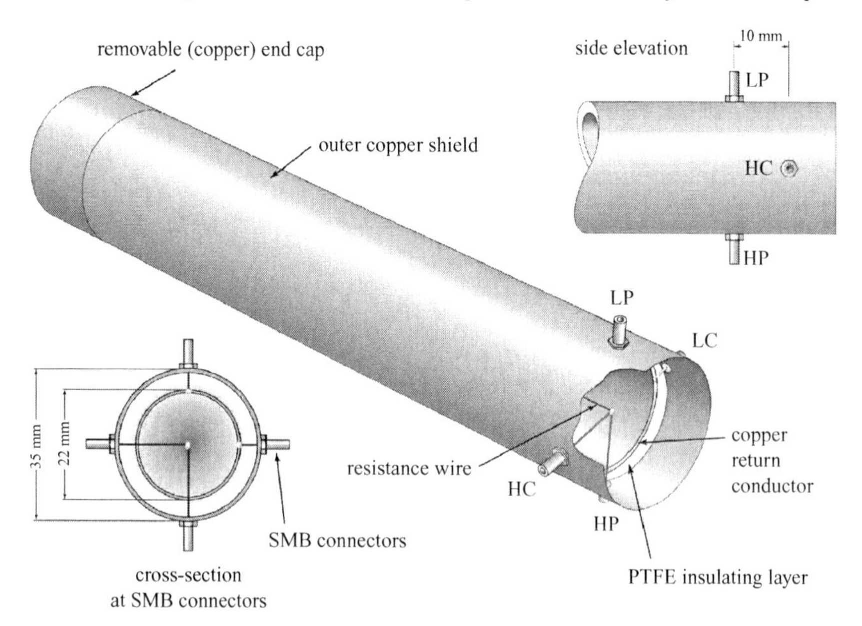

Figure 6.20 The high-frequency 4TP calculable coaxial resistance standard for broadband DC to 100 MHz operation (© IEEE 2005 [17])

location and the inner four-terminal-pair connections to the resistance wire and the coaxial copper return conductor are orthogonal in order to minimise the mutual inductance between the current and potential leads. The outer connections are to the copper shield, and their orthogonality ensures that no outer-conductor impedance is added to the component. This configuration significantly reduces the f^2 frequency dependence of the standard by minimising the mutual inductance between the leads that is one of the main components of uncertainty in the MHz range of frequencies. Supporting dielectric material is eliminated and there is only air between the resistance element and the copper return conductor; any dielectric other than air would make a finite additional contribution to its frequency dependence and complicate analytical calculations. If the greatest achievable accuracy is required, the effect on the frequency dependence of any oxide layers on the resistance wire element or on the inner surface of the coaxial return conductor also must be calculated. In addition, there may be a finite frequency dependence associated with the resistivity of the resistance element (although this is expected to be quite small for $f \ll f_p$, where f_p is the plasma frequency of the metal). End effects can also be virtually eliminated if the potential connections are displaced by a distance comparable with the diameter of the inner return conductor away from the current connections while maintaining the orthogonality of the four-terminal-pair connections (as illustrated in the side elevation in Figure 6.20).

The performance of an HF calculable resistance standard can be represented by a relatively simple model which contains only the main sources of frequency dependence, although the typical accuracy of the calculated frequency dependence at 100 MHz of either 100 Ω or 1 kΩ standards is only about ±0.1% (1σ). For most applications, such as calibration of commercial four-terminal-pair LCR meters and impedance analysers, this is perfectly adequate, but for some metrological applications, such as Johnson noise thermometry (where the 100-Ω sense resistor needs to be calibrated with about ±1 ppm uncertainty at frequencies up to 1 MHz) or quantum Hall resistance measurements to a few MHz, much more accurate analytical calculations and measurements of a HF calculable resistance standard will be needed in the future. This work is currently in progress and will be published in due course. The major complications in calculating the frequency dependence more accurately are the end effects at the end disc of the standard. Although this disc has a calculated DC resistance of only 0.1 μΩ and the eddy current losses are unlikely to change its resistance by more than a factor of 1000 from DC to 100 MHz (which would amount to a contribution of only 0.1 ppm, for a 100 Ω standard), the effect of the capacitive current distribution near the end disc and its inductance need to be accounted for more accurately.

The impedance $Z(\omega)$ of an HF calculable resistance standard can be described in terms of a pure series resistance R_S with a series inductance L_S and an overall shunt capacitance C_S. This model is found to be suitably appropriate given the geometrical design shown in Figure 6.20. The voltage between the return conductor and shield is zero, and consequently, the capacitive current through the supporting PTFE dielectric material is also zero. The impedance $Z(\omega)$ of the standard is

$$Z(\omega) = \frac{R_S + j\omega[L_S(1 - \omega^2 L_S C_S) - C_S R_S^2]}{1 + \omega^2[C_S^2 R_S^2 - 2L_S C_S + \omega^2 L_S^2 C_S^2]} \tag{6.32}$$

Typical values for a 100-Ω resistance standard, with the radius of the copper return conductor $r_1 = 10$ mm and wire radius of $r_w = 15$ μm and lumped parameters L_S and C_S are approximately 0.2 μH and 1.8 pF, respectively. Similarly, for a 1-kΩ resistance standard, these are typically 0.3 μH and 1.3 pF, respectively. Substituting these values into (6.32) shows that the 100 Ω standard has an apparent positive coefficient of frequency dependence, that is, a 30-ppm higher resistance at 1 MHz compared to its DC value. In contrast, the 1-kΩ resistance standard has a negative frequency-dependent coefficient, that is, −40 ppm at 1 MHz compared to its DC value. Furthermore, both resistance standards show a quadratic dependence on frequency, with about 10% uncertainty of the frequency-dependent coefficients.

The time constant τ is also an important physical quantity of the resistance standards and should be experimentally and theoretically accounted for in any precision impedance measurements. From (6.32)

$$\tau = \frac{L_S}{R_S} - C_S R_S \tag{6.33}$$

For the 100-Ω and 1-kΩ resistance standards, for example, $\tau \approx 2$ ns and $\tau \approx 1$ ns, respectively. Finally, the parallel equivalent circuit resistance of the standard is related to the series equivalent circuit resistance by $R_P = R_S(1 + \omega^2\tau^2)$.

6.5.4 A bifilar resistance standard

This standard, which is suitable for frequencies less than 100 kHz, consists of a resistance wire forming a single loop, of outward and returning current, surrounded by a cylindrical shield. Figure 6.21 shows an equivalent electrical model of the standard consisting of a uniform transmission line with shunt capacitance and leakage conductance to the shield and between the two wires of the loop. For the greatest accuracy, the standard should be terminated with four-terminal-pair connections. Its conductance is

$$G_{4TP} = \frac{1}{R}\left\{1 - \frac{R}{6}(G_0 - 2G_1) - \frac{\omega^2(L - 2M)^2}{R^2} - \frac{\omega^2 R^2}{720}\left[15C_0^2 - (C_0 - 4C_1)^2\right]\right\} \tag{6.34}$$

where R is the resistance of the standard, G_0 and G_1 are the conductances between the resistance wire and the shield and between the wires of the loop, respectively, M is the total mutual inductance between the two halves of the wire loop and C_0 and C_1 are the shunt capacitances between the wires and the shield as well as between the sides of the loop. The relative change in conductance, $\Delta G_{4TP} = 1 - G_{DC}/G_{4TP}$ where G_{DC} is the conductance at DC, at a frequency of 1 kHz is approximately 1×10^{-9} for a 1-kΩ resistance standard.

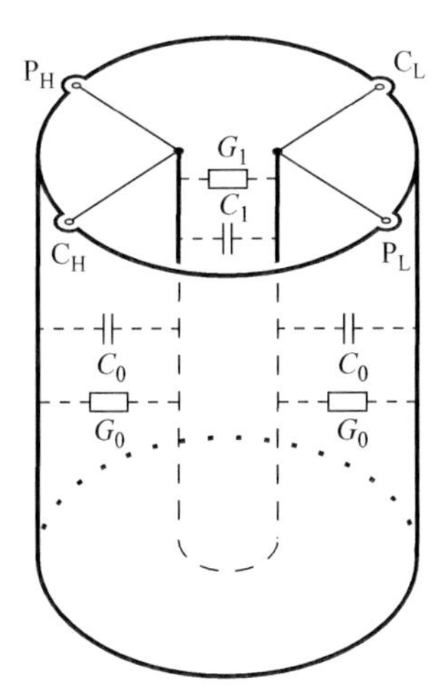

Figure 6.21 Equivalent circuit lumped parameter model of a bifilar resistance standard

6.5.5 Gibbings quadrifilar resistance standard

Initially, this standard [18] was devised for determining the ohm from the Thompson–Lampard calculable capacitor. This required a 10-kΩ resistance standard whose change in resistance between DC and 10^4 rad/s could be calculated to within ± 1 part in 10^7 uncertainty. Many national standard laboratories now employ this standard to relate the DC value of a resistor to its AC value at $\omega = 10^4$ rad/s. The DC value of a resistor is measured in terms of the quantum Hall resistance (see also section 6.6.1) and comparison with the quadrifilar standard at $\omega = 10^4$ rad/s enables the impedance of the standard to be directly compared to the impedance of a capacitance standard in a quadrature bridge (see sections 9.2.1–9.2.3). For a 10-kΩ value of the standard the change of resistance with frequency is dominated by second-order terms of the type $\omega^2 R^2 C^2$. In contrast, for resistance standards with lower values than 10 kΩ, the dominating effects are of the type $\omega^2 L^2/R^2$, due to mutual inductive coupling and to dielectric loss. For the intermediate nominal value of 1 kΩ eddy current losses that would be induced in the shield conductor are cancelled by the reversed quadrifilar configuration of the resistance wire shown in Figure 6.22.

An additional advantage of the reversed quadrifilar configuration is that it reduces the physical length of the standard, which is also advantageous for controlling its temperature. Gibbings in his original publication [18] reported both 100 Ω and 1 kΩ reversed quadrifilars whose change in resistance with frequency from DC to 10^4 rad/s was less than 1 or 2 parts in 10^8 with the same uncertainty. A 10:1 ratio bridge scaled these values to the required 10 kΩ.

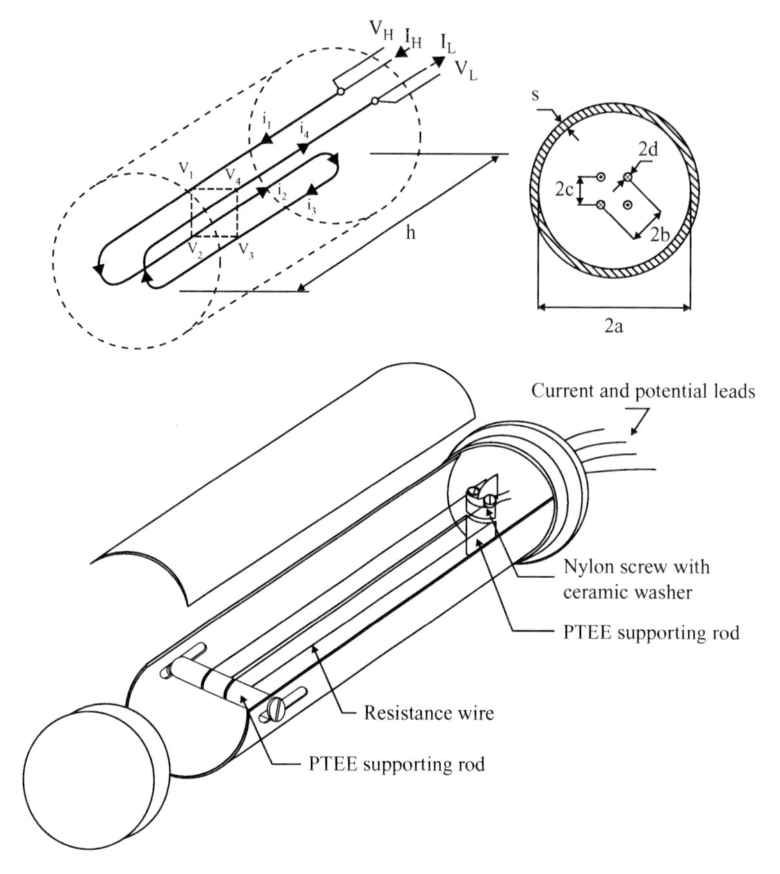

Figure 6.22 Gibbings 100 Ω reversed quadrifilar resistance standard whose resistance increases quadratically with frequency and at 8 kHz has a value 6.5×10⁻⁹ higher than DC (with τ = 16 ns)

The four-terminal-pair impedance of the reversed quadrifilar resistance standard can be defined as

$$Z_{4\mathrm{TP}}(\omega) = R_0[1 + \alpha(f) + j\omega\tau] \tag{6.35}$$

where R_0 is the value of the standard at DC, and the dissipative or in-phase component of the impedance $\alpha(f)$ is the summation of three contributions; β due to reactive impedances, χ due to eddy current losses induced in the shield conductor and γ due to skin effect in the resistance wires, thus

$$\alpha(f) = \beta + \chi + \gamma \tag{6.36}$$

such that

$$\beta = \left\{ \begin{array}{l} \frac{R_0}{6}(G_0 - 5G_1 - 3G_2) + \frac{\omega^2 R_0^2}{11520}\left[240C_0^2 - 15(C + 8C_1 + 8C_2)^2 - (C_0 + 16C_1)^2\right] \\ -\frac{\omega^2}{R_0^2}(L - 8M_1 + 4M_2)^2 \end{array} \right\}^{-1} \tag{6.37}$$

where G_0 and C_0 are the total leakage conductance and capacitance to the shield conductor, respectively, and the remaining symbols have the usual meaning with subscripts 1 and 2 referring to adjacent or diagonally opposite wires in the reversed quadrifilar resistance standard. The eddy current losses induced in the shield conductor are then given by

$$\chi = \frac{8\mu_0}{\pi} \cdot \frac{m(f)\omega h}{R_0} \left(\frac{1}{4 + m^2(f)} \right) \left(\frac{b}{a} \right)^4 \tag{6.38}$$

where $\mu_0 = 4\pi \times 10^{-7}$ H/m, h is the overall length of the resistance standard, a and b are shown in Figure 6.7 and $m(f)$ can be evaluated from

$$m(f) = \frac{\omega \mu_0 s a}{2\rho_S} \tag{6.39}$$

with s being the thickness of the shield conductor and ρ_S its resistivity. Lastly, the losses due to the skin effect in the resistance wire is given by

$$\gamma = \frac{\omega^2}{12} \left(\frac{\mu_0 \mu_r}{4\pi t} \right)^2 \left(1 - \frac{\omega^2}{15} \left(\frac{\mu_0 \mu_r}{4\pi t} \right)^2 \right) \tag{6.40}$$

where μ_r is the relative permeability of the wire material (assumed to be unity) and t is the overall resistance per unit length of the wire.

The time constant of the Gibbings reversed quadrifilar is given by

$$\tau = \left[\frac{R_0}{6} (5C_1 + 3C_2 - C_0) - \frac{1}{R_0} (L - 8M_1 + 4M_2) \right]^{-1} \tag{6.41}$$

6.5.6 Boháček and Wood octofilar resistance standard

Boháček and Wood [19] noted that the Gibbings model could readily be extended to realise a standard whose frequency dependence could be made smaller by a factor of approximately 3.5 over the frequency range 0.5–5 kHz. As the name suggests, an octofilar resistance standard has additional four turns of wire compared to that normally found in the quadrifilar design. This consequently means that the overall length of the octofilar standard can be half the length of a quadrifilar standard, if all the other parameters, in particular the resistivity of the resistance wire are kept the same. The reduced length of the standard also makes it relatively simpler to achieve good temperature control of the standard. The parallel four-terminal-pair resistance of the octofilar is then also given by (6.34), except in this case β is

$$\beta = \left\{ \begin{array}{l} \frac{R_0}{192}(17G_0 - 128G_1 - 144G_2 - 128G_3 - 48G_4) \\ \quad + \frac{\omega^2 R_0^2}{737280} \left[\begin{array}{l} 192C_0^2 - 4(C_0 + 32C_1 + 32C_3)^2 - \\ 60(C_0 + 16C_1 + 32C_2 + 16C_3)^2 - \\ 15(C_0 + 16C_1 + 16C_2 + 16C_3 + 16C_4)^2 \end{array} \right] \\ \quad + \frac{\omega^2}{R_0^2}(L - 16M_1 + 16M_2 - 16M_3 + 8M_4)^2 \end{array} \right\} \tag{6.42}$$

where the symbols and subscripts have the same meanings as in (6.37). Its time constant is

$$\tau = \left[\begin{array}{l} \frac{R_0}{192}(17C_0 - 128C_1 - 144C_2 - 128C_3 - 48C_4) \\ \quad + \frac{1}{R_0}(L - 16M_1 + 16M_2 - 16M_3 + 8M_4) \end{array} \right] \tag{6.43}$$

Similarly, the eddy current losses induced in the shield conductor are given by

$$\chi = \frac{32\mu_0}{\pi} \cdot \frac{m(f)\omega h}{R_0} \left(\frac{1}{16 + m^2(f)} \right) \left(\frac{b}{a} \right)^8 \tag{6.44}$$

where $m(f)$ is given by (6.39), and the remaining factor in (6.36) for the eddy current losses in the resistance wire is

$$\gamma = \frac{\omega^2}{192} \left(\frac{\mu_0 \mu_r r^2}{\rho_w} \right)^2 \tag{6.45}$$

where r_w and ρ_w are the radius and resistivity of the resistance wire, respectively.

The quadrifilar and octofilar resistance standards assisted in initially verifying the frequency independence of quantum Hall resistance devices (quantised at 12.906 or 6.45 kΩ) at frequencies ranging up to 5 kHz. Typically a 12.906-kΩ quadrifilar standard changes fractionally with frequency by approximately $+1 \times 10^{-7}$ to $+2 \times 10^{-7}$ at 5 kHz and has a time constant of 10–15 ns. An octofilar standard changes fractionally with frequency by $+0.55 \times 10^{-7}$ at 5 kHz and has a time constant of 10 ns. Both have an uncertainty of approximately $\pm 3 \times 10^{-8}$ (1σ) at 5 kHz.

6.5.7 HF secondary resistance standards

The preceding sections discuss standards of resistance whose frequency dependence is calculable from first principles. There is also a need for secondary standards whose frequency dependence, although not calculable, is nevertheless as small as possible. A simple and practical four-terminal-pair secondary impedance standard with applications in high-frequency metrology is shown in Figure 6.23. Four female BNC connectors are mounted equidistant along the lid of a 70 mm × 20 mm × 20 mm die-cast box so that their spacing allows a direct plug-in to the input ports of commonly available commercial impedance bridges and network analysers. The outer conductor of the L_C connector makes contact with the metal of

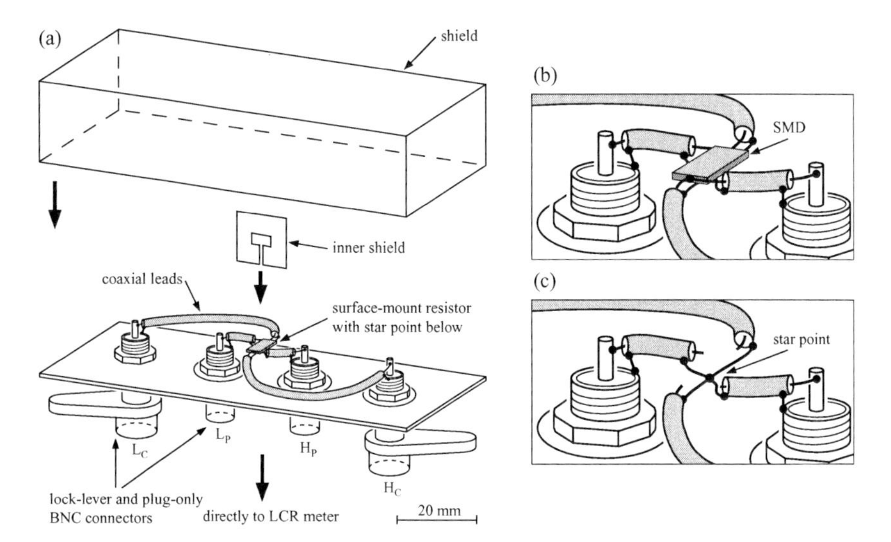

Figure 6.23 *(a) High-frequency secondary 4TP resistance standard using a (b) internal coaxial transmission lines (c) and surface-mount chip resistor. The details (b) and (c) show the coaxial leads before and after mounting the chip*

the box, and the other three connectors are insulated from it so that the box constitutes an electrical shield. The in-line configuration of the BNC connectors is converted to an orthogonal geometry of current and potential connections as shown, with the inner conductors connected to a surface-mount chip that can be a resistor, a capacitor or an inductor. These chips are widely available from numerous manufacturers of electronic components. The connections are made with coaxial cables whose outer conductors are brought together as a star connection adjacent to the chip. This arrangement minimises the mutual inductance between potential and current connections. For resistance or inductance standards, the chip penetrates a closely fitting hole in a planar shield connected to the star point to minimise direct capacitance from one end to the other. This reduces the phase angle and frequency dependence of the standard. Its frequency response can be calibrated using an HF calculable resistance, capacitance or inductance standard. This design of resistance standard is a good high-frequency transfer standard for calibrating high-frequency four-terminal-pair meters and impedance analysers. If the standard is connected to an instrument via short coaxial cables rather than being directly plugged in, a cable impedance correction to the measured results is normally required, and in some cases, cable loading corrections may also be needed if the measurement accuracy for a particular application demands it. These effects are typically of the order of 0.1% or less at 1 MHz.

An alternative construction based on four female SMB connectors soldered into a copper plate on the corners of a square to ensure an orthogonal geometry is illustrated in Figure 6.24. The resistance, capacitance or inductance chip component

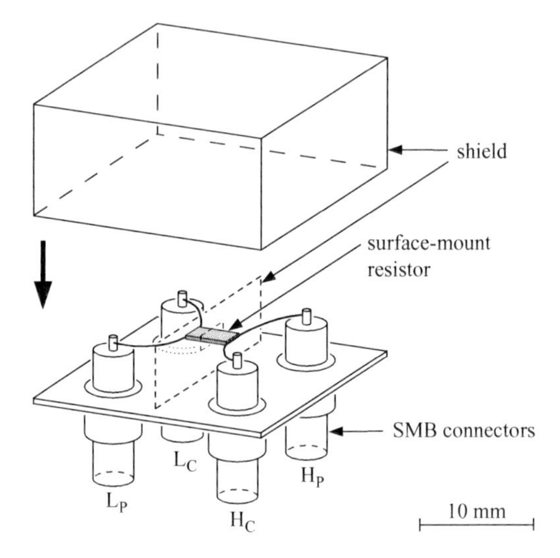

Figure 6.24 A connection geometry for minimising parasitic inductances of a chip component

is joined to the inner conductors of the SMB connectors in the orthogonal geometry as shown. For resistance or inductance standards constructed in this way, the chip component is also mounted through a closely fitting hole in a shield between the high and low SMB terminals. This minimises its shunt capacitance and, consequently, reduces its phase angle and frequency dependence. Surface-mount chip resistors are available in values ranging from 1 Ω to 1 MΩ, and can be obtained with values within 0.01% of nominal. The temperature coefficient of the surface-mount chip component is typically not more than 10 ppm/°C. Care has to be exercised as they usually are low-power components (around 100 mW). Their small size (2 mm × 1 mm × 0.5 mm) helps to keep the self-inductance of resistors or capacitors small, which is an advantage for high-frequency operation. Cables are needed to connect this type of standard to most commercial LCR meters, and this will reduce the accuracy of measurements unless appropriate cable loading corrections and cable corrections are applied.

6.5.8 HF parallel-plate capacitance standard

A design we have found particularly useful for high-frequency operation for a secondary, or transfer, capacitance standard is shown in Figure 6.25. The standard has four-terminal-pair connections and air as the dielectric. The four-terminal-pair connections are arranged orthogonally so that the current and potential paths have no mutual inductance (as discussed in more detail in section 6.5.9). This reduces the frequency dependence of the capacitance standard. In addition, since the standard has air as the dielectric its residual inductance (L_0) can be determined using a

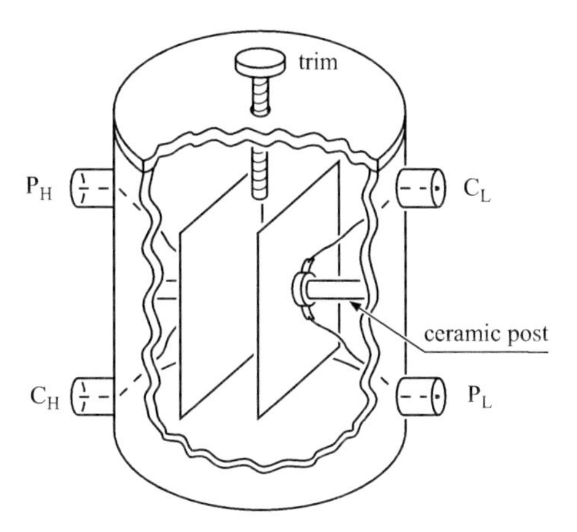

Figure 6.25 A high-frequency four-terminal-pair secondary capacitance standard

high-frequency LCR meter by measuring its resonance frequency f_0 (given by the expression in (6.46)). From its value C_0 at a low frequency, for example 1 kHz, its residual inductance can be determined. C_0 can readily be measured with an LCR meter. Therefore, the frequency dependence of the capacitance standard $C(f)$ at $f \ll f_0$ can be approximated to first order by

$$C(f) = C_0\left(1 + \omega^2 L_0 C_0\right) = C_0\left[1 + \left(\frac{f}{f_0}\right)^2\right] \tag{6.46}$$

In Figure 6.25, a screw is also shown on the top plate of the shield. This enables a small ΔC change in the value of the capacitance to be made by preventing a small amount of the electric field from the source plate reaching the detector plate and redirecting it to the shield. This arrangement has a negligible effect on the frequency dependence of the standard, and as such can be used to balance a high-frequency bridge system.

By changing the plate dimensions and separations, capacitors having the decade values of 0.1, 1 and 10 pF or other intermediate values can be constructed.

6.5.9 HF calculable coaxial capacitance standard

There has not been until recently a universal impedance device or standard that offers either negligible frequency dependence or a calculable frequency dependence over a broad range of frequencies. To resolve this problem, Awan and Kibble [17] reported a novel type of reference resistance standard (see section 6.5.3) in 2005 whose complete frequency dependence can, in principle, be calculated ab initio. In 2007, they then extended the key concepts behind the calculable resistance standard to realise air-dielectric standards of capacitance and inductance whose frequency dependence is also calculable. Taken together, the three *LCR* standards of impedance are underpinned by a *universal coaxial geometry* together with an orthogonal connection topology. The advantages of this geometry are discussed in section 6.5.3. Although the calculable standards of resistance, capacitance and inductance share a common geometry, they also have distinct features. For the high-frequency calculable coaxial capacitance standard, the capacitance is defined between the central copper conductor, inner coaxial return conductor and the two guards at either ends of the standard, as shown in Figure 6.26. In practice, the four-terminal-pair connections are made at one end of the standard for ease of construction. Awan and Kibble constructed calculable 4TP 10-pF capacitance standards [20] that were approximately 150 mm in length and had a shield diameter of 25 mm. They also reported its calculated frequency dependence up to 100 MHz, which they estimated as being 20 ± 2% higher than the value of the standard at 1 kHz. This was based on the calculated 54 nH inductance of the standard that also resulted in $f_0 = 217$ MHz. The time constant of the standard, given by $\tau = (L/R) - (CR)$, is 5.3 ns (assuming a dissipation factor of 10^{-6} rad or an equivalent series resistance R of 10 Ω). The calculations for a calculable geometry whose largest dimension is l will be valid up to a frequency corresponding to $l \ll \lambda/4$. This frequency is about 100 MHz using conventional constructional techniques, or possibly GHz and higher if micro- and nanomachined constructional techniques are employed. Some preliminary work using

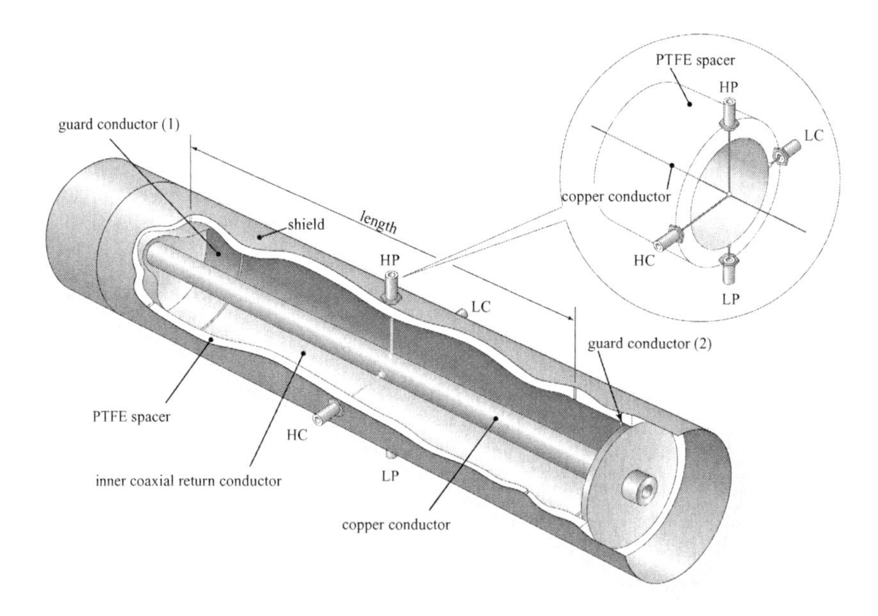

Figure 6.26 The ideal design outline of the high-frequency 4TP calculable coaxial capacitance standard for operation at frequencies up to 100 MHz. The inset diagram shows a closer view of the 4TP orthogonal connections (© IEEE 2007 [20])

micromachined coaxial transmission lines and capacitive devices has already been reported [21]. Further work is needed to optimise and characterise these structures before they can be usefully deployed in precision broadband electromagnetic metrology at the on-chip scale, covering the DC to millimetre wave parts of the spectrum.

Applications of a calculable coaxial capacitance standard include a reference standard for material/dielectric measurements, comparisons with finite element electromagnetic modelling of impedance devices and calibration of modern electronic LCR meters up to 100 MHz.

6.5.10 HF calculable coaxial inductance standard

This standard was described by Awan and Kibble [20]. The first 0.1 μH prototype was constructed as shown in Figure 6.27. It is similar to the HF calculable coaxial resistance standard shown in Figure 6.20. The central electrode is replaced with a copper wire. The calculated frequency dependence of the standard was reported as being 4 $\pm$ 0.4% higher than the 1-kHz value at a frequency of 100 MHz. Its resonance frequency was also estimated and found to be slightly over 500 MHz (using a calculated shunt capacitance of 0.997 pF). The estimated quality factor $Q = \omega L/R \ll 1$ since the inductance standard does not contain any magnetic or high-permeability material to increase its inductance. This renders its frequency dependence calculations much more complex with a higher uncertainty in its calculated value.

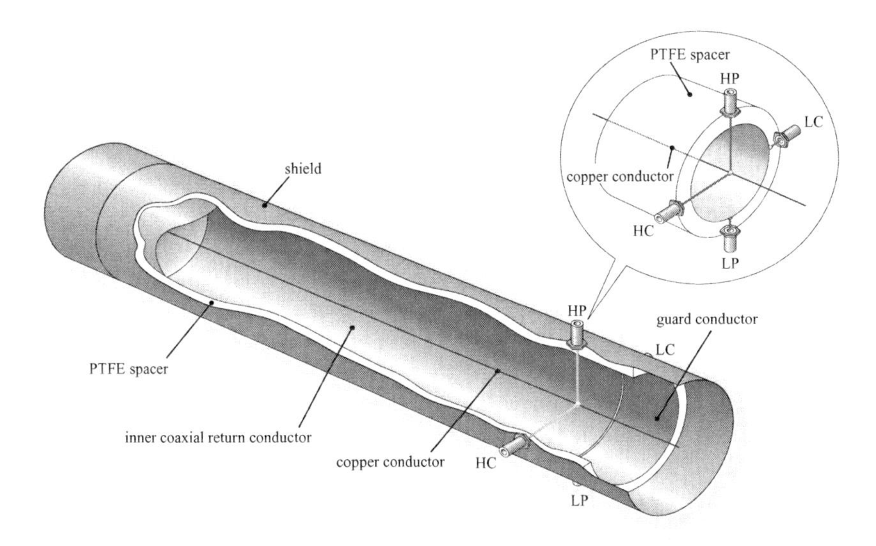

Figure 6.27 A high-frequency coaxial inductance standard of calculable frequency dependence [20]

The four-terminal-pair connector configuration contrasts with that of conventional two-terminal inductance standards, such as the Sullivan type discussed in section 5.3.2. The latter need an auxiliary measurement made with a shorting connector between the two terminals of the standard. This enables the inductance to be measured correctly as the change in inductance obtained with and without the standard being shorted, since inductance is a property that can only be defined and measured for a complete circuit or network. A shorting connector is not necessary for a calculable coaxial inductance standard, since the self-inductance of the standard is defined precisely as that of the current loop within the device.

6.5.11 A frequency-independent standard of impedance

From the previous discussions in this chapter, it is clear that an artefact standard, no matter how ingenious its design, can ever be completely frequency independent in its truest sense, even from DC to 100 MHz. This does not rule out the possibility of a quantum Hall resistance standard that might be a universal quantum standard of impedance in that its quantised resistance given by R_q remains the same from DC to perhaps a few THz (where it is theoretically expected to breakdown as the excitation frequency approaches the cyclotron frequency of the charge carriers, $\omega_c = eB/m^*$, where e is the electronic charge, B is the magnetic field and m^* is the effective electron mass). But there may be a design for an artefact standard whose frequency dependence could be quite small or negligible over a substantial portion of the DC to 100 MHz frequency range.

The main factors that influence the frequency dependence of the three *LCR* calculable coaxial impedance standards, discussed in sections 6.5.3, 6.5.9 and 6.5.10, are the residual (or sometimes known as parasitic) capacitance and inductance that combine to give a small but finite apparent change with frequency in the impedance of these standards. Thus, to realise a frequency-independent standard of resistance, a method is needed to match the characteristic impedance of the standard with its resistance over a broad range of frequencies. Figure 6.28 shows a possible design for a potentially frequency-independent resistance standard having a conical inner return conductor surrounding the main resistance wire element. The rest of the features of this standard are identical to that shown in Figure 6.20 for the HF calculable resistance standard, and one particularly important test would be to compare the frequency dependence of the two standards.

This standard has not actually been developed or tested, and the calculations needed to determine its frequency dependence will be significantly more complex than those for the HF calculable resistance standard (which are themselves already quite complex). The frequency independence is expected to arise from the decrease in capacitance, and a corresponding increase in inductance, of the standard that occurs as the length of the standard is traversed from the four-terminal-pair connections towards the end cap.

A typical value for practical realisation is approximately 500 Ω given by the characteristic impedance $Z_0 \approx \sqrt{L/C}$. This is approximate because the transmission line is not lossless. A value of 50 Ω is preferable but the diameter of such a standard is impractical. As is well known from RF and microwave engineering, an impedance-matched system, usually having a 50- or 75-Ω characteristic impedance,

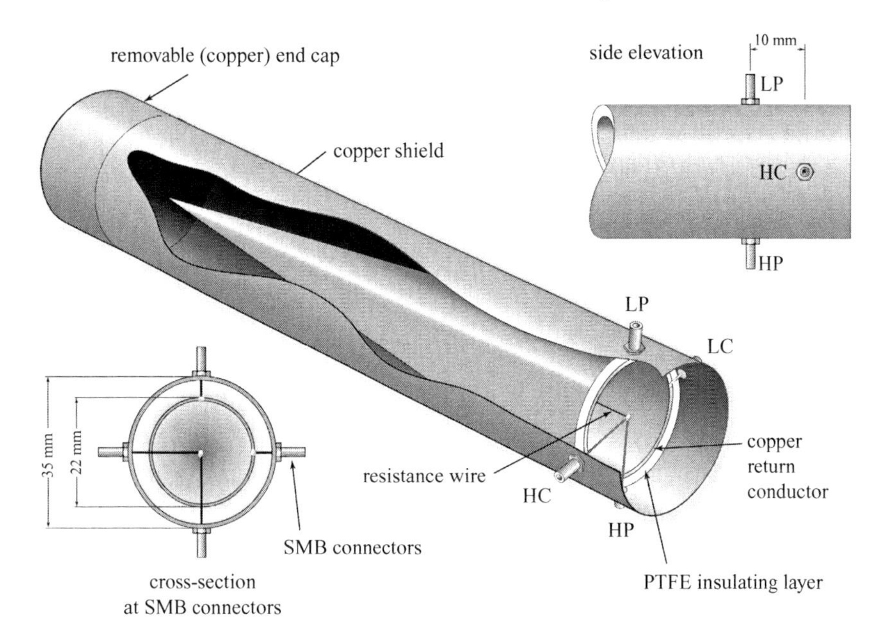

Figure 6.28 A possible design of a 500-Ω frequency-independent (DC to 100 MHz) resistance standard

does not reflect, or equivalently, display any significant dispersion of the transmitted microwaves. Therefore, the conical resistance standard shown in Figure 6.13 potentially provides convergence and unification of low-frequency coaxial bridge and microwave techniques. Clearly, a lot of work still remains to complete this convergence and unification process since there is a mismatch between a 500-Ω device and the 50-Ω systems normally employed at RF and microwave frequencies, but the techniques discussed in this chapter are a good starting point for combining both RF and microwave impedance metrology with traceability directly to quantum Hall resistance than to length standards, as is currently the case.

6.5.12 An ideal standard of impedance of calculable frequency dependence

In the preceding sections 6.5.3, 6.5.9 and 6.5.10, calculations of the frequency dependencies of the standards are limited in accuracy by the difficulty of accounting for effects in the neighbourhood of the transverse plane conductor connecting the inner and outer conductors at the end remote from the coaxial connectors. This limitation can be overcome (see section 6.5.3) by separating the current and potential terminals along the length of the resistive element at either end of the device by a distance L approximately equal to D (see Figure 6.29). This, crucially, is expected to result in virtually negligible *end effects* and leads to the desired ideality of an impedance standard with a calculable frequency dependence.

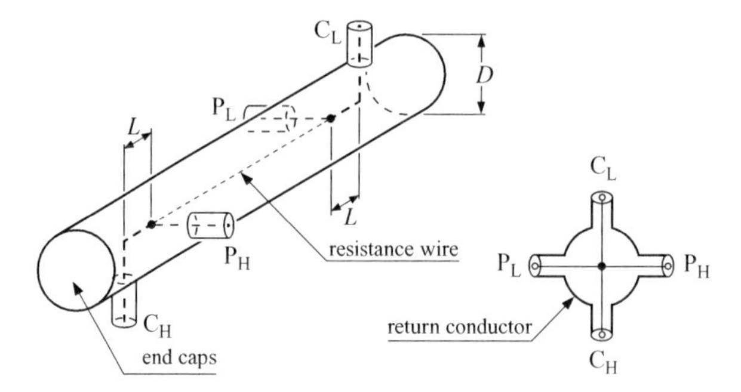

Figure 6.29 An ideal four-terminal-pair coaxial resistance standard of calculable frequency dependence

Although Figure 6.29 illustrates a resistance standard, the concept can equally be extended to apply to the capacitance and inductance standards. The central element is connected to the inner leads going to all four ports and the outer conductors of the ports are connected to the outer coaxial return conductor. As shown in the cross-sectional view, the four ports are arranged orthogonally in order to minimise mutual inductance between the current and potential leads.

The elimination of the end effects in this manner is analogous with guarding techniques but here guarding is achieved with minimal added complexity.

The analytical calculations of the frequency dependence of this standard are much more straightforward than for the high-frequency calculable resistance standard discussed in section 6.5.3 because there are no complexities arising from the presence of an end plate, although if the calculated frequency dependence of the standard is not required to better than 0.1%, both standards are expected to give the same performance. This new standard will only be needed if analytical calculational accuracy better than 0.1% is demanded for some application, such as in Johnson noise thermometry or for investigating quantum Hall devices at higher frequencies.

The disadvantage of this new design lies in the difficulty of assembling it without compromising the conducting homogeneity of the outer conducting tube near the connections by splitting the outer coaxial return conductor in half and mounting the resistance wire and its connections in the split cylinders. If this procedure can be done so that it has negligible effect, then the ideal standard shown in Figure 6.29 could be realised. Alternatively, the resistance wire, of predetermined value, could be joined to the four extended inner pins of the connectors and then be mounted in a relatively large continuous outer shield.

There are four main factors that will contribute to the frequency dependence from DC to 100 MHz of the standard; eddy current losses in the resistance wire and the outer return conductor and the shunt capacitance and inductance of the resistance wire and the return conductor. There are many other factors that cannot be calculated but that potentially also affect the frequency dependence of the standard, such as oxide layers and misalignment of the resistance wire from the centre of the return conductor, but in a carefully designed and constructed standard, these effects can be made negligible.

The complete analytical expressions for the eddy current losses and the series inductance in the resistance wire have been derived from first principles using Maxwell's equations. The resistance of the wire, of radius a, at a given frequency will be higher than its DC value, $R_{DC} = 1/\pi\alpha^2\sigma$, due to eddy currents and is

$$R_S^i = \frac{\zeta}{2\sqrt{2}\pi a\sigma}\left\{\frac{Bei_1(a\zeta)[Ber_0(a\zeta) - Bei_0(a\zeta)] - Ber_1(a\zeta)[Ber_0(a\zeta) + Bei_0(a\zeta)]}{[Bei_1(a\zeta)]^2 + [Ber_1(a\zeta)]^2}\right\} \tag{6.47}$$

where R_S^i is the series resistance of the inner resistance wire per unit length, $\zeta = \sqrt{2}/\delta = \sqrt{(\omega\mu\sigma)}$, δ is the skin depth, σ is the conductivity of the resistance wire and Ber_n and Bei_n are the complex Kelvin functions of order n. The expression in (6.47) is plotted in Figure 6.30, which shows the calculated frequency dependence of a practical 1-kΩ resistance wire in an HF calculable resistance standard. The key physical properties of this standard needed for calculating its electromagnetic properties are given in Table 6.2. The calculated frequency dependence due to eddy currents in the resistance wire at frequencies up to 100 MHz is plotted on the bottom of Figure 6.30 and shows that the apparent resistance of the wire increases quadratically with frequency. The total change in resistance, from DC to 100 MHz, is found to be higher by only 8.2 ppm. in contrast, the linear-log plot shown on the top of Figure 6.30 shows that the skin effect becomes significant as the frequency

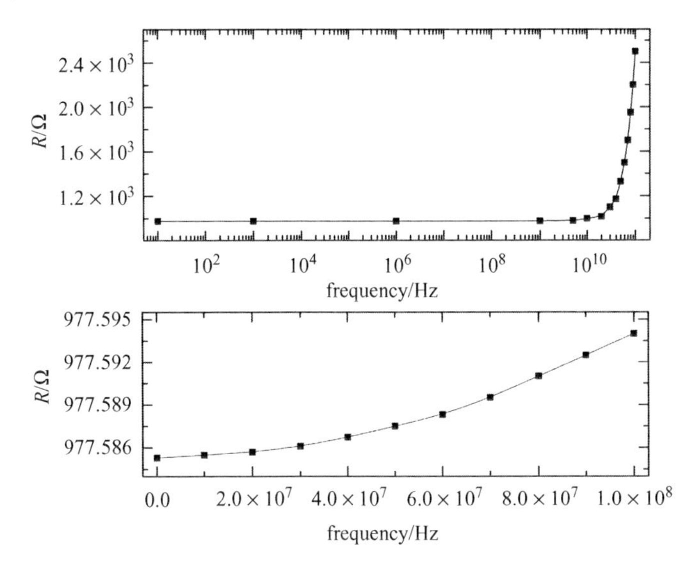

Figure 6.30 Calculated frequency dependence from equation 6.47 of the ideal 1-kΩ HF calculable resistance standard up to 100 MHz (bottom) and up to 100 GHz (top)

Table 6.2 The physical parameters of a 1-kΩ HF calculable resistance standard for calculating the results shown in Figure 6.30

Nominal value	ρ	*a*	*b*	*c*	Length
1 kΩ	1.33 $\mu\Omega$.m	0.00835 mm	10 mm	11 mm	161 mm

The resistivity of the resistance wire is ρ and its radius is a; b and c refer respectively to the inner and outer radii of the copper return conductor.

approaches 5 GHz, and the frequency dependence has a $\sqrt{f}$ dependence, as expected for eddy current losses in the skin-effect regime. Furthermore, neglecting radiation loss effects, calculations using the software Maple have also confirmed that as $f \to 0$, (6.44) gives the DC value of the standard, as shown in Figure 6.30.

The self-inductance of the 1-kΩ resistance wire is

$$L_S^i = \frac{\zeta}{2\sqrt{2}\pi a \sigma \omega}\left[\frac{Bei_1(a\zeta)[Ber_0(a\zeta)+Bei_0(a\zeta)]+Ber_1(a\zeta)[Ber_0(a\zeta)-Bei_0(a\zeta)]}{[Bei_1(a\zeta)]^2+[Ber_1(a\zeta)]^2}\right] \tag{6.48}$$

where L_S^i is the series self-inductance of the resistance wire per unit length, assuming that the current is returned and the circuit completed via a coaxial return conductor at infinity. Figure 6.31 shows the calculated frequency dependence of the self-inductance. The bottom graph shows that the inductance for the actual finite-sized standard is essentially constant at approximately 8 nH for frequencies up to

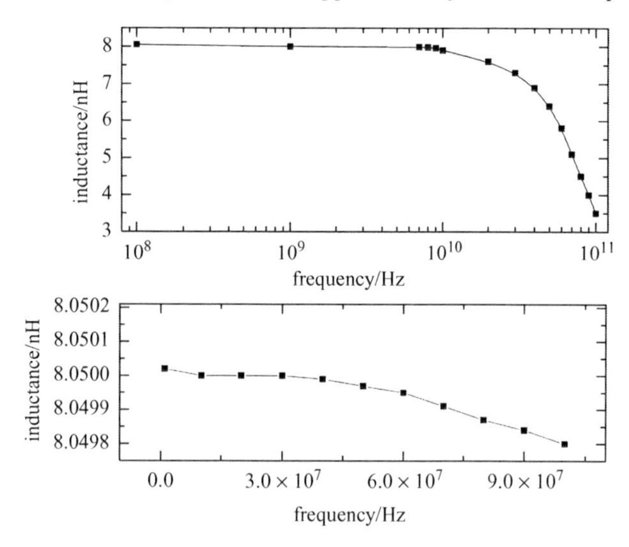

Figure 6.31 Calculated frequency dependence of the self-inductance of the 1-kΩ HF calculable resistance standard at frequencies up to 100 MHz (bottom) and 100 GHz (top). The erratic low-frequency behaviour is merely an artefact of the numerical calculations

100 MHz because the skin effect in the 16-μm diameter resistance wire is negligible up to 100 MHz. In contrast, for frequencies above 5 GHz, when the skin effect in the wire begins to become prominent, as shown in the top graph, the apparent self-inductance reduces with a dependence proportional to $1/\sqrt{f}$, as expected.

The section of coaxial return conductor that contributes to the actual defined, calculated and measured impedance is only that between the two planes located at H_P and L_P. That is, between these two planes is an *active* section of the coaxial return conductor that contributes to the defined impedance in terms of losses, inductance and shunt capacitance. This is because since in this ideal geometry end effects are negligible, the circumferential magnetic flux lines caused by the transported AC in the inner resistance wire that cut the outer coaxial return conductor only contribute to the measured impedance between the H_P and L_P connections, as required by the four-terminal-pair definition. Therefore, it is possible to calculate, again from first principles using Maxwell's equations, the frequency dependence of the eddy current losses and self-inductance of this section of the return conductor. In the case of eddy current losses,

$$R_S^0 = \frac{\zeta}{\sqrt{2}I_{pk}}\{\alpha[Ber_0(c\zeta) - Bei_0(c\zeta)] + \beta[Kei_0(c\zeta) - Ker_0(c\zeta)]\} \tag{6.49}$$

where R_S^0 is the series resistance of the outer coaxial return conductor per unit length. I_{pk} is the peak AC transport current in the inner and outer conductors and α and β are parameters, also involving the complex Kelvin functions defined in Appendix 4.

The corresponding frequency dependence of the self-inductance of the active part of the outer return conductor is

$$L_S^0 = \frac{\zeta}{\sqrt{2}\omega I_{pk}}[\alpha(Ber_0(c\zeta) + Bei_0(c\zeta)) - \beta(Ker_0(c\zeta) + Kei_0(c\zeta))] \tag{6.50}$$

where L_S^0 is the series self-inductance of the outer return conductor per unit length.

Although similar expressions to those of (6.47) to (6.50) can be found in the literature, they are often full of errors and do not give consistent results. Therefore, to the best of our knowledge, these are the first set of correct expressions that have also been robustly checked for consistency using Maple. By robust consistency we mean first that as the frequency $f \to 0$, the above expressions lead to the DC value of the resistance standard, as shown in Figure 6.30, and second, at frequencies where the skin effect is negligible, the frequency dependence of the resistance wire should be proportional to f^2 due to eddy current losses in the wire, as demonstrated in Figure 6.30 (lower graph). Finally, when the skin effect in the wire is present then the frequency dependence of the resistance should be proportional only to $\sqrt{f}$, which is also shown in Figure 6.30 (upper graph). Similar robustness checks have also been carried out on the calculated self-inductance of the inner resistance wire and the outer coaxial return conductor.

The influence of shunt capacitances on the apparent resistance of the standard is as expressed in equation 6.27 of section 6.5.2 when normalised by $\omega^2 R^2$.

6.6 Quantum Hall resistance

6.6.1 Properties of the quantum Hall effect (QHE) and its use as a DC resistance standard

The integer QHE discovered in 1980 by Klaus von Klitzing [22,23] is a macroscopic quantum effect and its extraordinary and fascinating precision makes it very valuable and important for resistance and impedance metrology. The QHE occurs in a so-called two-dimensional electron gas where the electrons of a thin planar layer in a semiconductor are confined to two-dimensional motion. This can be achieved with a GaAs heterostructure semiconductor device that has a doped layer designed in such a way that the electrons of the donor atoms accumulate in a very thin layer and so constitute the two-dimensional electron gas (Figure 6.32). A typical quantum Hall device manufactured from such a heterostructure is shown in Figures 6.33 and 6.34.

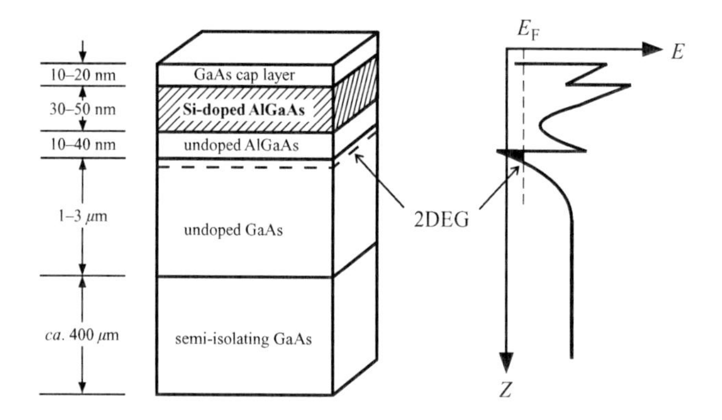

Figure 6.32 The layered structure of a quantum Hall device

Usually, the two-dimensional electron gas is formed as a rectangular bar with a current terminal at each short side and three potential contacts along each long side. The quantised state occurs only in a strong perpendicular magnetic field that forces the electrons to move in cyclotron orbits with quantised energy levels (the so-called Landau levels). To avoid a thermal breakdown of the quantised state, the temperature of the QHE device must be below a few Kelvin, and the current must be below a certain threshold (as discussed below). Under this condition and measured with DC, the Hall resistance defined as

$$R_H = \frac{U_H}{I}$$

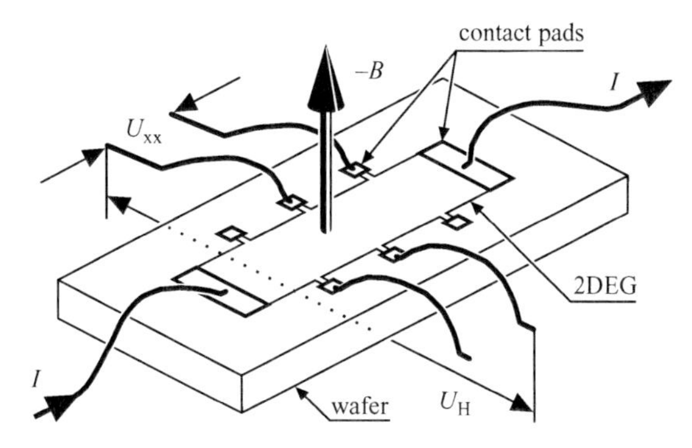

Figure 6.33 The terminals and layout of a quantum Hall bar

Figure 6.34 A photograph of a quantum Hall device mounted on a printed-circuit board, which is 16 mm in diameter and designed in the frame of the Euromet project No. 540

shows ranges of the applied magnetic induction B over which R_H is constant. That is, quantised steps, which are also termed plateaux, occur successively in a graph of R_H versus B according to

$$R_H = \frac{R_K}{i}$$

with $R_K = h/e^2$, the von Klitzing constant, and $i = 1, 2, 3, \ldots$ an integer quantum number. The electron density of the two-dimensional electron gas determines the magnetic induction (typically of the order of 10 T for the most commonly measured $i = 2$ plateau), and the mobility and the bath temperature determine the width of the plateaux.

The longitudinal resistance, defined as

$$R_{xx} = \frac{U_{xx}}{I}$$

always vanishes for a fully quantised state of the two-dimensional electron gas. The Hall resistance of a poor device can deviate from the quantised value, but this is always accompanied by a non-zero longitudinal resistance, and so the measurement of the longitudinal resistance is a key parameter for testing whether the quantisation of a particular device is complete. A typical measurement of the quantum Hall and longitudinal resistances as a function of magnetic induction is shown in Figure 6.35. From the various well-defined plateaux available for measurement, the quantum Hall resistance at the plateau $i = 2$ whose value of 12.906 kΩ is closest to a decade value, occurs in devices with a practicable electron mobility at a magnetic induction that is easily produced by a superconducting magnet. This plateau is particularly wide and is therefore commonly used in resistance metrology. The detailed requirements for reliable DC quantum Hall resistance measurements as well as for suitable devices are specified in the DC guidelines document [24,25] and are not repeated here.

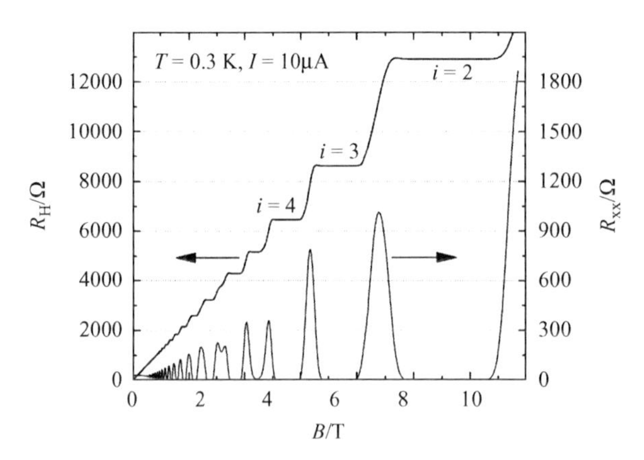

Figure 6.35 The quantum Hall and longitudinal resistance of a GaAs heterostructure

As far as it is known, the von Klitzing constant R_K depends only on two unchangeable fundamental constants, Planck's constant h and the charge of the electron e, in the combination h/e^2. It does not depend on any other parameter of our macroscopic world, such as the geometrical dimensions of the two-dimensional electron gas, the semiconductor material, time, temperature, barometric pressure or humidity. It does not require any precise geometrical adjustment or dimensional measurements, it does not depend on all the side effects that occur with conventional resistance artefacts, and it is neither subject to long-term drift nor affected by transportation, unlike conventional resistance standards. Therefore, the quantised resistance value can be reproduced with an impressive relative uncertainty of at least 10^{-10}, which is much better than all conventional resistance artefacts, and this extraordinary property is what makes quantum Hall resistance so valuable for resistance and impedance metrology.

The physics that governs this quantum effect is outlined in several other publications. Although some details are still not completely understood, for metrological purposes, the important facts are well supported by experiment and are sufficient to establish the utility of a quantum Hall device as an excellent DC resistance standard [23–25].

The only possibility for accurately measuring the SI value of the von Klitzing constant without assuming any relation to other fundamental constants is the measuring chain that links a calculable capacitor to the quantum Hall resistance (Figure 6.36). But its SI value can also be obtained from the measured value of the fine structure constant. SI resistance and capacitance can then be obtained from the inverted measurement chain of Figure 6.37. The individual measuring chains that have been realised up to now in various NMIs follow a scheme similar to either Figure 6.36 or Figure 6.37. Their implementations differ in some details, but they all require quadrature and 10:1 ratio bridges, a calculable AC–DC resistor to effect the AC–DC transfer (see sections 6.5.1–6.5.4), and either a DC resistance ratio bridge or a cryogenic current comparator. All of these components together with calculable capacitors are described in this book.

Now the alternative of a direct AC measurement of the quantum Hall resistance is not only possible but offers an even better uncertainty via the shorter measurement chain of Figure 6.38 [26].

Following a recommendation of the Comité International des Poids et Mesures that has been in force since 1 January 1990, a conventional value, $R_{K-90} = 25\,812.807\ \Omega$, has been chosen as the best representation of the SI value of the von Klitzing constant. The choice was made from a consensus mean value of the results that had been obtained from calculable capacitors. R_{K-90} is a defined, exact quantity and the resulting DC resistance unit, the Ω_{90}, has no uncertainty according to this agreement. It is not exactly the same as the SI unit and in SI units the Ω_{90} has a relative uncertainty of 2×10^{-7}.

A second numerical value of the von Klitzing constant recommended by the CODATA committee [2] assumes the relation $R_K = h/e^2$ to be exact, an assumption that is justified by theory as well as by experimental universality tests. These have been verified with a very high precision of about 10^{-11} that R_K does not depend on any other quantity. The numerical SI value of R_K with a relative uncertainty of 6.8×10^{-10} can thus be inferred from an experimental value of the fine structure constant, $\alpha = (\mu_0 c^2/2)e^2/h$. The constants μ_0 (permittivity of vacuum) and c (speed of light in vacuum) contribute no extra uncertainty as their values are defined to be exact in SI units. The SI units of resistance, capacitance and inductance can also be

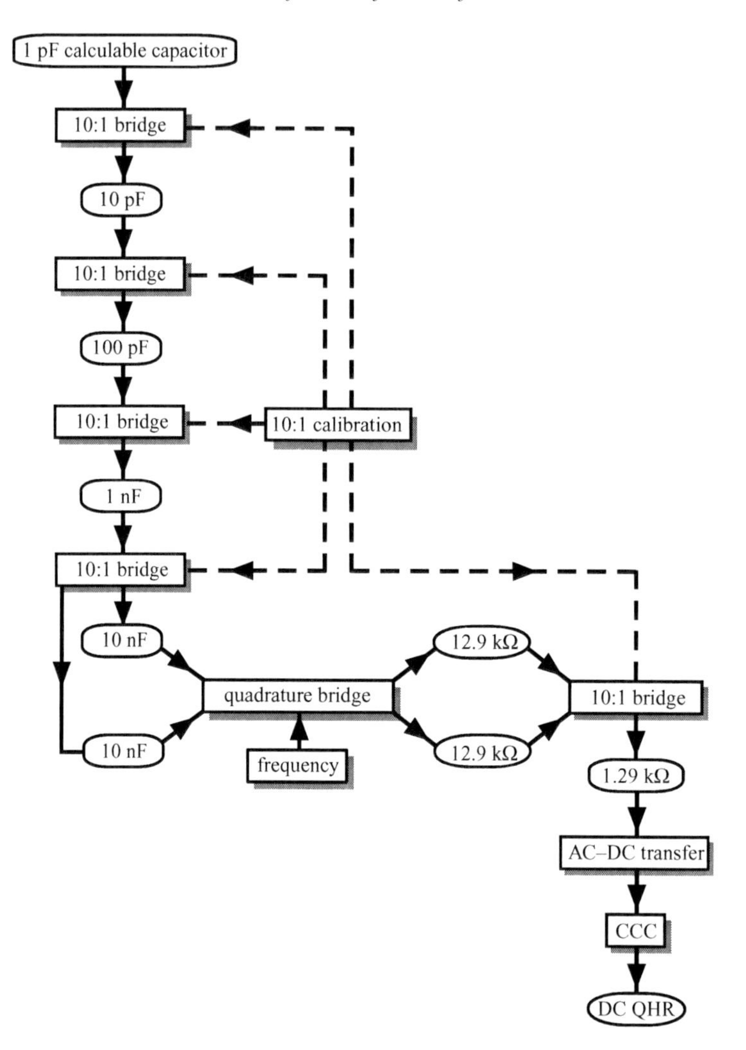

Figure 6.36 A measuring chain from a calculable capacitor to the DC quantum Hall resistance

realised from this value of R_K. Its very low uncertainty neither limits the accuracy of any resistance or impedance calibration nor can it be made better by any present-day calculable capacitor. Nevertheless, resistance and impedance standards so far are required to be calibrated in terms of R_{K-90}.

6.6.2 *The properties and the equivalent circuit of a quantum Hall device*

As noted in the previous section, in the quantised state, the Hall resistance measured with a Hall bar having current terminals at either end and potential terminals in the

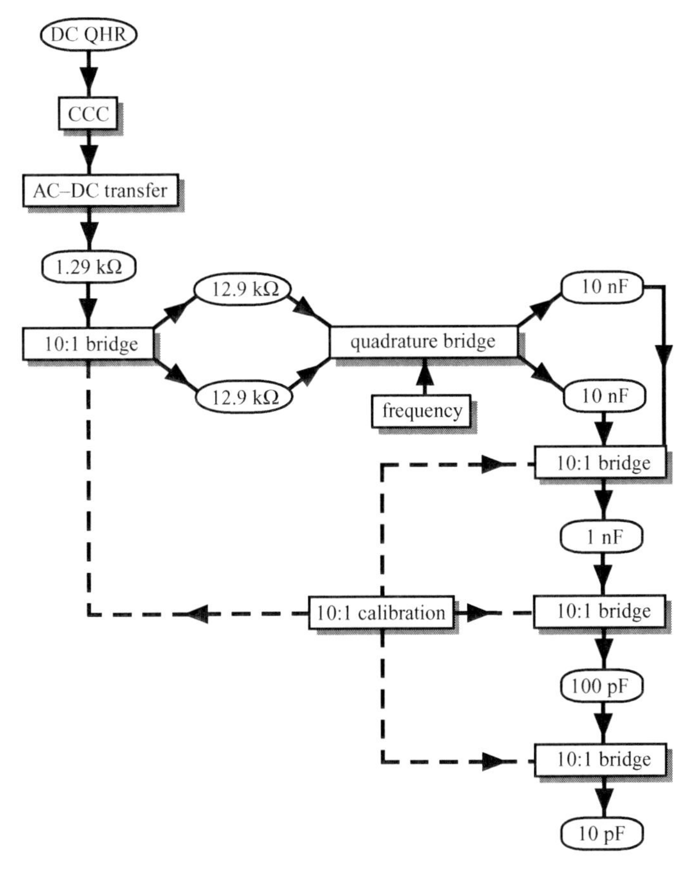

Figure 6.37 A measuring chain from the DC quantum Hall resistance to a 10-pF capacitance standard

middle of the device and at either side of it is an integer fraction $R_{\mathrm{H}} = R_{\mathrm{K}}/i$ of the von Klitzing constant, with $i = 1, 2, 3, \ldots$, and the longitudinal resistance is zero. In the following, we consider only ideal QHE devices, which means, for example, that the dimensions of the two-dimensional electron gas are sufficiently large and that the contact resistances and the Hall current are sufficiently low, as specified in the *DC guidelines* of the QHE [24,25]. Contact with the two-dimensional electron gas can be made by diffusing gold or tin into it. Gold contacts are usually laid down through a lithographic mask and are geometrically quite precise. Tin in the form of sub-millimetre diameter balls can also be diffused in by heating, and good contacts made this way can be of very low resistance but are inevitably geometrically less precise.

An ideal quantum Hall device has some basic and unique properties:

1. The two-terminal resistance measured between *any* pair of terminals is, apart from the resistance of the measuring leads and the particular contact resistances involved, equal to the quantised Hall resistance value.

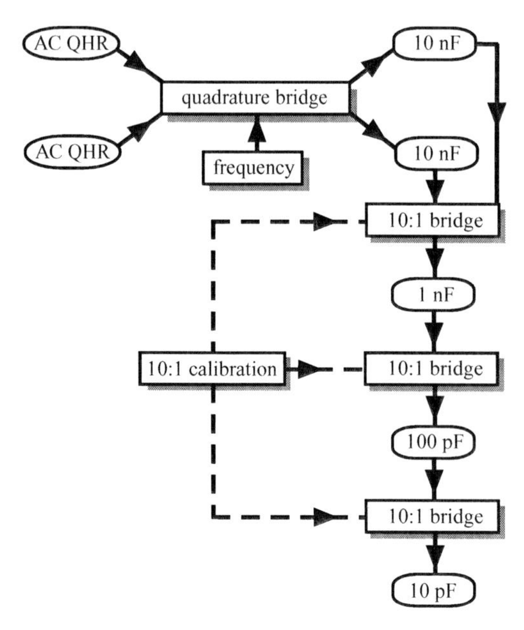

Figure 6.38 A measuring chain from an AC measurement of quantum Hall resistance to a 10-pF capacitance standard

2. When two different potentials are applied to *any* pair of terminals of the quantum Hall device so that a current flows through the device, the two device edges from the respective current entry and exit terminal in a clockwise direction as seen against the direction of the magnetic induction are at the same potential as the particular current terminal.
3. When a four-terminal measurement is carried out, the Hall voltage between *any* pair of potential terminals on opposite sides of the current flow gives the same quantised value.
4. The equipotential effect described under (2) of the list has a further consequence. In general, the metallic current terminal pads in contact with the edge of the two-dimensional electron gas have a finite width. A current entering or exiting the two-dimensional electron gas causes the equipotential effect also along these contact borders but can enter or exit the two-dimensional electron gas only in a region having a potential difference between it and the contact. This means that the currents entering or exiting regions automatically bunch to the most extreme counterclockwise corner of the contact border as seen against the direction of the magnetic induction. The full potential difference occurs at each of these corners. Also the whole energy dissipation I^2R_H occurs there and can be made visible, for example, by imaging of cyclotron emission [27] whereas the rest of the two-dimensional electron gas is noise- and dissipation-less. Therefore, the regions at the corners are often called 'hot spots'.

After the hot electrons have entered the two-dimensional electron gas through the hot spot and assuming that the distance between the hot spot and the defining potential terminals is large enough, the hot electrons thermalise to the quantised state and are unaffected by all the complicated physics of the hot spot. The electrons flowing through the two-dimensional electron gas need a finite time for this transition, and this makes clear that a hot spot has a finite dimension (of the order of some 10–100 μm). It is not a singularity although it is often termed so in the literature. The same situation in reverse applies to the hot spot at the opposite corner of the device where the electrons accumulate before they leave the two-dimensional electron gas by tunnelling into the current contact. If the voltage across the device is too large, avalanche effects of ballistic electrons hinder the development of the quantised state. This 'breakdown' typically happens at a few hundreds of microamperes and appears as a sudden and dramatic increase of the longitudinal resistance by several orders of magnitude [28].

If the current contacts are split into several parallel subcontacts with appropriate series resistances (i.e. multiples of R_H), the current can be shared between them and the hot spot is divided into several hot spots (each with less energy dissipation), but in practice, this approach gives little relevant improvement of the device performance.

The fact that the effective point of contact lies in the extreme corner of each current contact applies also to the potential contacts. As a result, the direction in which the Hall voltage is measured differs a little from the geometrical direction of the potential contacts (Figure 6.33). Even if the potential contacts are exactly opposite to each other, the longitudinal resistivity is superposed on the quantum Hall resistance. For precision lithographic contacts, the proportionality parameter (the so-called *s*-parameter) often is equal to the width of the potential arms divided by the width of the two-dimensional electron gas and has a typical value of -0.1, but other and positive values have been observed. The *s*-parameter of tin-ball contacts usually has larger, negative or positive values because of their unavoidable geometrical imperfection. A negative *s*-parameter produces a weak minimum at the high-field side of the quantum Hall plateaux, and a positive *s*-parameter produces a weak maximum at the low-field side of the quantum Hall plateaux. These effects occur because as the magnetic induction is changed to move away from the plateau centre, R_{xx} increases. The Hall resistance further from the plateau centre also deviates from the quantised value R_H and at a greater rate than R_{xx}. As a result, R_{xx} causes a structure in a limited range of magnetic induction at the edges of the Hall plateau. Further, if the longitudinal resistance is not zero across the whole central plateau region, the shape of the quantum Hall plateaux changes accordingly.

5. When a small current I' flows through a potential terminal (e.g. a capacitive AC current, a thermoelectric DC current, or a deliberately injected current to compensate for some unwanted voltage drop in the QHE device), it comes from the preceding current terminal in counterclockwise direction as seen

against the direction of the magnetic induction and causes a potential drop $I'R_H$ between the ends of the edge joining the current and potential terminals.

Because a QHE device behaves as a linear network, the effect of currents simultaneously flowing through different terminals can be calculated according to the principle of superposition. AC QHE measurements often make use of this property, as will be shown later on.

In an AC measurement, a conductor connected and bonded to a high-potential contact of a quantum Hall device should not be left unconnected at its other end because the associated capacitive current would be drawn through the quantum Hall resistance device and the Hall voltage of that current would falsely contribute to the measured Hall voltage. Therefore, open-ended conductors are not permitted in AC quantum Hall resistance measurements. Moreover, when a DC quantum Hall resistance measurement is carried out and the current direction is reversed to eliminate thermal emfs, the capacitance of an open-ended high-potential conductor is recharged, and due to dielectric relaxation, the effect of charging may last several minutes before a reliable Hall resistance can be measured.

To facilitate mathematical analysis and to support the interpretation of experimental results, a phenomenological equivalent circuit of a quantum Hall resistance device with four terminals has been suggested by Ricketts and Kemeny [29]. Their equivalent circuit is shown on the left-hand side of Figure 6.39. It consists of linear circuit elements that are ideal resistors and current-driven voltage sources and is constructed symmetrically in such a way that – neglecting the resistance of the leads and contacts – the two-terminal DC resistance between *any* two terminals gives the quantised resistance value R_H (property (1) from the list above).

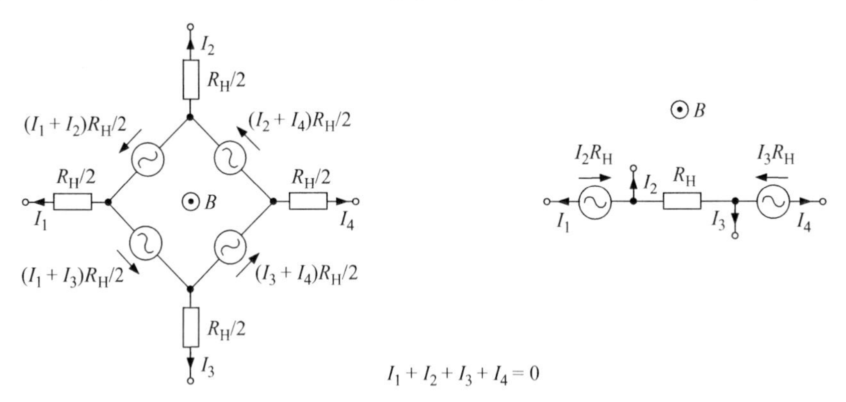

Figure 6.39 The equivalent circuit of Ricketts and Kemeny and an equivalent, simpler circuit

The alternative equivalent circuit shown on the right of Figure 6.39 can be derived from the Ricketts–Kemeny scheme using network theory and is thus fully equivalent to it [30]. It requires fewer components and is thus more suitable for mathematical analysis as well as aiding intuitive understanding. To describe a QHE device with two current terminals and two (or three) potential terminals at each side of the device, this alternative scheme can be easily extended (Figure 6.41b) and enables the basic properties of a QHE device listed above to be clearly visualised. For the same situation, the Ricketts–Kemeny scheme has to be joined up three times in series but the many components and relations make the analysis quite complicated. Furthermore, the alternative scheme, but not the Ricketts–Kemeny scheme, permits longitudinal resistances that can be different at either side of the device to be added to the equivalent circuit.

In all equivalent circuits of QHE devices that are shown in this book, the voltage sources are for convenience represented as alternating voltage sources but of course, all considerations also apply to DC.

6.6.3 Device handling

QHE devices should be cooled down slowly from room temperature to 4 K or below at a constant rate of not more than about 10 K/min. They should be cooled in the dark, with no applied voltages or currents, with the connections to them short-circuited, and be shielded from radiofrequency radiation. Once cooled, the measuring leads should be handled carefully to prevent the device from being exposed to static electrical charge. The measuring current should be turned on and off smoothly as otherwise a large current pulse could flow through the device and affect it, probably because of frozen-in charge carriers. Frozen-in charge carriers (usually electrons) can affect the internal capacitances and their associated loss factors resulting in irregular features appearing in AC plateaux and causing error in AC measurements. The magnitudes of these features are proportional to frequency, so they do not appear in DC measurements, but they nevertheless affect the desirable spatial orthogonality of current path through the device and the potential across it (the DC s-parameter), and therefore may affect the quantised DC value. It is also good practice to short-circuit the leads of a quantum Hall resistance device when no measurement is being carried out and when the magnetic field is rapidly ramped up or down. Devices intended for AC measurement should be exposed to a sufficiently good vacuum with mild baking at a maximum of 60 °C for about 48 hours to remove electrically lossy water films before cooling.

These effects of misshandling are cumulative, but fortunately nature is kind in that a device can always be returned to its original good state by simply warming it up to room temperature and cooling it again.

6.7 QHE measured with AC

The QHE measured with DC has been well established since 1990 for making precise DC resistance calibrations. Because of this great success, Melcher et al. suggested [31] that quantum Hall resistance could be measured with AC at frequencies in the kHz range and the same quantum effect could then be used for AC impedance calibrations.

A measurement of the quantum Hall resistance with alternating current is practically the same as the QHE measured with direct current, and the commonly used phrase 'AC QHE' does not imply a new physical phenomena – it is only a short form of 'QHE measured with AC'. One small effect of AC measurement arises because the localised electrons that are essential for the finite width of the plateaux oscillate cyclically in the alternating electric Hall field. This gives rise to an intrinsic frequency dependence that might be relevant in the GHz frequency range. In the kHz frequency range, it is smaller than 10^{-9} R_H/kHz and may also be negligible for frequencies up to 1 MHz because the achievable uncertainties of actual measurements are larger than for audiofrequency measurements.

Furthermore, while in a DC measurement of quantum Hall resistance, the capacitances associated with a quantum Hall device are charged only during a short time interval after switching the Hall current on or off or after every reversal of the current direction, they are continuously charged and recharged in the case of an AC quantum Hall resistance measurement. These capacitances have associated current- and frequency-dependent AC losses that vary across a quantum Hall plateau. As a result, in the early measurements, the quantum Hall plateaux measured with AC were not as flat as those measured with DC, the AC quantum Hall resistance differed from the quantised DC value, and both effects increased linearly with frequency and current. These effects limited the accuracy of AC quantum Hall resistance measurements before special precautions were developed to compensate or remove the effects of AC loss currents flowing through the device.

To avoid these problems that are less than 1 ppm/kHz but are nevertheless significant, a few NMIs have traced their impedance standards to DC quantum Hall resistance measurements by employing a special resistance artefact that has a calculable difference between its value measured with AC and with DC (its AC–DC difference) to effect the transfer between AC and DC. A possible measuring chain is shown in Figure 6.36 or Figure 6.37. Other measuring chains using standards of other or non-decade nominal values are also possible. Since then all the early difficulties of AC quantum Hall resistance measurements appear to be understood and solved (as will be reported in sections 6.7.4 and 6.7.5), and it is possible to measure the QHE directly and precisely with alternating current. The capacitance unit, the farad, can now be realised

- by a short measuring chain (Figure 6.38),
- from an unchangeable and highly reproducible cryogenic quantum effect,
- with good resolution because of a very low thermal noise power,
- without the need for any artefact of calculable frequency dependence that might suffer from other limitations,
- and therefore with a remarkably low relative uncertainty that is less than 1×10^{-8}.

As inductance is usually derived from capacitance and AC resistance, the three impedance units ohm, farad and henry, as well as the ohm defined for DC can be derived from the *same* quantum effect. This is an advantage for the consistency of the SI.

6.7.1 Multiple-series connection scheme

The conventional four-terminal-pair measurement scheme cannot be applied straightforwardly to an AC quantum Hall resistance measurement. Unlike all conventional four-terminal-pair standards, the potential terminals of a quantum Hall device do not have a low-impedance connection to the associated current terminal. Consequently, capacitances and inductances associated with the quantum Hall device and its connections and surroundings give rise to serious frequency-dependent systematic errors. A 'multiple-series' connection scheme [32,33] avoids all these difficulties in an elegant manner and thus became the standard connection scheme for AC measurements of quantum Hall resistance. This scheme requires that the inner conductors of coaxial leads from all equipotential contacts on either side of the quantum Hall device are connected to two respective star points. An example is shown in Figure 6.40. The coaxial star points are usually outside the cryostat and at room temperature. The two star points are the internal defining points of a two-terminal-pair device but, as will be shown later, with the advantage that the cable corrections are numerically the same as, but are of opposite sign to the simpler ones of a four-terminal-pair standard.

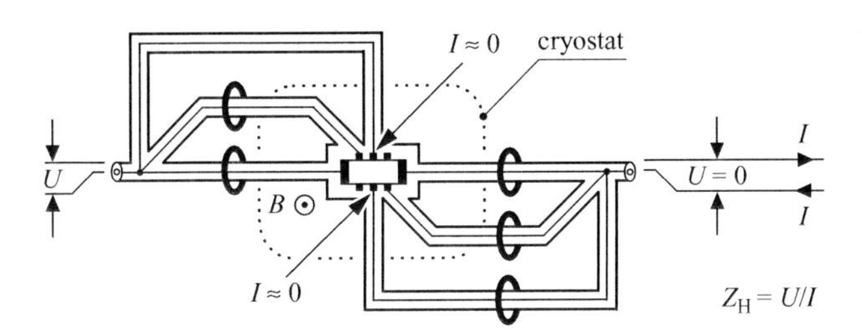

Figure 6.40 A triple-series connection

The multiple-series connection scheme is not restricted to AC quantum Hall resistance measurements. It can also be used for DC quantum Hall resistance measurements. Because shielded leads are also advantageous for DC measurements (e.g. they reduce the requirements for the insulation resistance of the leads) and because the multiple-series connection scheme also eliminates unwanted DC effects such as the slow relaxation of charge storage in the dielectric insulation of the measuring leads, one and the same quantum Hall device in one and the same cryomagnetic system can be used for both DC and AC measurements.

As an example, Figure 6.42 shows the equivalent circuit of a triple-series connection scheme. To explain the multiple-series connection scheme, we start with a two-terminal measurement at the two current terminals. The measuring current undergoes unwanted voltage drops at the lead and contact resistances to the current contacts so that the Hall potential at each edge of the quantum Hall device differs from the potential applied to the room temperature end of its corresponding current lead. If now another lead is connected from each star point to the next potential terminal around the device, as drawn in Figure 6.42 the small potential difference between each edge and the respective star point causes a small current to be drawn through the potential lead. This current causes an additional Hall voltage that compensates for the small voltage drop along the current lead. The compensation is not complete because this small current causes an even smaller voltage drop down the added lead but the process can be repeated by another connection to the next potential terminal. When one additional lead after another is connected from each star point to the successive potential terminals down either side of the device, the currents in them become successively smaller. If the number of leads is large enough and the lead and contact resistances are small enough, the Hall voltage between the two final potential contacts on either side of the device becomes practically the same as the voltage applied to the two accessible star points outside the cryostat, which is the purpose of the multiple-series connection scheme. This compensating effect can be analysed with the aid of the equivalent circuit of Figure 6.42 [30].

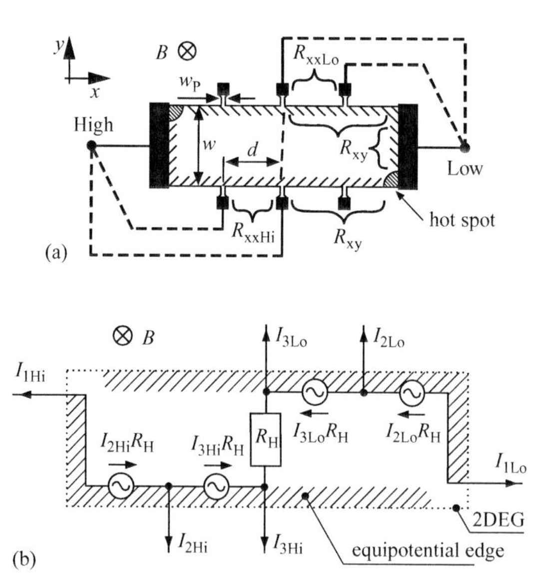

Figure 6.41 (a) The relevant device edges of a six-terminal device (b) The equivalent circuit (for simplicity without the longitudinal resistances)

The relevant lead impedances and contact resistances are shown explicitly in Figure 6.42. For simplicity, the longitudinal resistances down each side of the device, which should be zero if it is fully quantised, have been neglected in the following analysis. The multiple leads at the high- and the low side are labelled n and m, respectively. The apparent impedance Z_H as defined at the high and low star points is calculated with respect to the quantum Hall resistance R_H and yields to a good approximation:

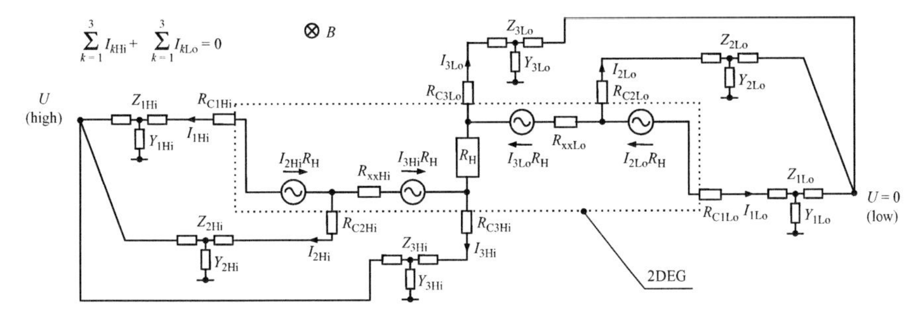

Figure 6.42 The equivalent circuit of a quantum Hall device with a triple-series connection scheme ($n = m = 3$) including the contact resistances and the lead impedances and admittances. For simplicity, the outer conductors of the coaxial leads are not shown

$$Z_{\mathrm{H}} = R_{\mathrm{H}}\left[1 + \frac{Z_{n\mathrm{Hi}}Y_{n\mathrm{Hi}}}{2} + \prod_{k=1}^{n}\left(\frac{\tilde{Z}_{k\mathrm{Hi}}}{R_{\mathrm{H}}}\right)\right] \times \left[1 + \frac{Z_{1\mathrm{Lo}}Y_{1\mathrm{Lo}}}{2} + \prod_{k=1}^{m}\left(\frac{\tilde{Z}_{k\mathrm{Lo}}}{R_{\mathrm{H}}}\right)\right] \tag{6.51}$$

$$\text{with } Z_k = R_k + j\omega\, L_k \tag{6.52}$$

$$Y_k = \omega C_k(j + \tan\delta_k) \tag{6.53}$$

$$\tilde{Z}_k = \frac{Z_{\mathrm{C}k} + Z_k}{(1 + Z_k Y_k)} \approx R_{\mathrm{C}k} + R_k \ll R_{\mathrm{H}} \tag{6.54}$$

The first indices (k, l, m or n) number the leads and the second indices (Hi or Lo) indicate the high- and the low-potential side of the quantum Hall resistance device, respectively. Z_{H} and R_{H} with the single index H stand for the Hall impedance and the Hall resistance, respectively.

Each multiple lead has been represented as a T-network: the lead impedance Z_k is split into two equal halves (see Figure 6.42) consisting of the series resistance R_k and the series inductance L_k. Y_k is the admittance of the lead consisting of the capacitance C_k with its associated loss factor tan δ_k. For simplicity, the indices Hi or Lo have been omitted in (6.51) to (6.54). $\tilde{Z}_k$ in (6.51) and (6.54) is called the 'connecting impedance'. The term $(1 + Z_kY_k)$ in (6.54) is in any practical case equal to 1 and the imaginary component of $\tilde{Z}_k$ is usually small so that $\tilde{Z}_k$ is approximately equal to the resistance of the contact, R_{Ck}, plus the series resistances of the particular lead (see (6.54)). The terms $Z_kY_k/2$ in (6.51) are independent of the number of multiple leads, and they are identical with the usual quadratic cable correction terms as they apply to any conventional two-terminal-pair impedance.

The product terms in (6.51) are specific to the multiple-series connection scheme. In the case of a two-terminal-pair connection scheme ($n = m = 1$), (6.51) is just the usual two-terminal-pair formula for the resistance R_H in series with the connecting impedances $\tilde{Z}_k$. For every additional multiple-series lead, the potential difference between each star point and the respective edge of the quantum Hall device is reduced by an additional factor of $\tilde{Z}_k/R_H$, as described by the product terms in (6.51). If the number of leads is large enough and the connecting impedances are small enough, the product terms become negligibly small. The cable correction now corresponds to that of a four-terminal-pair device (but with opposite sign), and, apart from this cable effect, the Hall voltage between the two final potential contacts becomes practically the same as the voltage at the two star-point junctions.

To assess a particular connection scheme, the numerical values of the product terms in (6.51) for a typical connecting impedance of a few ohms, $\tilde{Z}_k/R_H$ are a few times 10^{-4}. Then the error of a double-series connection scheme is a few times 10^{-8}, which is not sufficiently small for precise AC quantum Hall resistance measurements. Furthermore, as open leads are not allowed, only one potential contact at each edge of the quantum Hall device can be provided with a lead so that it is not possible to measure its longitudinal resistance. In contrast, triple- and quadruple-series connection schemes sufficiently eliminate the effect of the lead and contact resistances, and two or three potential terminals are available at each side of the quantum Hall device for measuring longitudinal resistances.

Finally, a quantum Hall device with a multiple-series connection scheme (triple-series or higher) can provide a self-contained combining network (see section 5.6). The currents flowing through each defining Hall potential contact are automatically adjusted to a value which, while not zero, is much smaller than what is needed for an effective combining network like a current source or a Kelvin arm (see section 5.6.1). Therefore, *a quantum Hall resistance device with a multiple-series connection scheme can provide a self-contained combining network.* This simplifies both coaxial bridge networks and their balancing procedures. Examples are discussed in section 9.3.

6.7.2 A device holder and coaxial leads

For AC quantum Hall resistance measurement, the quantum Hall device should be mounted into a complete metallic shield to achieve a well-defined impedance without any stray capacitive current to the cryostat (see Figure 6.43). This shield

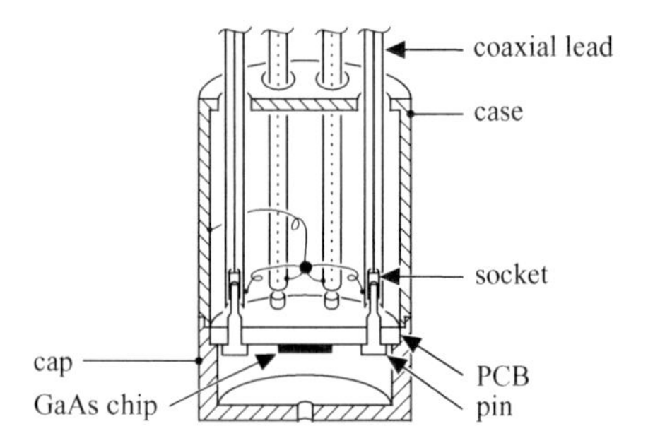

Figure 6.43 A shielded device holder

containing the quantum Hall resistance device is usually mounted at the lower end of a probe that can be lowered to the centre of a superconducting magnet coil inside a cryostat, as shown in Figure 6.44. A pumped ^{4}He system with a base temperature of 1.5 K is sufficient for ordinary applications, but a ^{3}He system having a base temperature of 0.3 K produces wider Hall plateaux and allows a wider scope for research investigations. The coaxial multiple leads have to be fed through the probe and the shield to the quantum Hall resistance device. Coaxial leads having a low thermal conductivity minimise the thermal load on the cryosystem. The outer conductors of the coaxial leads and the shield of the quantum Hall resistance device

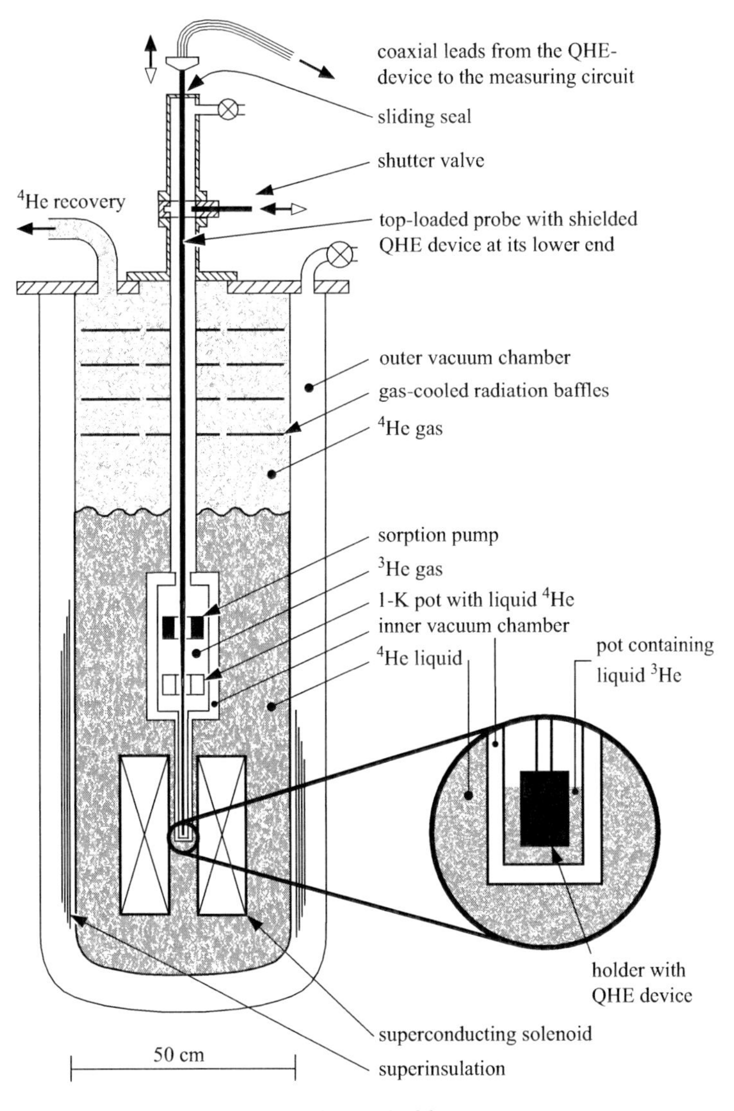

Figure 6.44 A device holder in a cryostat

should be electrically isolated from the cryostat. If two shielded quantum Hall resistance devices insulated and isolated from one another can be mounted side-by-side into the same cryomagnetic system, they can be measured in a quadrature bridge (see section 9.3.7) to make a direct realisation of the SI farad.

To equalise the return currents in the outer conductors from the multiple-series junctions outside the cryostat up to a point that is as close to the quantum Hall device as possible, the coaxial leads should be fed through the shield of each quantum Hall device. Their outer conductors as well as the device shield should be connected to an internal star point close to the quantum Hall resistance device (see Figure 6.43). If this is not done, the open-loop inductance and capacitance of any remaining length of unequalised inner conductors will cause a large unwanted quadratic frequency dependence. The four-terminal resistance of the outer-conductor network is transformed by the equalisers to be in series with the quantum Hall resistance, but it can be kept negligibly small with a star-point connection. Its four-terminal resistance can be measured with a four-wire micro-ohm meter. If this resistance is not negligibly small, a correction must be made for it.

The quantum Hall device is usually mounted on a printed-circuit board or a ceramic carrier. Both types of carriers are provided with a certain number of pins that can be plugged into the coaxial miniature sockets of the mounting system to allow an easy exchange of quantum Hall devices. The printed-circuit board may carry metal pads that act as gates or shields [34,38], and usually it is printed with line conductors that reach from the pins to pads close to the quantum Hall device so that connections can be made to the device with bonding wires. The bonding wires should be as short as possible and the quantum Hall device should be fixed to the printed-circuit board with glue to avoid AC-driven oscillations of the bonding wires and the quantum Hall device in the magnetic field. As a further precaution, the current contacts of the quantum Hall resistance device can be provided with two parallel bonding wires of different lengths and therefore different resonance frequencies. The effect of a resonant frequency in one bonding wire will be shorted out by the other. As there is capacitance between quantum Hall device and surrounding metals, the glue as well as the printed-circuit board material should have a sufficiently low loss factor and a sufficiently high isolation resistance. If ungated quantum Hall devices are used, the conductors on the printed-circuit board and the star point in the case should not be too close to the quantum Hall device, otherwise they act as unwanted gates. A standardised mounting system has been developed in the frame of the EUROMET project No. 540 to allow the exchange of quantum Hall devices between different national standard laboratories (see also Figure 6.34).

The coaxial leads require a minimum isolation resistance between the inner and outer conductors. To obtain a desired relative uncertainty of 1×10^{-9} a moderate isolation resistance of about 2 GΩ is sufficient for the defining leads of the coaxial connections that have a typical inner conductor resistance of 2 Ω. For the other cables even less will suffice. This is one of the useful advantages of a coaxial network over a four-wire measurement that applies to AC as well as to DC measurements. In a four-wire measurement, the isolation resistances between the high- and low-potential line conductors on the printed-circuit board as well as between the quantum Hall device and a split-gate or double shields are a shunt resistance in parallel to the quantum Hall resistance. Therefore, these isolation resistances must be larger than 10 TΩ. Isolation resistances can be checked with a DC tera-ohm meter.

The outer conductors of the coaxial leads and the shield of the quantum Hall device must provide sufficient shielding to avoid any spurious crosstalk between the inner conductors of different leads or between each inner conductor and any other device, for example, the cryomagnetic system. The shielding can be tested with a high-resolution ($\approx$0.1 fF) capacitance bridge.

In general, any precise impedance standard requires a specific correction for the impedance and admittance of the defining cables (section 5.4), and this also applies to the quantum Hall resistance. To measure the cable parameters with a precision LCR meter, two dummy printed-circuit boards without a quantum Hall device should be used: one printed-circuit board with open terminals for the measurement of the lead admittance, and a second printed-circuit board with short-circuited terminals for the measurement of the lead impedance. The cable parameters should be measured in the cryomagnetic system under the same conditions as for the actual quantum Hall resistance measurement.

The coaxial leads connected to the quantum Hall device are usually composed of sections having a different impedance and admittance per meter (e.g. a section of room temperature cable from the measuring bridge to the head of the cryostat and a section of a thin cryogenic cable inside the cryostat) and the combination requires a more complex cable correction. For defining cables consisting of two sections, the cable correction is given in section 3.2.6. It is determined by the total lead series impedance and shunt admittance if the different sections have the same characteristic impedance.

6.7.3 *Active equalisers*

AC quantum Hall resistance measurements usually require long coaxial leads for the multiple-series connection scheme because the QHE occurs only at cryogenic temperatures in a cryostat but the device is connected to measuring bridges at room temperature. To keep the heat transport into the liquid helium bath of the device small, the coaxial leads are usually thin, and thus the long and thin leads have comparatively large resistances, typically about 2 Ω. This greatly reduces the efficiency of the current equalisers involving these cables in their meshes and increases the net current in these cables and its effect on the main balance. For equalisers with the usual 12–20 turns, the effect on the main balance is too large to be accurately evaluated and corrected (as described in section 3.2.2). Active current equalisers are recommended in connection with the ac quantum Hall resistance because their lead capacitances are not as large as for passive current equalisers of enhanced number of turns.

In a coaxial bridge network employing a quantum Hall device, it might not be sufficient to use active equalisers only in the multiple leads of the device. For example, the equalisers might be arranged in such a way that the *two* defining leads of the quantum Hall resistance do not carry an equaliser. However, these two leads are part of other meshes in the bridge network and the corresponding equalisers experience the large impedance of the long leads. This usually gives rise to significant net currents. If the resulting effect on the main balance (evaluated as described in section 3.2.2) is too large, these equalisers should also be replaced by active equalisers or by passive increased-efficiency equalisers. It is also possible to arrange the equalisers in the bridge network in such a way that only one lead of the quantum Hall device does not carry an equaliser. It then might be sufficient to use active equalisers only in the multiple leads to the device but not in the rest of the bridge network. However, the residual effect of all the equalisers in the network on the main balance must still be evaluated.

6.7.4 *Capacitive model of ungated and split-gated quantum Hall devices*

If no special precautions are taken, various internal and external capacitances draw a current through a quantum Hall device, and this gives rise to additional Hall voltages that contribute to the quantum Hall impedance Z_{H} as well as to the longitudinal impedance Z_{xx} (see also section 6.6.2) [35]:

$$Z_{\mathrm{H}} = R_{\mathrm{H}} + \sum_{n} \omega R_{\mathrm{H}}^{2} [C_{n}^{\mathrm{High}} (j + \tan \delta_{n}^{\mathrm{High}}) - C_{n}^{\mathrm{Low}} (j + \tan \delta_{n}^{\mathrm{Low}})]$$

$$Z_{\mathrm{xx}}^{\mathrm{ab}} = R_{\mathrm{xx}}^{\mathrm{ab}} + \sum_{n} \omega R_{\mathrm{H}}^{2} C_{n}^{\mathrm{ab}} (j + \tan \delta_{n}^{\mathrm{ab}})$$

The capacitances of the relevant edges as shown in Figure 6.41(a) and the associated loss factors are labelled by the index n. We wish to discuss the situation in general and so the meanings of C_1, C_2, etc. depend on the particular quantum Hall resistance device and the metals, gates or shields in its surroundings. The contributions to Z_{H} originating from the high- and low-potential edge of the two-dimensional electron gas have opposite sign and the relevant edge regions reach from the particular 'downstream' defining potential contact to the low-current terminal (Figure 6.41a). For simplicity, those additional correction terms of Z_{H} that arise from the multiple-series connection scheme (section 6.7.1) are not explicitly listed here again but, of course, have to be added. The contributions to Z_{xx} measured at two contacts a and b originate only from the edge region between those contacts. R_{xx} denotes the AC longitudinal resistance that would be found if there were no capacitive effects, and that in the kHz frequency range is practically equal to the DC longitudinal resistance (i.e. for an ideal quantum Hall device it is zero in the plateau centre and steeply increases at the plateau edges).

So far, three different contributions to the AC losses have been identified [37]. The main contribution stems from losses incurred by electrons in the two-dimensional electron gas caused by internal capacitances in it as well as by external capacitances between it and surrounding conductors. The loss mechanism is polarisation and this explains the linear frequency dependence of the measured Hall resistance [39]. The loss factor increases linearly with current because both the polarising electrical Hall field and the number of polarisable electrons in the two-dimensional electron gas increase with current. Consequently, the AC losses of the two-dimensional electron gas appear as a linear frequency and current dependence of AC quantum Hall resistance measurements. Further, both the capacitance and the associated loss factor measured as a function of the magnetic field increase from the plateau centre towards the plateau edges, and this explains the curved shape of the plateaux observed for both the quantum Hall and the longitudinal impedances.

A second, much smaller, contribution has been attributed to lossy dielectric materials located between the two-dimensional electron gas and surrounding conductors. Relevant conductors could be the line conductors on a printed-circuit board device carrier, gates or shields and lossy dielectric materials could be, for example, the GaAs substrate and the device carrier.

The third small loss mechanism has been attributed to adsorbates on the surface above the two-dimensional electron gas. In both cases, the relevant loss mechanism is polarisation, which is approximately frequency independent, and this, when combined with the pre-factor ω in the equations given above, results in a linear frequency dependence of the quantum Hall and longitudinal impedances. As the number of polarisable dipoles does not depend on the magnitude of the Hall current, the corresponding contribution is current independent. The surface adsorbates can be removed by standard vacuum outgassing techniques before the QHE device is cooled down in the cryostat, and the other AC losses can be eliminated by gates or shields, at least to within some limit as discussed below.

According to the current and frequency dependence of the individual loss mechanisms, the current and frequency dependence of the quantum Hall and longitudinal resistance can be empirically written as

$$\mathrm{Re}(Z_{\mathrm{H}}) = R_{\mathrm{H}} + fk' + fIc'(+\,\text{higher order terms})$$

$$\mathrm{Re}(Z_{\mathrm{xx}}) = R_{\mathrm{xx}} + fk + fIc.$$

where k and k' are current-independent frequency coefficients and c and c' current coefficients. This is the current and frequency dependence that has been empirically known since the very first AC quantum Hall measurements but was not understood for a long time.

6.7.5 *Ungated quantum Hall devices*

Conductors should not be located too close to ungated devices to avoid an unnecessary increase of external capacitances beyond the inevitable capacitance between the device and its shield. For shield dimensions much larger than the dimensions of the device, the capacitance to the shield becomes independent of the shield dimensions. Assuming an approximate relation $k'/k \approx c'/c$ and a device that satisfies the DC guidelines, the deviation of the AC quantum Hall resistance from the quantised DC value is approximately proportional to the AC longitudinal resistance. This proportionality can be measured and the AC quantum Hall resistance can be extrapolated to that which would be found if the longitudinal resistances were zero [40].

6.7.6 *Split-gated quantum Hall devices*

Another approach is based on subjecting the device to electric fields from conductors near its edges. These conductors are termed as 'split gate', and can be realised either as a split back-gate where the device is mounted on two conducting surfaces separated by a small gap along the centre line of a device (Figure 6.45) or two side-gates [35,37].

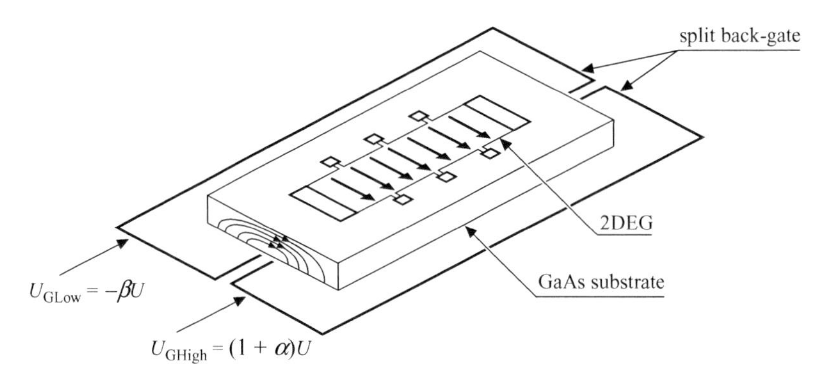

Figure 6.45 A quantum Hall device mounted on a split back-gate

Applying adjustable AC voltages of the same frequency as the measuring frequency to each half gate generates an electric AC field that penetrates into the dielectric 2DEG and biases the capacitances of the QHE device. This effect can be described by two parameters λ_{High} and λ_{Low}:

$$Z_{\mathrm{H}} = R_{\mathrm{H}} + \omega R_{\mathrm{H}}^2 \left[C_1^{\mathrm{High}} \left(j + \tan \delta_1^{\mathrm{High}} \right) \lambda_{\mathrm{High}} - C_1^{\mathrm{Low}} \left(j + \tan \delta_1^{\mathrm{Low}} \right) \lambda_{\mathrm{Low}} \right]$$

$$Z_{\mathrm{xx}}^{\mathrm{High}} = R_{\mathrm{xx}}^{\mathrm{High}} + \omega R_{\mathrm{H}}^2 C_1^{\mathrm{High}} \left(j + \tan \delta_1^{\mathrm{High}} \right) \lambda_{\mathrm{High}}$$

$$Z_{\mathrm{xx}}^{\mathrm{Low}} = R_{\mathrm{xx}}^{\mathrm{Low}} + \omega R_{\mathrm{H}}^2 C_1^{\mathrm{Low}} \left(j + \tan \delta_1^{\mathrm{Low}} \right) \lambda_{\mathrm{Low}}$$

Only the dominating internal capacitive effects are considered here. It is found experimentally that the effect of a split gate on both Z_{H} and Z_{xx} can be described by two multiplying factors rather than by sum terms and that each of these parameters can be adjusted to be positive, negative or zero. If the component of the external electric AC field in the plane of the 2DEG is equal to the internal Hall field, the internal capacitances no longer load the quantum Hall device, that is, both parameters λ_{High} and λ_{Low} are zero. Then, the main contribution to the frequency and current dependence of the AC quantum Hall resistance and the AC longitudinal resistances along both sides of the device vanish simultaneously. The criterion for adjusting the gate voltages is the attainment of a zero AC longitudinal resistance at both sides of the device (assuming again that the device satisfies the DC guidelines).

Apart from this special setting of the gate voltages, many other solutions exist where the capacitive contributions to Z_H at both sides of device are not individually zero but mutually cancel each other. These solutions lie on a line in a two-dimensional gate-voltage diagram (and the special solution discussed above is a point on that line. The criterion for adjusting the gate voltages is a zero current coefficient of the AC quantum Hall resistance [34,35]. This approach is easier to realise but it results in slightly curved Hall plateau if the capacitive effects at both sides of the device have a different shape as a function of the magnetic field (an effect that does not limit this approach), whereas the first criterion results in perfectly flat plateaux (which is neither a necessary nor a sufficient criterion).

Both the extrapolation and the split-gate approach are only approximate solutions because residual frequency dependences of the order of a few parts in 10^8/kHz remain. In the case of the extrapolation method, the AC longitudinal resistance shows a strictly linear current dependence whereas the current dependence of the AC quantum Hall resistance shows a non-linear saturation at higher currents. Further, the internal and external capacitances contribute to the AC longitudinal and quantum Hall resistance in a different way. Both effects limit this approach. The split-gate approach is limited by the external capacitive AC losses in the GaAs substrate that are not eliminated. However, the residual frequency dependences of both methods are of the order of a few parts in 10^8/kHz and are thus comparable to the best calculable capacitors presently in operation. But, this is not the end of the story.

6.7.7 Double-shielded device

More than ten times better and, for all practical purposes, perfect results can be obtained by a third approach. This approach considers where the relevant currents flow and from this information, how to achieve that unwanted currents do not affect the measurement. The strategy is not to eliminate the unwanted capacitances or to use materials having lower loss factors but to handle all the currents in such a way that the defining condition of the particular impedance standard is precisely met. This is not only a simple and elegant strategy but also the most precise and sensible solution of the problem.

The double-shield approach is based on two conducting shielding boxes with a narrow gap in between them. The device is located inside the boxes in such a way that the defining Hall potential contacts are centred in the gap (Figure 6.46) [36,41].

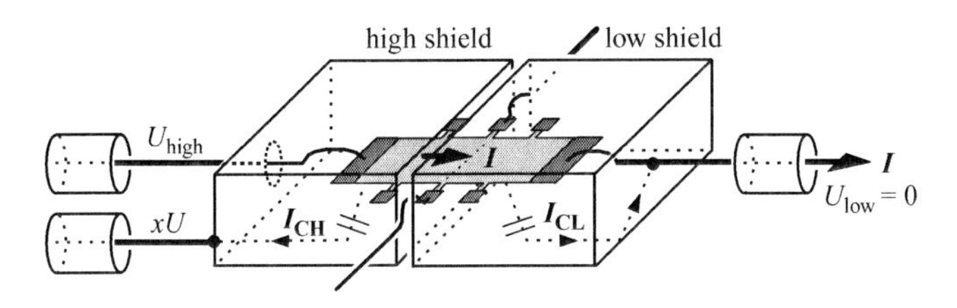

Figure 6.46 A double-shield device

The low shield is connected to the low-current terminal so that the capacitive loss currents are picked up and added to the measured current, whereas the shield around the high-current terminal is connected to the outer conductor network so that the capacitive stray currents from within it are intercepted and routed back to the source so that they do not falsify the measurement.

As an additional detail, an adjustable potential xU has to be applied to the high shield in order to electronically adjust the position of the shield gap relative to the defining Hall-potential contacts and in this way to eliminate the residual AC losses in the two-dimensional electron gas due to an imperfect mounting of the device. The criterion for adjusting the high-shield potential is, exactly as for the split-gate approach, the vanishing of the current coefficient of the AC quantum Hall resistance. This double-shield approach also eliminates the effect of the AC losses of adsorbates as well as the AC losses in the substrate that is the limitation of the split-gate approach. The purpose of the double-shield method is to precisely meet the defining conditions of an AC impedance standard, and this simple basic principle is the reason of its success. Altogether, this approach allows the realisation of a quantum standard of impedance that is more accurate than all conventional impedance artefacts. Residual parasitic capacitances can be measured using a scaled-up room temperature model of the device [41] and their effect on the phase angle calculated so that the device constitutes a phase angle standard of as low an uncertainty as can be obtained by other methods (see section 9.2.1).

To make use of such a precise quantum Hall impedance standard and to realise a traceability chain like those shown in Figure 6.37 or Figure 6.38, we now need appropriate coaxial bridges. Because a quantum Hall device can provide a self-contained combining network, the design of the bridges can be simplified. They are described in sections 9.3.6 and 9.3.7.

References

1. Thompson A.M., Lampard D.G. 'A new theorem in electrostatics and its application to calculable standards of capacitance'. *Nature.* 1956;**177**:188
2. Mohr P.J., Taylor B.N., Newell D.B. '2008 CODATA recommended values of the fundamental physical constants'. *Rev. Mod. Phys.* 2006;**80**:633–730
3. Lampard D.G., Cutkosky R.D. 'Some results on the cross-capacitances per unit length of cylindrical three-terminal capacitors with thin dielectric films on their electrodes'. *J. IEE.* 1960;**107c**:112–19
4. So E., Shields J.Q. 'Losses in electrode surface films in gas dielectric capacitors'. *IEEE Trans. Instrum. Meas.* 1979;**28**:279–84
5. Shields J.Q. 'Phase angle characteristics of cross-capacitors'. *IEEE Trans. Instrum. Meas.* 1972;**21**:365–68
6. Rayner G.H. 'The time-constant of carbon composition resistors'. *Br. J. Appl. Phys.* 1958;**9**:240–42
7. Engheta N., Salandrino A., Alú A. 'Circuit elements at optical frequencies: nanoinductors, nanocapacitors and nanoresistors'. *Phys. Rev. Letts.* 2005; **95**:095504

8. Cutkosky R.D. 'Four-terminal-pair networks as precision admittance and impedance standards'. *IEEE Trans. Commun. Electron.* 1964;**80**(70):19–22
9. Solymar L., Shamonina E. *Waves in Metamaterials*. Oxford University Press, Great Clarendon Street, Oxford, OX2 6DP, UK, 2009
10. Shields J.Q. 'Voltage dependence of precision air capacitors'. *J. Res. NBS.* 1965;**69c**:265–74
11. Kibble B.P., Rayner G.H. *Coaxial AC Bridges*. Bristol: Adam Hilger Ltd.; 1984. (Presently available from NPL, Teddington, TW11 0LW, U.K. www.npl.co.uk)
12. Kibble B.P., Schurr J. 'A novel double-shielding technique for AC quantum Hall measurement'. *Metrologia.* 2008;**45**(5):L25–27
13. Campbell A. 'On a standard of mutual inductance'. *Proc. Roy. Soc. A.* 1907;**79**(532):428–35
14. Haddad R.J. 'A resistor calculable from DC to $\omega = 10^5$ rad.s^{-1}'. M.Sc. thesis, School of Engineering and Applied Science, George Washington University, 1969
15. Grover F.W. *Inductance Calculations: Working Formulas and Tables*. Dover Phoenix, 1973
16. Delahaye F., Goebel R. 'Evaluation of the frequency dependence of the resistance and capacitance standards in the BIPM quadrature bridge'. *IEEE Trans. Instrum. Meas.* 2005;**54**(2):533–37
17. Awan S.A., Kibble B.P. 'Towards accurate measurement of the frequency dependence of capacitance and resistance standards up to 10 MHz'. *IEEE Trans. Instrum. Meas.* 2005;**54**(2):516–20
18. Gibbings D.H.L. 'A design for resistors of calculable AC/DC resistance ratio'. *Proc. IEE.* 1963;**110**:335–47
19. Bohacek J., Wood B.M. 'Octifilar resistors with calculable frequency dependence'. *Metrologia.* 2001;**38**(3):241–48
20. Awan S.A., Kibble B.P. 'A universal geometry for calculable frequency response coefficient of LCR standards and new 10 MHz resistance and 1.6 MHz quadrature bridge systems'. *IEEE Trans. Instrum. Meas.* 2007;**56**(2):221–25
21. Reilly S.P., Leach R.K., Cuenat A., Awan S.A., Lowe M. *Overview of MEMS Sensors and the Metrology Requirements for Their Manufacture*. NPL report, DEPC-EM 008, section on 'MEMS Devices for Electrical Measurements', 2006, pp. 25–31
22. Von Klitzing K., Dorda G., Pepper M. 'New method for high-accuracy determination of the fine structure constant based on quantized Hall resistance'. *Phys. Rev. Lett.* 1980;**45**:494–97
23. Jeckelmann B., Jeanneret B. 'The quantum Hall effect as an electrical resistance standard'. *Rep. Prog. Phys.* 2001;**64**:1603–55
24. Delahaye F. 'Technical guidelines for reliable measurements of the quantized Hall resistance'. *Metrologia.* 1989;**26**:63–68
25. Delahaye F., Jeckelmann B. 'Revised technical guidelines for reliable DC measurements of the quantized Hall resistance'. *Metrologia.* 2003;**40**:217–23

26. Schurr J., Bürkel V., Kibble B.P. 'Realizing the farad from two AC quantum Hall resistances'. *Metrologia.* 2009;**46**:619–28
27. Ikushima K., Sakuma H., Yoshimura Y., Komiyama S., Ueda T., Hirakawa K. 'THz imaging of cyclotron emission in quantum Hall conductors'. *Physica E.* 2006;**34**:22–26
28. Nachtwei G. 'Breakdown of the quantum Hall effect'. *Physica E.* 1999; **4**:79–101
29. Ricketts B.W., Kemeny P.C. 'Quantum Hall effect devices as circuit elements'. *J. Appl. D: Appl. Phys.* 1988;**21**:483
30. Schurr J., Ahlers F.J., Hein G., Melcher J., Pierz K., Overney F., Wood B.M. 'AC longitudinal and contact resistance measurements of quantum Hall devices'. *Metrologia.* 2006;**43**:163–73
31. Melcher J., Warnecke P., Hanke R. 'Comparison of precision AC and DC measurements with the quantised Hall resistance'. *IEEE Trans. Instrum. Meas.* 1993;**42**:292–94
32. Delahaye F. 'Accurate AC measurements of the quantized Hall resistance from 1 Hz to 1,6 kHz'. *Metrologia.* 1994/95;**31**:367–73
33. Delahaye F. 'Series and parallel connection of multiterminal quantum Hall-effect devices'. *J. Appl. Phys.* 1993;**73**:7914–20
34. Delahaye F., Kibble B.P., Zarka A. 'Controlling AC losses in quantum Hall effect devices'. *Metrologia.* 2000;**37**(6):659–70
35. Schurr J., Ahlers F.J., Hein G., Pierz K. 'The AC quantum Hall effect as a primary standard of impedance'. *Metrologia.* 2007;**44**(1):15–23
36. Kibble B.P., Schurr J. 'A novel double-shielding technique for AC quantum Hall measurement'. *Metrologia.* 2008;**45**(5):L25–27
37. Schurr J., Kibble B.P., Hein G., Pierz K. 'Controlling losses with gates and shields to perfect a quantum Hall impedance standard'. *IEEE Trans. Instrum. Meas.* 2008;**58**:973–77
38. Overney F., Jeanneret B., Jeckelmann B. 'Effects of metallic gates on AC measurements of the quantum Hall resistance'. *IEEE Trans. Instrum. Meas.* 2003;**52**:574–78
39. Jeanneret B., Overney F. 'Phenomenological model for frequency-related dissipation in the quantized Hall resistance'. *IEEE Trans. Instrum. Meas.* 2007;**56**:431–34
40. Overney F., Jeanneret B., Jeckelmann B., Wood B.M., Schurr J. 'The quantized Hall resistance: towards a primary standard of impedance'. *Metrologia.* 2006;**43**:409–13
41. Schurr J., Kucera J., Pierz K., Kibble B.P. 'The quantum Hall impedance standard'. *Metrologia.* 2011;**48**:47–57

Chapter 7
Transformers

Transformers couple circuits without the need for any direct conducting connection between them, although often a single connection is added to make the potential of one winding definite with respect to another. The magnetic flux caused by the current in one winding, usually designated the primary winding, threads the other windings, and the aim of a good transformer design is to make this flux coupling as complete as possible. This is usually achieved by linking the windings with a high-permeability magnetic core to provide an easy path for the magnetic flux.

In instrument transformers, we describe the cores as toroids wound from extremely high-permeability material, formed into a very thin ribbon to avoid flux exclusion by eddy currents at frequencies up to 1 MHz. To enhance the completeness of flux coupling further, the windings are usually tightly coupled, for example, being the separate strands of a tightly spun rope appropriately connected in series or parallel.

Enclosing the primary winding around its toroidal core in a toroidal shield, as shown in Figure 5.4 ensures that the entire magnetic flux of the primary winding threads equally, with extremely high accuracy of parts in 10^9, two or more secondary windings. The ratios of the voltages induced in secondary windings are then quite accurately equal to the ratio of the number of turns of these windings.

Transformers can be divided into two categories. In one, the purpose is to transmit electrical power from one circuit to another, commonly at a different voltage. Unless they are autotransformers having only a single tapped winding, they may also provide isolation of one network from another so that there is no current flow between the networks. The design of isolating transformers is such that they transmit power efficiently, solely by the magnetic coupling, and a high precision of voltage or current ratio between the primary winding and the secondary windings is not an important consideration.

In this chapter, we are concerned with the other category of transformer that may deliver some power, but the design objective is rather to approach an ideal transformer. In an ideal transformer the emf induced in each and every turn is equal and the voltage developed by a winding is very close to the sum of these induced emfs, i.e. to one part in a million or better. Isolation may also be required. In the case of current transformers, the ratio of the currents in the windings is the important quantity, and the accompanying voltages are only of incidental interest.

Sometimes the two windings are, in fact, continuous and a tapping point is provided, as in an autotransformer, and sometimes a continuous winding is divided into ten or more sections by nine or more taps.

Other uses for transformers in measuring and sensing circuits include impedance matching and noise matching (see section 2.1.3).

7.1 General considerations

The easiest and most accurate way to compare impedance standards is in terms of the voltage or current ratio of transformer windings, because the ratio between output taps of a properly constructed transformer is not affected, at least to parts in 10^8, by aging or moderate changes in its environmental conditions (position, temperature, pressure, humidity, etc.). Moreover, if an accuracy of a part in a million is satisfactory, the ratio of the transformer can often be taken to be equal to the ratio of the number of turns of the windings. When lower uncertainty is required, a method of calibrating the transformer is needed.

We first discuss some general considerations that limit the performance of transformers, then we describe how to construct and design transformers to minimise these effects, and finally, we discuss calibration methods. This final part can be omitted on a first reading.

7.1.1 The causes of departure from an ideal transformer

It is often convenient when analysing the role of a transformer in a circuit to construct its equivalent circuit from a combination of an ideal transformer and circuit elements, which represent the imperfections in the characteristics of the actual transformer. In an ideal transformer, (1) there are no energy losses either in the windings or in the core; (2) the only magnetic flux is that which threads equally all the turns of the ratio windings; (3) flux from a current in any of the ratio windings threads completely the other ratio windings and (4) there are no significant capacitances within or between the ratio windings or to the magnetic core or any surrounding screens.

In the following account of various transformers and the circuits in which they are employed, we shall be considering the causes of departure from an ideal transformer and how they can be minimised and their effects eliminated in some circuits.

The departures of a transformer from ideal characteristics arise from several causes.

The magnetic material of the core absorbs, as well as stores, energy when it is magnetised; there is a magnetic power loss, and because magnetisation characteristics are non-linear, the power loss cannot be strictly represented by current flowing through a simple resistance. Eddy currents induced in the core material give rise to additional power loss.

The windings of the transformer possess resistance, which, at higher frequencies, will be increased by skin-effect eddy currents both within the conductors and as induced in adjacent conductors. There will also be capacitance between turns and between layers of a winding; it is shown later that these distributed capacitances can be represented by a single 'lumped' capacitance between the ends of a winding. In addition, some small proportion of the flux linking one winding will follow a path that does not link another winding. This small amount is usually called leakage flux and is the main cause of leakage inductance, which affects the ratio of loaded ratio windings. The design and construction of transformers used for sensing and measurement purposes aim at minimising the effects of some or all of these imperfections.

The core material should have a high stored magnetic energy compared with the energy lost per cycle. Within the restrictions imposed by other considerations, the wire for the windings should be thick to minimise the resistance and self-inductance of the wire itself; the latter contributes to the leakage inductance. The capacitances within a winding can be reduced by spacing the individual turns and layers, but the advantages gained are offset by increased leakage flux and higher winding resistances resulting from the use of finer wire for the windings. In addition, it may be necessary, for example, to consider the dielectric loss of the wire insulation and other quantities of secondary importance.

Although a transformer may depart considerably from the ideal, it will still constitute a precise and stable ratio device provided that the causes of departure do not change. One cause of variability is the magnetic state of the core; this is considered in the next section.

7.1.2 The magnetic core

Until recently, the core material was always one of the mu-metal or permalloy class of nickel–iron alloys having a composition of 75% Ni and 25% Fe; the alloy with the highest permeability is known as super mu-metal in the United Kingdom or supermalloy in the United States and is used where its higher cost is not an important factor. Like all magnetic alloys, its magnetic properties are far from linear. The relative permeability of typical mu-metal alloys is shown in the graph of Figure 7.1. It increases by a factor of about 4 from its initial value at very low flux densities up to a maximum that can be as high as 200 000 at a flux density of 0.2 T.

A solid magnetic core can be used only if the magnetic flux does not change, because when magnetic flux in the core changes, an emf is induced in the core in planes perpendicular to the direction of the flux, and in accordance with Lenz's law, eddy currents are set up in the core such that the flux changes are opposed (Figure 7.2a). The emfs and hence the eddy currents are proportional to the rate of change of magnetic flux. This effect occurs in any piece of metal in a varying magnetic flux, but the effect is enhanced in magnetic materials by their higher permeability. For a given uniform flux density B, the only way of reducing the magnetic flux $\int BdS$ is to reduce dS, where S is the cross section perpendicular to B. This is usually done by building up a magnetic core from thin sheets or stampings or turns of a ribbon (Figure 7.2b) insulated from one another, the sheets being in the direction of the flux. Since the induced emf is $d\int BdS/dt$, the larger the rate of change, the smaller the $\int dS$ required to reduce the induced emfs and eddy currents.

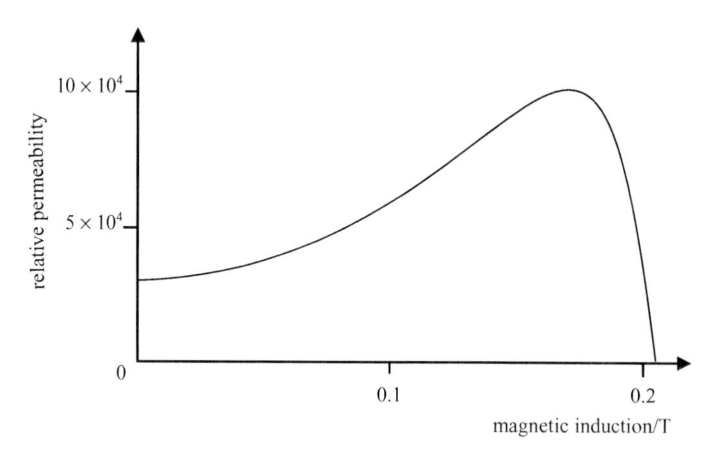

Figure 7.1 The permeability of magnetic core material as a function of flux density

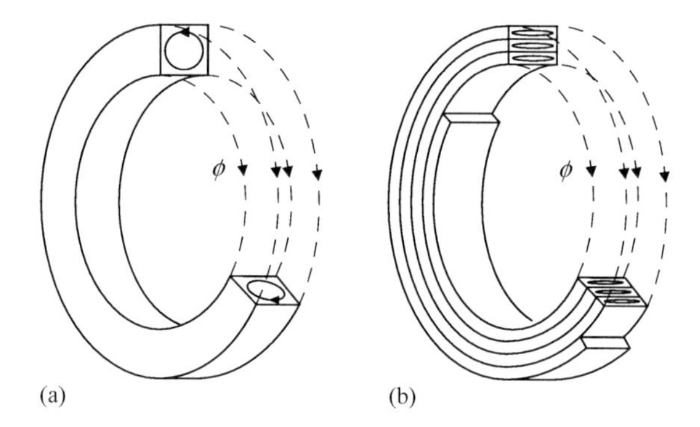

Figure 7.2 Eddy currents induced by flux changes in (a) a solid magnetic core and (b) a ribbon-wound core

That is, higher frequencies require thinner sheets. The Vitroperm and Ultraperm materials, from which the more recent cores are made, are created as amorphous (i.e. non-crystalline or glassy) ribbons only a few micrometres thick by extremely rapid cooling of the molten material on a cold rotating cylinder. The thinness of the ribbon, together with the high resistivity of the material, ensures even greater freedom from eddy current effects than the older mu-metal cores. Cores wound from the thinnest amorphous metal tapes are usable up to megahertz frequencies. Unfortunately, the space factor (the proportion of the total cross section occupied by magnetic material) becomes worse for thinner ribbons as the necessary insulation between adjacent turns of ribbon becomes a larger proportion of the total core volume.

If the flux density is raised appreciably above that which produces maximum permeability, the material saturates, and at the peaks of a large applied sinusoidal waveform, the permeability is very low. As a result, the input impedance of a winding falls at these excitation peaks so that it is no longer much larger than the output impedance of a practical source. Excessive voltage drop occurs in the source, the waveform is clipped, and the proportion of harmonics generated in the current and voltage waveforms associated with the transformer is increased considerably.

The permeability and the power loss at a given flux density change with time and depend on the past history of the material. A DC current in a winding or mechanical and magnetic shocks are also likely to change the permeability and increase the loss and flux leakage between windings, altering the precise voltage ratios of the transformer, but the magnetic properties of a transformer core of this material can always be restored to a fairly well-defined state when required by demagnetising the material. A transformer core can be demagnetised by an alternating current in a winding whose peak value is large enough to thoroughly saturate the core and then smoothly decreasing the current to zero in a time long compared to its period. Using a much lower frequency than the normal operating frequency of the transformer will lower the reactance of the core and therefore the driving voltage needed to produce a saturating current. Unfortunately, following mechanical shock or demagnetisation, the permeability decreases with time, and this spontaneous change renders the demagnetised condition somewhat uncertain.

Eddy currents may be regarded as reducing the effective permeability of the magnetic core and increasing the total power lost via the electrical resistance of the current path during the alternating magnetisation.

Toroidal cores are universally used in precise ratio transformers because of their symmetry. However, this symmetry is not quite perfect because they are made by winding a ribbon of material as shown in Figure 7.3. The ribbon has a beginning on the inside of the toroid and an end on the outside, and a small proportion of the alternating flux, instead of being confined to the interior of the toroid, will jump from one end of the ribbon to the other, as illustrated in the figure. Also, another

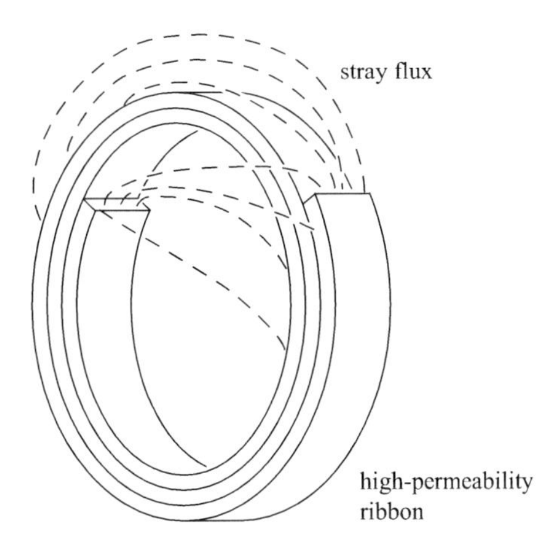

Figure 7.3 Flux leakage from a ribbon-wound toroidal magnetic core

small proportion loops from outside to inside of the core on both sides of it. This looping flux is greater near the inner and outer ends of the spiral strip. Effects arising from these stray fluxes can be, for all practical purposes, eliminated by enclosing the core and the winding energising it in a toroidal screen of mu-metal.

7.1.3 The windings; the effect of leakage inductances, capacitances and resistances

We now consider these three effects, all of which may be to some degree undesirable in a particular device and whose individual minimisation are unfortunately, in practice, incompatible.

If the conductors comprising two windings could be brought into exact geometrical coincidence in space, all the flux that links one winding also links the other. Since it is impossible to get two conductors to spatially coincide, a small proportion of the flux generated by a current flowing in one winding will fail to thread the other winding. This small proportion causes a small inductance, termed the leakage inductance, of the windings. This leakage inductance manifests itself as an internal inductance L of the windings, which alters their voltage ratio when current is drawn from one or more of the windings by a load.

Current I drawn by a load Z causes a flux ϕ, of which a small amount $\delta\phi$ does not thread the second winding, as shown in Figure 7.4.

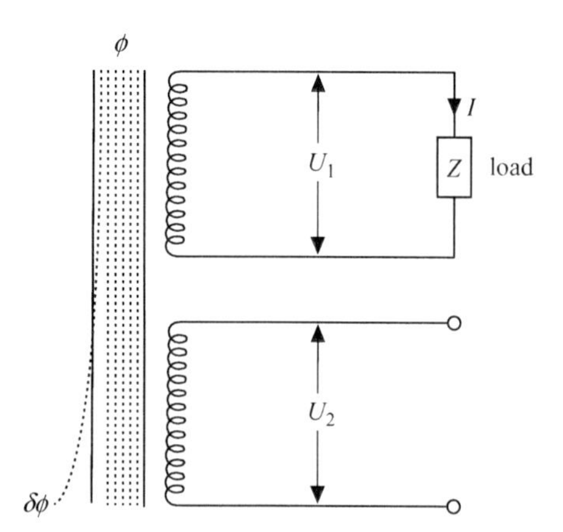

Figure 7.4 The effect of leakage flux on a transformer

$\delta\phi$ gives rise to a missing emf–$d(\delta\phi)/dt = L_1 dI/dt$ and modifies the ratio U_1/U_2 by this amount, so L_1 is a real physical inductance in series with the winding. The resistance of the winding and the loss of L_1 can be represented as a lumped resistor R_1 in series with the winding, so that we have the equivalent circuit of Figure 7.5 where the windings are ideal, that is, perfectly coupled and of zero resistance.

In a transformer where the windings are wound very closely together to minimise this effect and spaced a distance away from the core or magnetic screens by insulation materials, etc., the presence or absence of the core or magnetic

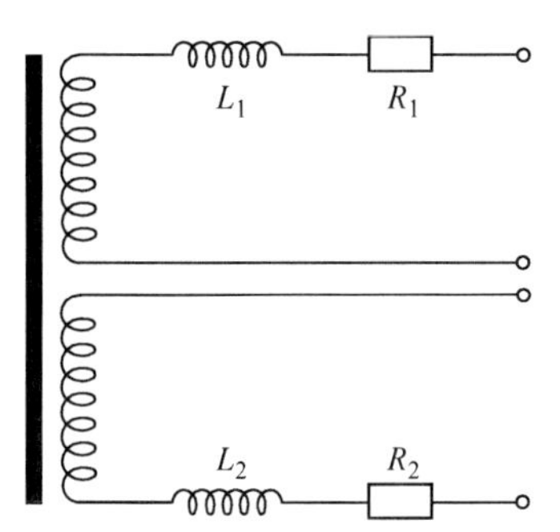

Figure 7.5 The equivalent circuit of a transformer having leakage flux and winding resistances

screens will not greatly affect the magnitude of the leakage inductance because the flux in the core or screens threads both windings. Also the leakage inductance and capacitance of windings do not depend much on the core, so that it is again a useful approximation in considering them to think of the device as air-cored. Any magnetically permeable material interposed between the windings for whatever purpose will, of course, greatly increase the leakage inductance.

The apparent resistance of a winding is also not affected much by the presence of the core, so that it is nearly the same as if the winding were air cored. The low-frequency and DC resistance value will be augmented at higher frequencies by skin- and proximity-effect eddy currents within the conductor, particularly if the winding is closely packed or multi-layered.

In a voltage ratio transformer, the potential of successive turns of the windings increases in a regular manner until the full output voltage is reached across the complete winding. Between all pairs of turns, capacitance currents will flow as shown in Figure 7.6.

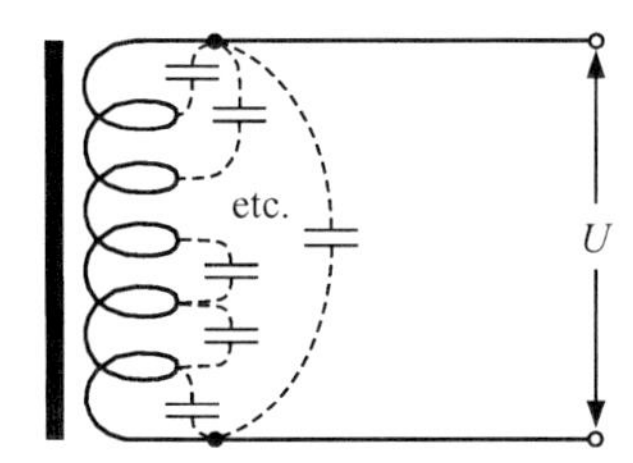

Figure 7.6 The internal capacitances of a transformer winding

The net effect of these distributed capacitances can be represented by a single lumped capacitance. Each component capacitance δC, by virtue of transformer action, is equivalent to a capacitance $\delta C\rho^2$ across the winding, where ρ is the fraction of the voltage across δC compared with the total voltage across the winding. Alternatively, a string of capacitances in series are equivalent to a single capacitance $\delta C/n$ across the winding. These relations are valid for all frequencies for which the phase of the current throughout the winding is constant; in practice, this even applies to frequencies well above self-resonance.

Since L and R also arise from distributed phenomena, they must be shown on the winding side of this distributed shunt capacitance. Again, the winding itself is now assumed to be free of self-capacitance effects. The equivalent circuit is as shown in Figure 7.7.

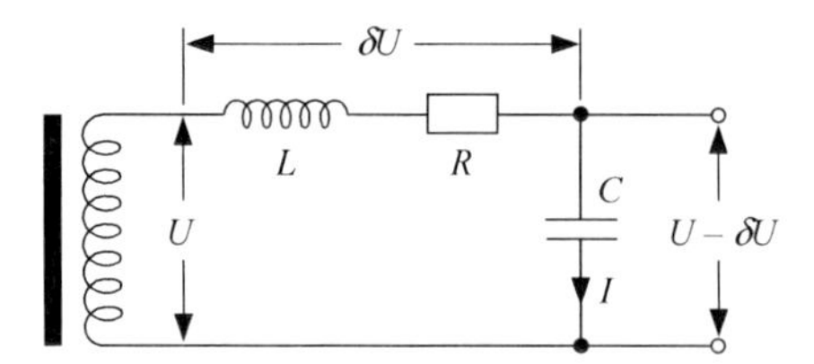

Figure 7.7 The equivalent circuit of a transformer winding including the effects of leakage inductance L, resistance R and internal capacitance C

The current I drawn by the self-capacitance alters the output voltage U of the winding to $U - \delta U$. As

$$
\begin{aligned}
I &\approx j\omega CU \\
\delta U &= I(R + j\omega L) = j\omega CU(R + j\omega L) \\
&= (-\omega^2 LC + j\omega CR)U
\end{aligned}
$$

The real part of δU is negative, and hence, the in-phase component of voltage is increased either by the presence of self-capacitance or by an external capacitive load connected across the output terminals. L, R and C are best regarded as parameters describing the behaviour of the winding.

Cutkosky [1] has pointed out that the deliberate introduction of a small impedance z in common between the two windings, as shown in Figure 7.8, can be advantageous. The output impedance of one section can be decreased at the expense of another, and a better overall compromise can be reached. The current

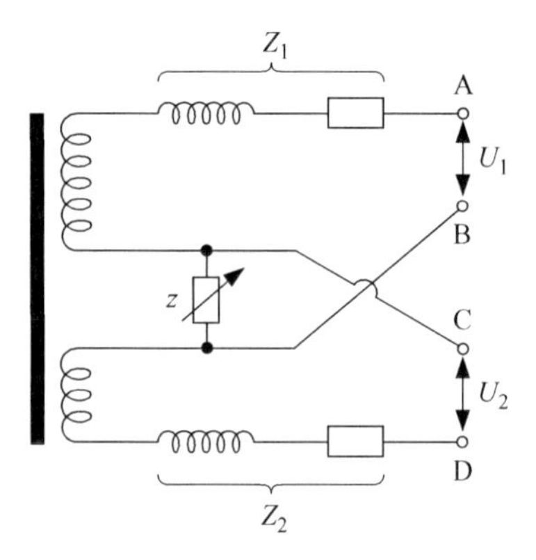

Figure 7.8 Connecting a small resistance between the windings

drawn by a load between terminals A and B flows through z, thus altering the potential of terminal C, so that the ratio of U_1/U_2 is immune to this loading if $z = -Z_1/(U_1/U_2 + 1)$. In the case of a 10:−1 ratio, $z = Z_1/9$.

7.1.4 Representation of a non-ideal transformer: the effect of loading on its ratio windings

There are two kinds of transformer used in bridge networks. These two kinds may formally be termed two- and three-winding transformers. Two-winding transformers are usually known as inductive voltage dividers (IVDs) or their dual, inductive current dividers, and are discussed in sections 7.3.1 and 7.3.2. Three-winding transformers, usually known as voltage (or current) ratio transformers, are discussed in sections 7.3.7–7.3.11. The essential difference between these two kinds is that the energising current from a source flows through some or all of the windings of a two-winding transformer, causing small potential differences across their leakage inductances and winding resistances, but in a three-winding transformer, the energising current flows in a separate primary winding. The inherent accuracy of the latter arrangement is therefore greater.

An equivalent circuit of a two-winding transformer is drawn in Figure 7.9.

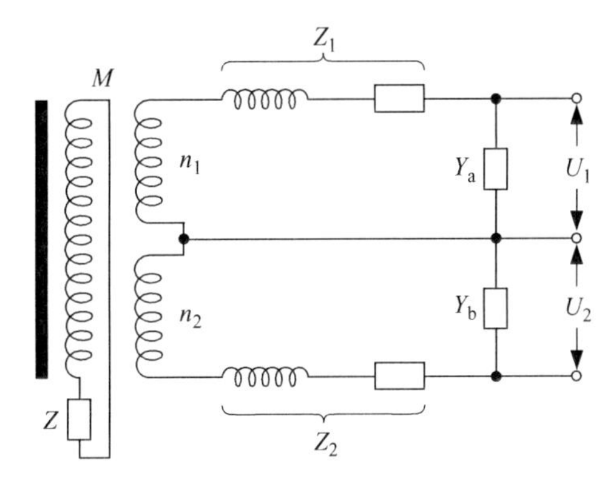

Figure 7.9 Equivalent circuit of a two-winding transformer

A source may be connected across either winding, or if the windings have a common connection as shown, it may be connected across both. The winding and load Z drawn to the immediate right of the core *do not represent an actual winding and load.* They represent the extra impedance resulting from loss mechanisms presented to the driven winding, which might be the one with n_1 turns or the one with n_2 turns or both in series. Z_1 and Z_2 represent the series resistances and leakage inductances, and Y_a and Y_b, the internal shunt admittances of the windings. Analysis, neglecting product and higher terms, of the small quantities $z_{1,2}$ yields

$$\frac{U_1}{U_2} = \rho\left[1 + \frac{Z_1 - \rho Z_2}{(1+\rho)Z} + \frac{(Y_b - \rho Y_a)(Z_1 + \rho^2 Z_2)}{\rho(1+\rho)}\right] \tag{7.1}$$

where $\rho = n_1/n_2$. Equation (7.1) shows how Z_1, Z_2, Y_a and Y_b cause the voltage ratio U_1/U_2 to depart from the nominal ratio of the winding turns.

The equivalent circuit of a three-winding transformer (which has a separate energising winding), when measuring a ratio of admittances Y_1/Y_2, is shown in Figure 7.10. It shows the effect that loading produces on the transformer ratio.

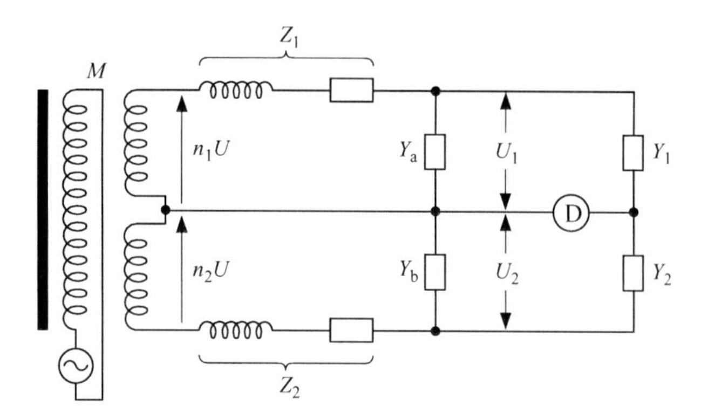

Figure 7.10 The equivalent circuit of a three-winding transformer

If the flux induced in the core by the primary winding in turn induces an emf U per turn in the secondary windings, the ratio of emfs U_1 and U_2 applied to the admittances Y_1 and Y_2 to be compared is

$$\frac{U_1}{U_2} = \left(\frac{n_1 U}{n_2 U}\right) \frac{[1 + Z_2(Y_2 + Y_b)]}{[1 + Z_1(Y_1 + Y_a)]} = \frac{Y_2}{Y_1} \tag{7.2}$$

Equation (7.2) shows that the ratio of the voltages applied to the admittances Y_1 and Y_2 is the turns ratio n_1/n_2 modified by the (usually small) product terms $Z_1(Y_1 + Y_a)$ and $Z_2(Y_2 + Y_b)$.

Z_1 and Z_2 are typically 10 μH in series with 0.1 Ω, so that when comparing admittances Y_1 and Y_2 of 10^{-3} S at an angular frequency of 10^4 rad/s, the corrections to the ratio U_1/U_2 contained in the square brackets of (7.2) are of the order of 10^{-4} and must be accounted for, or otherwise eliminated, in accurate work.

Simply connecting the energising source across Y_1 and Y_2 will eliminate their effect, since the current through them is drawn from the source rather than the transformer. The effect of Y_a and Y_b is not entirely eliminated, as they represented the distributed admittances of the windings.

The error in the voltage ratio of a transformer caused by the loading of the shunt admittances $Y_1 + Y_a$ and $Y_2 + Y_b$ presented by the measuring circuit to its output ports can be readily found and accounted for by measuring this shunt admittance, increasing it temporarily by a known ratio with another shunt admittance and extrapolating back to the zero load condition. The internal impedances Z_1 and Z_2 of the transformer winding, which cause the loading errors, can also be calculated using (7.2).

7.1.5 The two-stage principle

If an only approximately correct flux in a core has been created by a primary or magnetising winding, the deficiency can be made up by a much smaller flux in a second core. This latter flux is produced by a winding energised by the emf caused by the discrepancy between the exact flux required and the smaller flux provided by the magnetising winding. As applied to voltage transformers, whose equivalent circuit is shown in Figure 7.11, it is desired that the total flux should correspond exactly to the source emf U_1 defined at the input terminals, but in the ordinary single-stage design, the flux corresponds to an emf smaller by an amount δU_1 because of the voltage drop the primary current suffers in flowing through the impedance associated with the primary windings. This impedance is composed of the resistance r of the primary winding plus the leakage inductance l between primary and secondary windings.

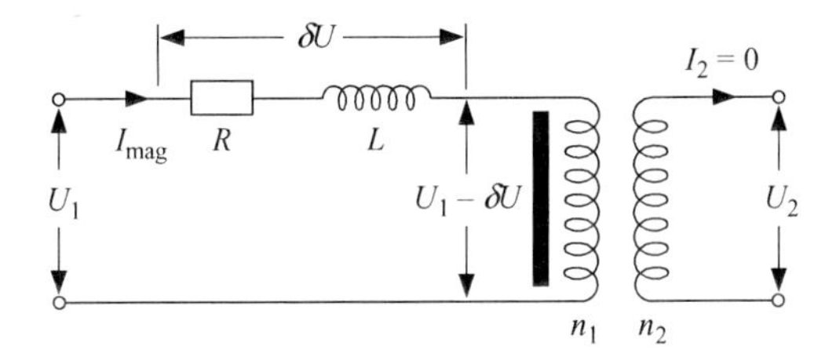

Figure 7.11 The equivalent circuit of the primary of a voltage transformer

Hence, this emf $U_1 - \delta U$ is equal to the rate of change of flux in the transformer core.

$$\delta U = I_{mag}(R + j\omega L)$$

And the emf appearing across the output terminals is

$$U_2 = \frac{(U_1 - \delta U)n_2}{n_1}$$

instead of the ideal value $U_2 = U_1 n_2/n_1$.

A second core added and wound as shown in Figure 7.12 is driven by the lost emf δU, which, to a good approximation, can be transformed to $\delta U n_2/n_1$ and added

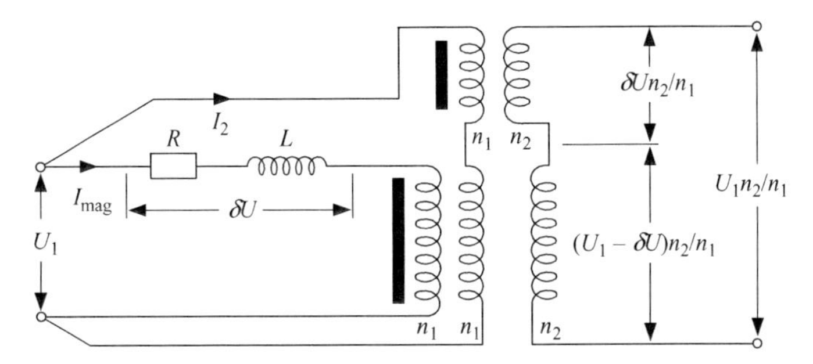

Figure 7.12 A second core added to a voltage transformer

to the emf in the secondary winding to give the corrected voltage

$$\frac{(U_1 - \delta U)n_2}{n_1} + \frac{\delta U n_2}{n_1}$$

which equals $U_1 n_2/n_1$, as was required.

An approximate analysis of the device can be made as follows, with reference to Figure 7.13.

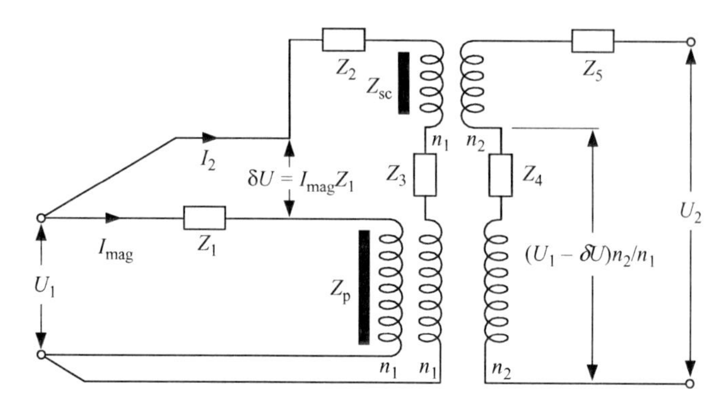

Figure 7.13 Parasitic impedances of a transformer having a second core and winding

$$\begin{aligned} I_2 &= \frac{\delta U}{Z_2 + Z_3 + Z_{SC}} = \frac{I_{mag} Z_1}{Z_2 + Z_3 + Z_{SC}} \\ &= \frac{U_1 Z_1}{(Z_1 + Z_P)(Z_2 + Z_3 + Z_{SC})} \end{aligned}$$

where Z_P and Z_{SC} are the impedances of the wound first and second cores, and $Z_1 \ldots, Z_5$ are composed of the resistances and leakage inductances of the five windings.

$$\begin{aligned} U_2 &= \frac{(U_1 - \delta U)n_2}{n_1} + \frac{[\delta U - I_2(Z_2 + Z_3)]n_2}{n_1} \\ &= U_1\left[1 - \frac{Z_1(Z_2 + Z_3)}{(Z_1 + Z_P)(Z_2 + Z_3 + Z_{SC})}\right]\frac{n_2}{n_1} \end{aligned} \tag{7.3}$$

I_2 is small because of the high impedance Z_{SC} of the lossy inductance of the second core.

Hence, since $Z_1 \ll Z_P$ and $Z_2 + Z_3 \ll Z_{SC}$, the actual voltage ratio U_2/U_1 differs from the turns ratio n_2/n_1 by approximately the product of two terms, Z_1/Z_P and $(Z_2 + Z_3)/Z_{SC}$.

In practice, a separate auxiliary transformer can be avoided by the construction shown in Figure 7.14. Remember that a winding threads *all* cores and their

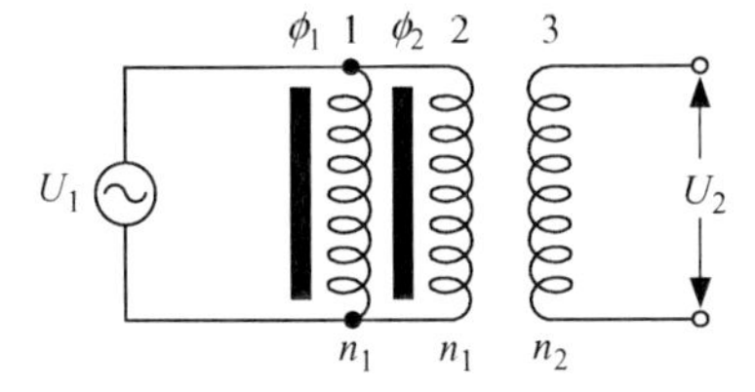

Figure 7.14 Avoiding the need for a separate auxiliary transformer

windings drawn to the left of it, so that winding 1 is wound only around core 1 and windings 2 and 3 are wound around both cores 1 and 2. By virtue of being wound around core 1, winding 2 senses the flux ϕ_1 and behaves like the second winding on the main transformer of the previous scheme, and by virtue of being wound around core 2, it can generate flux ϕ_2 in core 2. Since the emfs in winding 2 from ϕ_1 and ϕ_2 are additive, we can represent the device, as in the previous figure, as two windings in series, each with its individual series impedance Z_3 and Z_2.

Similarly, winding 3 is around both cores, and it can be represented as the secondary windings of both main and auxiliary transformers in series, each with its individual impedance Z_4 and Z_5.

Examples of this general concept of two staging will be discussed in the descriptions of particular devices later in this chapter.

The voltage across a second core of a voltage ratio transformer is in general higher by a few per cent than the voltage across the first-stage winding. This is because of the self-resonance induced by the self-capacitance of the winding in conjunction with its large inductance. The flux caused in the second core by the self-resonant current is easily opposed by an external adjustable current source connected across the winding, and in this way, the overall voltage of the potential windings, which are threaded by this flux, can be altered without altering their voltage ratio (see section 7.3.8).

7.1.6 Electrical screens between windings

The capacitive current between two windings can be reduced to zero by providing two screens between them, one screen being connected to each winding. Both screens are arranged to be at the same potential but are not directly connected within the transformer, as shown in Figure 1.19. As in section 1.1.2, the current through any residual capacitance between the screens can be diverted through a low-impedance single conductor so that it does not affect non-isolated sensitive circuitry (e.g. a phase-sensitive detector) in the measurement network supplied by the transformer. Addition of this conductor creates a mesh, which needs to be equalised by an equaliser in the input cable to the sensitive circuitry.

The screens must have overlapped breaks in them so that they do not produce a shorted turn on the transformer. The correct way to arrange the breaks is illustrated in Figure 1.27.

7.2 Constructional techniques

7.2.1 Design of transformer windings

A toroidal core for a transformer imposes some practical constructional limitations on the windings. The simple windings, which can be put on by a toroidal winding machine, do not usually produce an optimum transformer design, and the best compromise for a given purpose will usually necessitate winding by hand.

In this section, we describe techniques that help to ensure that a toroidal magnetic field is created by the energising winding.

Any transformer winding should be uniformly spread around the toroid so that its two ends are adjacent, as shown in Figure 7.15. Such a winding will be reasonably unaffected by external flux ϕ in its plane because the flux will divide approximately equally between the two sides of the toroid and in doing so will induce equal and opposite cancelling voltages in the two halves of the winding.

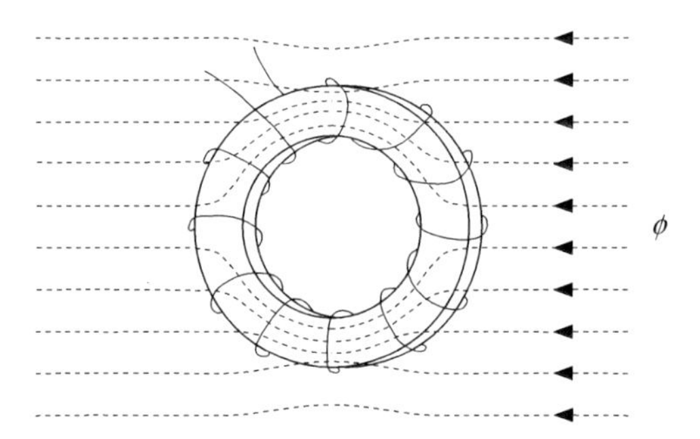

Figure 7.15 An uniformly wound toroid

In progressing around the toroid, however, the winding forms a loop in the plane of the toroid of a size equal to its mean diameter (Figure 7.16). The loop area may be decreased by carrying the conductor from the end of the winding round the toroid in the opposite direction to the winding progression so that it finishes adjacent to the beginning of the winding. This is called an anti-progression turn.

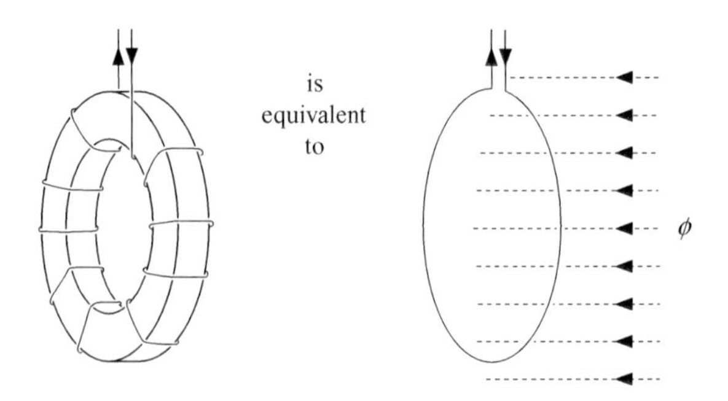

Figure 7.16 A toroidal winding progressing around a core

A small disadvantage of an anti-progression turn as sketched in Figure 7.17 is that it does not follow the mean path of the toroid. Therefore, flux in the toroidal plane may still couple into the winding, or an energised transformer core may produce flux in this plane. Thus, an anti-progression turn is only an approximation to a perfect solution to the problem because it evidently does not occupy the same physical volume as the actual winding, and therefore, some flux can still thread the combination.

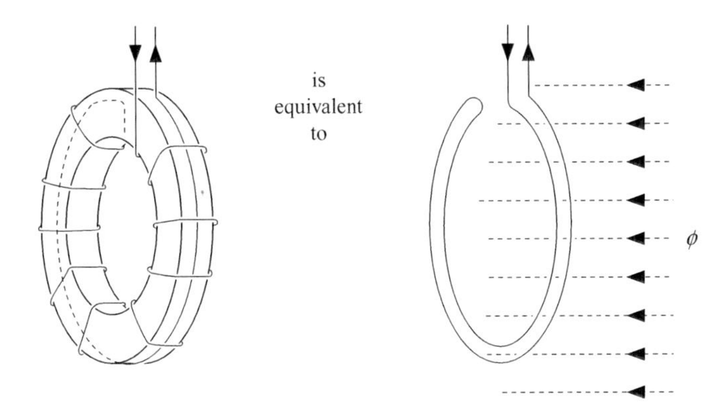

Figure 7.17 A transformer winding with an anti-progression turn

A better solution is to use the 'Ayrton-Perry' or 'bootlace' technique as sketched in Figure 7.18. Half the number of turns required are put on the toroid in the usual way, and the rest are wound on top of them, winding in the same sense through the core but in the opposite direction around it until the beginning of the winding is reached.

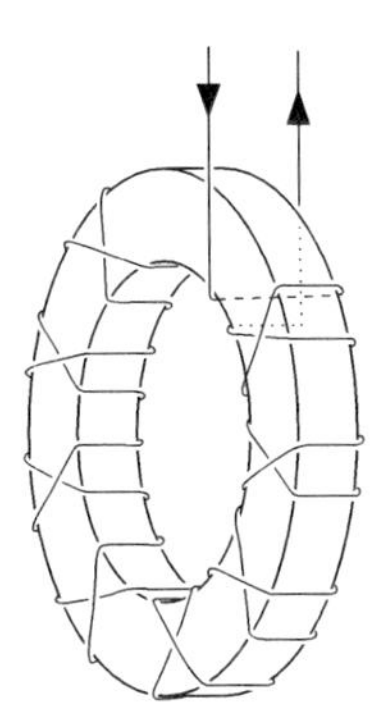

Figure 7.18 A 'bootlace' technique for transformer winding

A still more complex winding in which the *progression* of application of the turns is as indicated in Figure 7.19 has the additional advantage of decreasing the self-capacitance of the winding considerably.

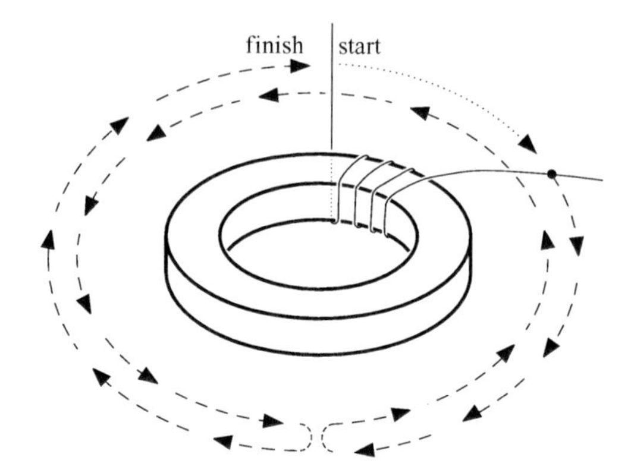

Figure 7.19 Progression of a bootlace winding to minimise self-capacitance

Windings of the kinds we have described, which do not have significant emfs induced in them by external fluxes, are termed 'astatic'.

In some of the following sections, the pictorial illustrations of various windings are not shown with an astatic feature for reasons of clarity. They should be understood to be astatically wound in practice.

The circuit shown in Figure 7.20 is a rapid way of counting a large number of turns wound on a core. A small number of turns m (typically 10), which can be quickly and reliably counted, is temporarily wound and connected in series-aiding (i.e. they thread the core in the same sense) with the n turns to be counted. Because of topology, n and m are exact integers corresponding to the number of passages of the conductor through the hole of the toroid.

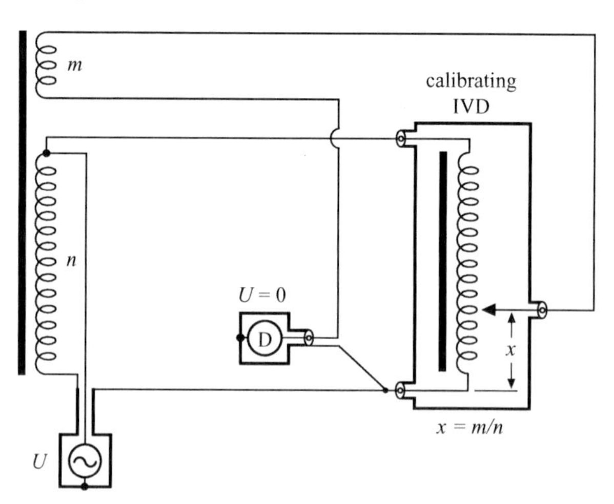

Figure 7.20 A circuit for counting the number of turns of a winding on a core

The windings are connected to a multi-decade IVD (see section 7.3.1) via a detector as shown, and a generator is connected across the whole circuit. The circuit is not coaxial, but no great care needs be taken about stray capacitances because this circuit of back-to-back transformers has a low impedance and because the detector is positioned in the low-potential side of the circuit. Also, because the greatest part of the magnetic flux associated with the circuit is within the wound core, the open meshes of this modest-accuracy circuit need only be minimised by sensible conductor routing.

The dial settings x of the IVD are adjusted until the in-phase component of the detector is nulled. The quadrature component caused by parasitic capacitances is small and irrelevant. Then $n = m/x$, and n can be calculated. The result will not be exactly an integer because of the finite resolution of the IVD and inaccuracies of the circuit, but it can be verified by repeating the procedure with a different value of m.

A simpler technique, which is adequate if the number of turns, or the turns ratio, is not too extreme, is to simply measure the ratio of the voltages with an AC voltmeter, as shown in Figure 7.21.

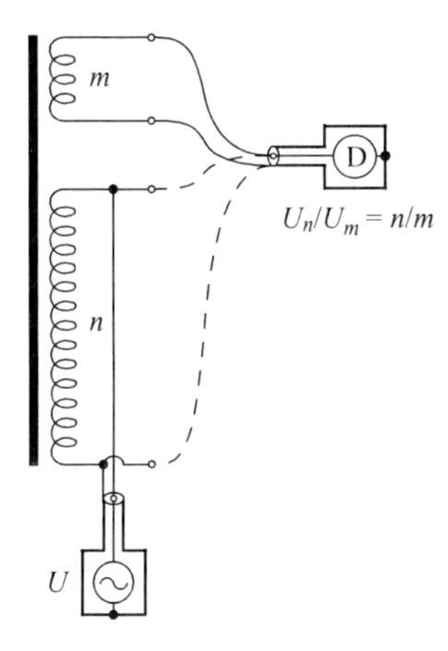

Figure 7.21 The ratio of numbers of turns measured using an AC voltmeter

7.2.2 Techniques for minimising the effect of leakage inductance, winding resistance and the capacitances of ratio windings

In this section, we describe winding techniques that optimise the performance of ratio windings.

It is first necessary to recall the concept of 'leakage inductance' (section 7.1.3). This small inductance arises from flux caused by current flowing in one section of a ratio winding, which does *not* thread the other sections. In a rope construction, described in sections 7.2.4 and 7.2.5, each strand of the rope is a section of the winding. Hence, the output impedance of a winding has an inductive component arising from these leakage fluxes, as well as each winding possessing resistance. To attain a voltage or current ratio as nearly equal to the turns ratio as possible, capacitance between sections of the windings should be minimised.

The requirements are unfortunately mutually exclusive; minimum leakage inductance requires that the separate ratio windings are as close together as possible, turn by turn, to one another, and this inevitably increases the capacitance between sections. The in-phase ratio error of a voltage or current ratio transformer is affected by the load the distributed capacitance offers to the impedance of the leakage inductance. That is, the effect is proportional to the product of the leakage inductance and the capacitance, so that if a winding design increases one and decreases the other, the ratio error tends to remain much the same.

It is usually desirable however to minimise the effect of output impedance on winding ratios to make the influence of external shunt capacitive loading of cables, etc. as small as possible. This means that leakage inductance should be decreased at the expense of inter-strand capacitance, and for the quadrature component, the resistance of the windings should be as small as possible. This is achieved in various ways by the winding designs described in sections 7.2.3–7.2.5.

If the windings are made as individual successive single layers on a toroid with interposed layers of insulation, the self-capacitance of each layer of winding will be small and the capacitance between layers will be reduced. The remaining capacitive currents can be further reduced by arranging that those parts of the windings at similar potentials are adjacent. The effect of inter-winding capacitance can be reduced to zero by appropriate screens placed between windings, as noted in section 1.1.5.

If high accuracy of the ratio of voltage windings *and* small leakage inductance are required, it is sometimes permissible to allow the capacitances between two windings to increase, but to make the principal capacitances symmetrical, so that they do not affect the ratio. This is achieved in the 'ordered rope' winding described in section 7.2.5.

7.2.3 Bifilar winding

The simplest possible windings are the bifilar windings of a $1{:}-1$ ratio transformer. If the winding is constructed symmetrically, the resulting $1{:}-1$ voltage ratio is usually closer to the 1:1 ratio of the turns than for the winding ratio of other transformers. To achieve this, a pair of conductors, using wire from the same reel to ensure equality of resistance and insulation properties, is twisted together so tightly that they lie in contact throughout their length, as shown in Figure 7.22a. This ensures symmetry of capacitance, resistance and leakage inductance along their length, so that the induced voltages are affected identically and their ratio is accurate. Because the windings are twisted, any small differences of emf induced in the pair by non-toroidal stray fluxes will sum to zero to a very close degree over the full length of the winding.

Figure 7.22b has been derived from the pictorial representation of Figure 7.22a by conceptually slitting the core in the vicinity of the zero tap and opening it out to a straight line. We can now imagine the windings unravelled by pulling the $+U$ point up and the $-U$ point down so that the zero points of the windings coincide, as shown in Figure 7.23.

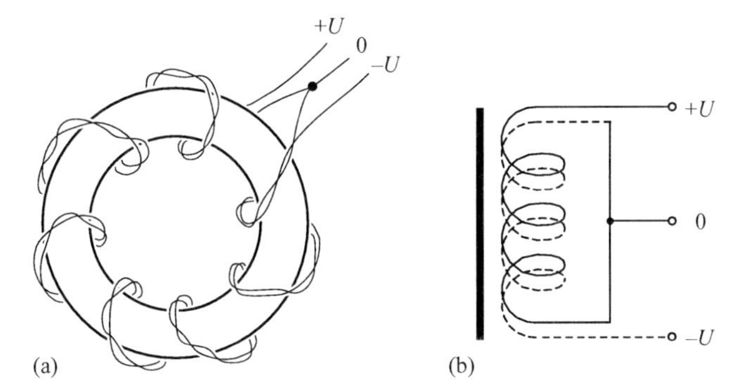

Figure 7.22 A bifilar winding: (a) pictorial representation and (b) diagrammatic representation

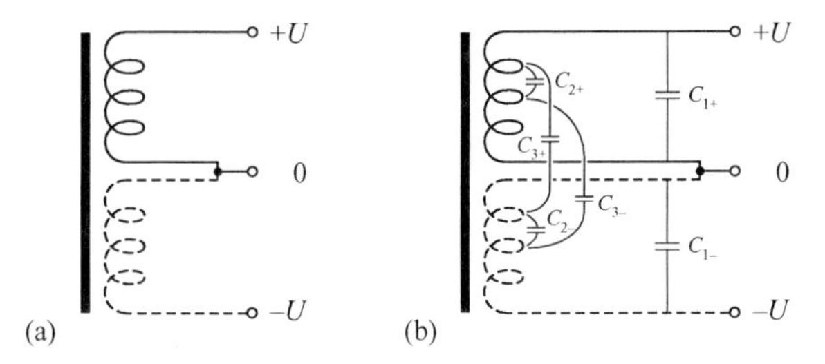

Figure 7.23 (a) Schematic representation of a bifilar winding. (b) Capacitances between the conductors of a bifilar winding

Near the $+U$ and the $-U$ ends of the windings, equal capacitive loads C_{1+} and C_{1-} to zero exist. Further capacitances symmetrical about zero occur throughout the winding, so that the distributed capacitive loadings on the $+U$ and $-U$ sections are the same. For example, for $C_{2+} = C_{2-}$, the voltage changes due to them are the same and the ratio of the output voltages remains accurately $+1{:}-1$.

We have assumed that a rope consisting of the twisted pair of conductors is wound on the toroid, so that capacitances between turns of the rope are small compared to those between the twisted conductors. Twisting will also have an averaging effect on the small between-turns capacitances. Moreover, they are largest between those adjacent turns that have the smallest potential differences, so that the capacitive current is small. Also, symmetry ensures that capacitances are equal for the two windings; that is, $C_{3+} = C_{3-}$, etc.

It is, therefore, not surprising that bifilar wound transformers can readily be made with $+1{:}-1$ ratio errors of the order of 1 in 10^9 at audio frequencies where core permeabilities are high. Their ratio errors remain within 1 in 10^6 up to 1 MHz, provided cores suited to these higher frequencies are used.

7.2.4 Rope winding having randomly arranged strands

A less perfect solution, but one that yields other voltage ratios and is easier to implement, is to take the conductors that are to form the ratio windings, lay them into a tight parallel bundle, twist one end to wind the complete set into a rope and wind the complete rope a sufficient number of times (10–100) around the core, as sketched in Figure 7.24. At the ends of the rope the conductors are then joined and can be connected to switches or connectors to form the ratio taps of the completed transformer.

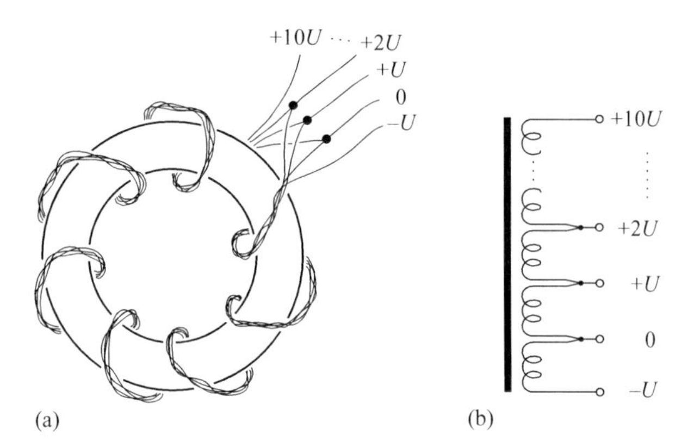

Figure 7.24 (a) A rope winding on a transformer core. (b) A schematic representation

Thus, a rope of 11 strands can be connected to form a +10:−1 ratio transformer. For the same reasons as for binary winding, the conductors should be as close together as possible, and twisted, but not as in an ordinary rope where preferred strands lie in the centre and others around the outside. Rather, the position of the strands in the rope should be randomised by periodic interweaving before twisting so that inter-winding capacitances and stray flux coupling are also randomised.

When the individual strands are connected in series to form a tapped winding, pairs of strands with relatively large inter-strand capacitances should be connected so that there is only a small voltage between them.

The departure from the nominal values of ratios equal to the ratios of numbers of turns produced in this way is typically a few parts in 10^7 at frequencies around 1 kHz. So long as the conductors are prevented from physically altering their position in the completed transformer and the magnetic state of the core is kept reasonably constant, the ratios are stable and can be calibrated with an uncertainty of the order of 1 in 10^9 (see section 7.4).

7.2.5 Ordered rope winding

An ordered rope winding is the very opposite of a random rope winding. The sections of voltage division are conductors spun around a central core, which can either be a plastic of low dielectric constant or another insulated conductor of greater cross section, as shown in Figure 7.25.

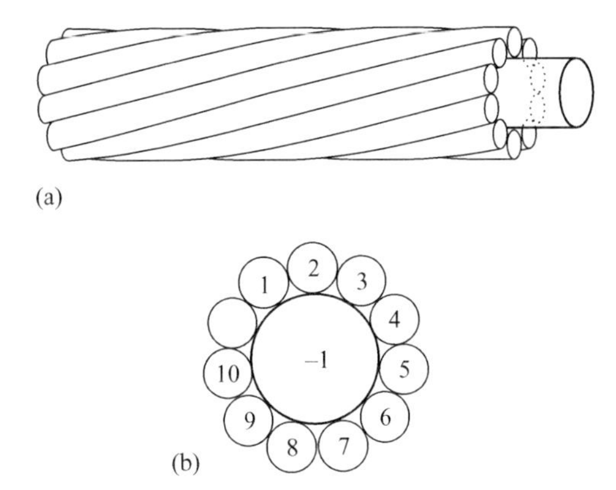

Figure 7.25 (a) An ordered rope of conductors for a transformer ratio winding. (b) A cross section

When this construction is achieved by spinning the strands of conductors as a single layer around a dielectric core of appropriate diameter, the capacitances between the adjacent strands are equal. Their symmetry ensures that they have little effect on the voltage ratios of the winding. The capacitance between the first and last strands could be merely a load on the generator connected across the winding as a whole or could be reduced by interposing a dummy strand of low dielectric constant plastic. Other capacitances between non-adjacent strands are much smaller. Because leakage inductances are also quite small, this winding technique is an appropriate compromise if the ratio errors are required to be small up to higher frequencies.

The technique can be employed to achieve two separate advantages. When employed in an IVD (with the addition of a dummy strand), the inter-strand capacitances form a capacitive divider in parallel with the inductive voltage division. Since the capacitances are approximately equal, their effect on voltage division is small, and the device retains good accuracy up to higher frequencies of the order of 100 kHz.

For use in a $10{:}-1$ voltage ratio transformer, the rope is spun around a conductor of greater cross section. The leakage fluxes between the spun 0–10 sections and the central conductor, which is the 0 to -1 section, are very small. A transformer made with such a rope can therefore be expected to exhibit small output leakage inductances and to have a very stable ratio. The inter-strand capacitances are now higher and not evenly distributed, being predominately between the 0 to -1 central conductor and the other sections, and so the transformer will give best performance at lower frequencies, below about 10 kHz.

If good performance is needed above 10 kHz, it might be better to wind an IVD with an ordered 11-strand rope (with a dummy strand) and to permanently connect the end of the first section where it connects to the second section as a zero-output tap.

7.2.6 Magnetic and electric screens

It is desirable to confine the flux produced by a winding to the interior of a toroidal space bounded by the outer surface of a winding. Therefore the winding should approximate a uniform current sheet, but a practical winding consists of a limited number of turns of wire of finite cross section, which allow flux leakage. Non-toroidal flux will also result from core imperfections, as described in section 7.1.2.

Both the effects of stray flux and capacitive currents can be reduced to negligible proportions by toroidal conducting screens of high-permeability mu-metal. Where a demountable joint is needed, it must be provided with a large overlap so that the reluctance of the air gap is small enough. Soft soldering can be used to give electrical continuity. Figure 1.10b shows a section through a toroidal shield; connections to the wound toroid in the annular space enter through the small extension on the left. An insulating sleeve prevents a disastrous complete conducting turn threading the core.

Shields can also be formed as washers and annular strips from a bendable form of mu-metal, which can be cut with scissors; although this is not as efficient as seam-welded and heat-treated thick sheet material, the resulting shield can be sufficiently good to be useful.

Screens made of thick copper have also been used. These rely for their magnetic effect on the skin depth phenomenon; eddy currents induced in them tend to prevent the passage of unwanted flux. The skin depth at 1 kHz is about 2 mm, so this technique is only useful at higher frequencies.

Sometimes the unwanted emergence of magnetic flux is not an important effect in the device concerned, and a purely electrical screen having an overlapping insulated gap suffices. This can be made of any conducting material, and conducting paint applied to a layer of an insulating tape is often employed. Whatever the material used, thought must be given to the possible threading of the core by the capacitive current flowing to and within the screens.

Consider a screen around a core energised with a winding, as shown in cross section in Figure 1.10a. Because the screen is a single turn around the energised core, there will be a corresponding potential difference between its ends, and a capacitive current will flow between them and through the screen. Therefore, a small, but finite, current threads the core, and its effect may be significant in the performance of the complete device. In the case of a magnetic screen, the amount of overlap is a compromise between a near-perfect magnetic screening path and the large capacitance of a big overlap. This capacitance permits a capacitive current to flow, driven by the emf induced in a single turn of the screen.

7.2.7 Testing the attainment of a nearly toroidal field

We have emphasised that a major objective when constructing accurate transformers is to ensure as far as possible a toroidal magnetic flux by means of windings properly distributed around the core and by using surrounding toroidal magnetic screens where necessary. Then even if the magnetic state of the core is affected by partial magnetisation or magnetic shock, the total flux still threads all ratio windings external to the toroid, and their ratio accuracy is unaffected.

It is useful to be able to test whether a nearly toroidal field has been attained, and a test is easily carried out by winding two equal coils, A and B, in opposing senses and connecting them to a detector as shown in Figure 7.26. T is the wound, energised toroid being tested, which may also have a surrounding toroidal magnetic screen.

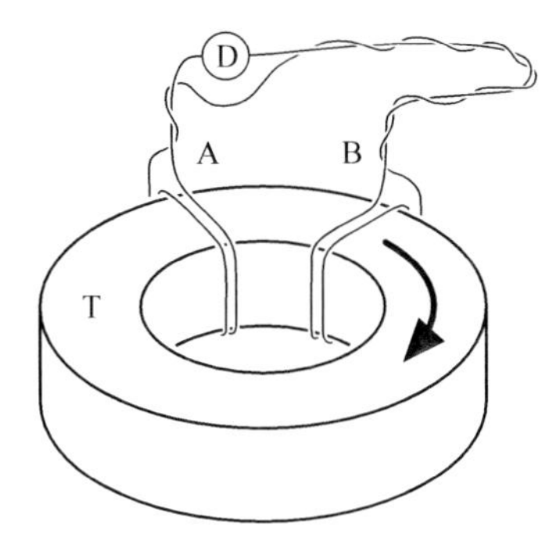

Figure 7.26 Testing the perfection of a nearly toroidal field

First, the coils A and B are brought together as closely as possible. Coil A is kept fixed with respect to the toroid, whilst coil B is moved around to various parts of the circumference. If the excited toroid assembly has a totally toroidal flux, no voltage will be registered by the detector, and if the detector does register a voltage, its ratio to that induced in a single turn of coil A is a measure of the imperfection.

7.2.8 Connections to the output ports

The basic winding techniques for the secondary windings have already been described, but the method of bringing out taps to the coaxial connectors requires some consideration if the highest possible accuracy is to be attained.

A good design is to mount the coaxial output connectors and the connector to the primary winding on one side of the enclosing magnetic box screen, but insulated from it. Thick wires from the outer conductor of the ratio winding connectors are routed alongside the tap connections to a point well within the volume of the box where they are joined together and to the zero tap as illustrated in Figure 7.27. In this way, unwanted small mutual inductances between the connectors to the taps and self-inductances from only an inner conductor to a connector passing through a hole in a magnetic sheet are avoided. The purpose of this method of mounting the output sockets is to ensure that the potentials of the outer conductors are all well defined and small and are not affected by the flow of current returning to the transformer through the outer conductors of other taps. This would happen if a multiplicity of other paths were provided by mounting the outer connectors directly and uninsulated on a conducting panel. The voltage of every tap is well defined, being that between the inner and outer conductor of its coaxial output socket. The connectors to the first (magnetising)-stage winding are also insulated from the box and their outer conductors connected together, as shown.

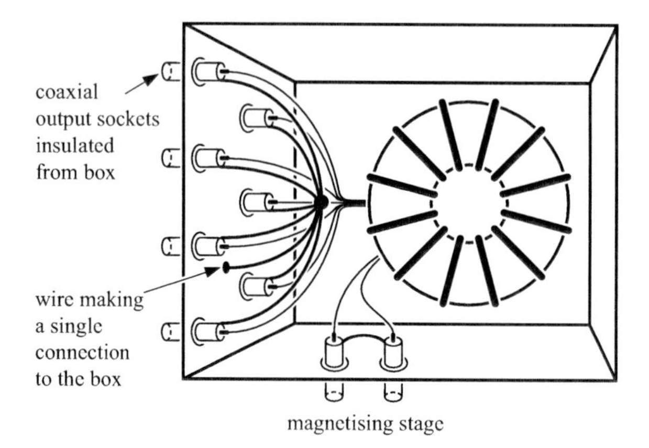

Figure 7.27 Connections to the output ports of a transformer

7.3 Types of transformers

The foregoing general principles have been applied to the construction of several kinds of transformers which have been found useful in accurate impedance-comparing networks.

7.3.1 Inductive voltage dividers

An IVD is an autotransformer in that it has no separate primary winding. A single, tapped winding fulfils the roles of both primary and secondary windings. It usually has several decades of output subdivision to provide a fine resolution of its voltage ratio.

These devices are of the greatest utility as they provide a voltage whose magnitude is related to the source voltage by an accurate and stable ratio, which can be finely adjusted as required and whose phase angle, if little current is drawn, departs from that of the source by as little as 1 μrad.

An ordered rope winding (see section 7.2.5) is best for all frequencies; for low and intermediate frequencies and the later decades of subdivision, a random rope winding is reasonable.

Since we are presupposing a construction in which all the flux threads all the windings, the voltages across the turns intervals n_1 and n_2 of Figure 7.28 must be such that $U_2/U_{in} = n_2/n_1$. Hence, U_2 can be altered in finite steps by switching in different turns to alter the ratio of integers n_2/n_1. Multi-outputs from paralleled switching are possible, and often two independent voltage outputs are provided in this way.

As there is a practical limit to the number of turns of reasonably sized wire that can be used (10^2–10^3 is usual), the fineness of the steps, which is set by the ratio of numbers of complete turns, is insufficient for most applications. Hence, some method of achieving what is, in effect, a fraction of a complete turn is required.

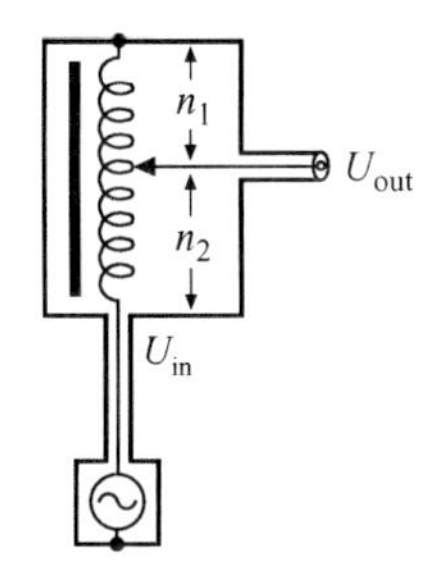

Figure 7.28 A simple IVD

Suppose another winding of m_1 turns is wound on the core and connected to a winding of m_2 turns on the core of a second transformer as shown in Figure 7.29.

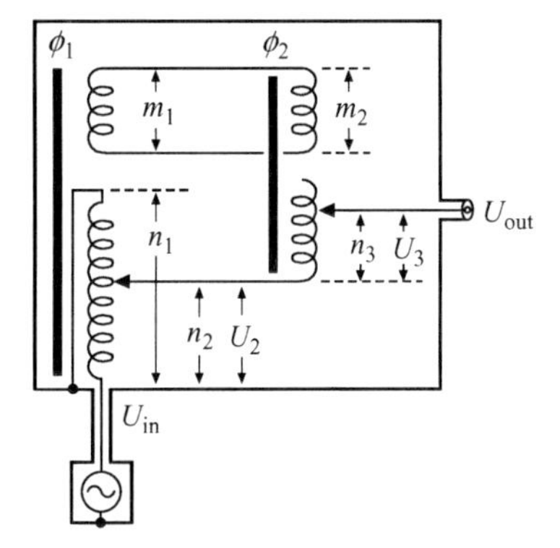

Figure 7.29 Further subdivision of the input voltage of an IVD

Only a small current flows because of the large inductance of the m_2 turns on the second transformer; and the flux linkages in cores 1 and 2 are almost equal – that is, $m_1\phi_1 = m_2\phi_2$. If n_3 turns are wound also on core 2, the emf induced in these turns is, to a good approximation,

$$U_3 = \frac{n_3 d\phi_2}{dt} = n_3\left(\frac{m_1}{m_2}\right)\frac{d\phi_1}{dt} = \left(\frac{n_3}{n_2}\right)\left(\frac{m_1}{m_2}\right)U_2$$

since $n_2 d\phi_1/dt = U_2$ and $U_2 = (n_2/n_1)U_{in}$

The winding of n_3 turns can be put in series with the winding of n_2 turns to achieve a total output voltage

$$U_{out} = \left(\frac{n_2}{n_1} + \frac{n_3 m_1}{n_1 m_2}\right)U_{in} \tag{7.4}$$

The procedure can be iterated to further cores to provide further subdivision as required.

Various modifications of this basic concept are employed to simplify construction or improve accuracy.

(i) The n_3 turns may be a tapped portion of the m_2 turns. Switch selection of taps results in the circuit illustrated in Figure 7.30.

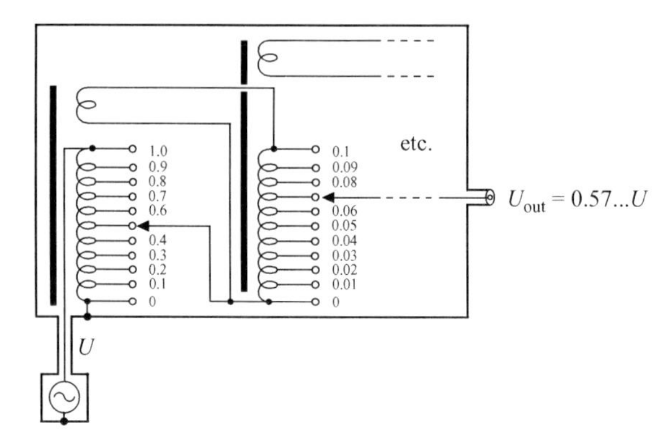

Figure 7.30 Switched selection of the taps of an IVD

(ii) Economical use of cores can be achieved by combining each successive pair of decades onto one core, that is, putting ten sections of $10p$ turns with taps between the sections on the core together with ten sections of p turns similarly tapped. The circuit is then as shown in Figure 7.31.

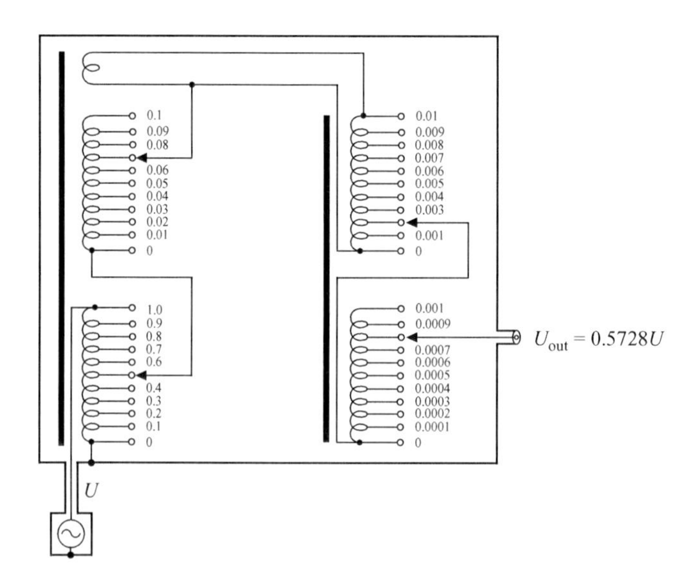

Figure 7.31 Minimising the number of cores needed for an IVD

An IVD is mostly used as a stable AC potential or current divider, for which a low output impedance is not of the utmost importance. Hence, a rope method of construction with a winding technique that limits the response to external flux is suitable. The first decade of the above example would consist of a rope of ten strands wound $10p$ times on the core with a 'bootlace' or non-progressive winding; the ends of appropriate strands are connected together at the fixed contacts of the first-decade switch.

Either the ends of the strands can be labelled before the strands are twisted up into a rope, or opposite ends of the same strand can be identified with an AC continuity tester after the rope has been twisted up and wound on the core. (A DC continuity tester could leave the core in a magnetically saturated condition even if it supplied only a fraction of an ampere turn.) Sometimes an 11th strand is connected across the next-decade switch (as shown by the dotted wire in Figure 7.30); we will describe its purpose later.

The second decade of Figure 7.31 is astatically wound with p turns of an 11-strand rope, 10 strands of which are connected to the second switch in a similar fashion to the connections to the first switch. The 11th strand connects across the $10q$ turns of the 10- or 11-strand rope of the next transformer assembly, which also has q turns of a 10-strand rope. The switches are connected as shown to form a complete four-decade divider. The construction technique can be iterated in an obvious fashion by adding further two-decade cores to make a divider of $2n$ decades. Four-, six- and eight-decade dividers are commercially produced in this way.

The input impedance of the device may be increased by increasing p up to the point where the comparatively large self-capacitance of the rope winding resonates with its inductance at the frequency ω at which the divider is to be used. A typical value would be $p = 5$ at 1.6 kHz ($\omega = 10^4$). Above the resonance frequency, the input impedance decreases again. The first core, therefore, has a large ratio of $\mu_r A/l$, where A is the cross-sectional area; l, the mean magnetic path length; and μ_r, the relative permeability of the core material, to maximise the inductance for a given p. Also, if B_{max} is the maximum flux in the core for which μ_r is still reasonably high (just below the flux at which the core material saturates), $B_{max}A$ must be sufficient enough so that $100pA\omega B_{max}$ exceeds the highest source voltage to be applied to the divider. Typically, if $A = 10^{-4}$ m^2, $B_{max} = 0.2$ T, $\omega = 10^4$ rad/s, $p = 5$ and $U_{max} = 100$ V. If minimum quadrature component of the output voltage is desired rather than high input impedance, it is better to use the minimum number of turns required to support the intended voltage rating. The input impedance of the ratio winding may be increased in another way, by the two-staged technique described in section 7.1.5. No inter-winding screens are necessary in these devices, and the internal capacitive loads alter the output voltage by typically less than one part per million of the input voltage at moderate frequencies. Some insulating spacing between the ropes forming successive decades on the same core is sufficient to reduce the error in the output voltage to 0.1 ppm or less of the input.

The transient conditions that occur as the switches are operated can be troublesome. It is preferable that the switches are of the 'make-before-break' type, that is, contact is made with the next position before it is broken on the present position. One section of the winding is thus momentarily shorted, imposing a transient low-impedance load on the source but not greatly affecting the output voltage. This is in contrast to the other possible situation of 'break-before-make' switching, where the circuit to the succeeding core is momentarily open and the collapse of flux in the core produces a large momentary emf at the output.

A better arrangement can be made and is worthwhile for the first decades where these transient effects are greatest. By means of two-pole switching with 'make-before-break' contacts on one pole and 'break-before-make' contacts on the other, first, a resistor of the order of 10 Ω can be switched across the taps, then the output contact is moved, and finally, the resistor is switched out.

In the design described above, each succeeding core is energised by a separate winding, which is a strand of the p turns per section. A similar arrangement can be made for the second winding on a core; although it might seem unnecessary as the second winding is already correctly energised by virtue of threading the same core as the first winding, it is advantageous for two reasons. First, particularly on the first core of a divider, the magnetising current flows through the windings comprising the first decade of division but in the absence of the extra connected strand, not through the second-decade windings. The voltage drop caused by this current traversing the leakage inductance and winding resistance then causes a lack of equality of potential across one section of the first decade to that across the whole of the second decade. This situation is alleviated by connecting an extra strand of the first-winding rope across the entire second winding. Second, the leakage inductance that exists between the first-decade rope and the second-decade rope is greatly reduced by coupling them together in this way. This results in a reduced output impedance of the complete divider.

If we consider now the connection between core assemblies, then the second assembly will draw its magnetising current from the first. The resistance and leakage inductance of the energised winding on the second core will cause a loss of emf across it because the magnetising current must flow through this impedance. Therefore, the total voltage across a succeeding core assembly will not equal the voltage difference between taps on the winding preceding it. Thus, small unwanted discontinuities will exist in the output voltage of the complete divider as a function of switch settings.

These discontinuities can be overcome, although the accuracy is not improved, by the switching shown in Figure 7.32, which arranges that succeeding windings are energised directly by the emf between taps.

Sometimes when using IVDs as adjustable voltage sources if their output is significantly loaded, the changes in output impedance as the ratio switches select various sections of the windings cause discontinuities. This is particularly troublesome when attempting fine adjustment using the lowest dial settings because there can be gaps in the range of output voltages. This defect is overcome by the circuit of Figure 7.32b, where the two decades of finest subdivision of output voltage are obtained from a 100:1 step-down transformer supplied by separate switches from the core two decades nearer the input. The impedance of the windings and their variations supplying the input to this transformer is thereby reduced by $(100)^2$, and this, together with the small and constant impedance of the transformer secondary winding, greatly reduces the output impedance of the last two decades.

The switch positions show a selected output voltage of 0.372026 times the input voltage.

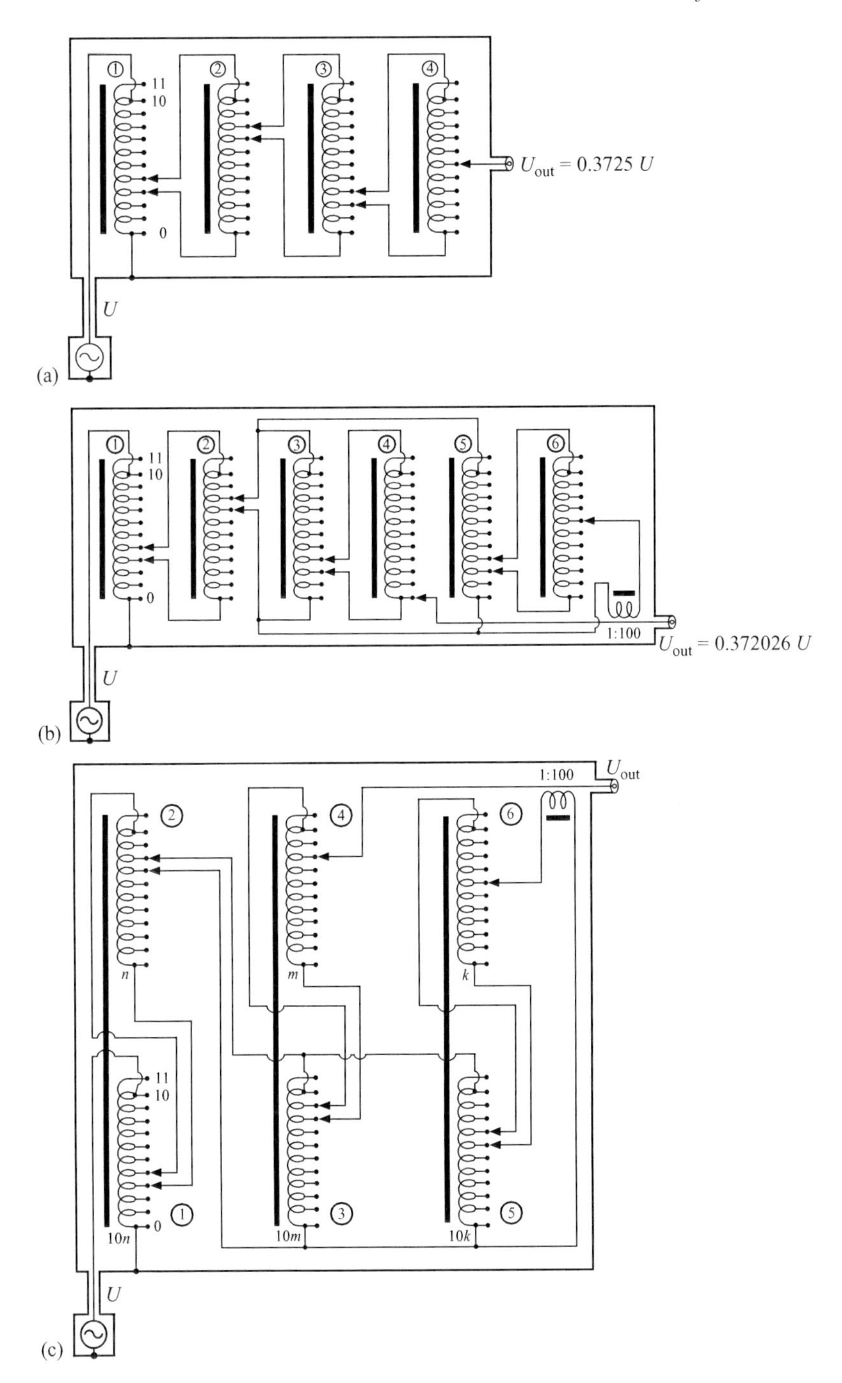

Figure 7.32 (a) Two-tap switching of an IVD. (b) The two final decades are via a step-down transformer. (c) Reducing the number of cores

As before, by winding a rope of $10p$ turns and a rope of p turns, adjacent pairs of cores can, in fact, be combined as one core with a saving of space and magnetic material, as in Figure 7.32c.

We shall use the formal representations of an IVD drawn in Figure 7.33a. Sometimes a single device containing some common circuitry, which provides two independent outputs, will be convenient, and this will be represented as shown in Figure 7.33b.

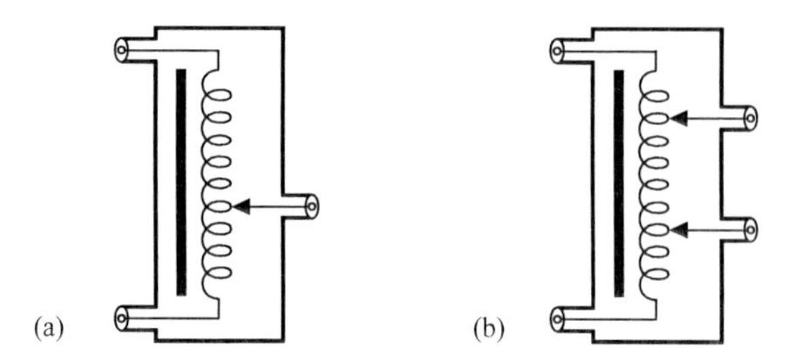

Figure 7.33 Formal representation of (a) an IVD having a single output and (b) an IVD having two independent outputs

The two-stage principle can provide further improvement in accuracy, as described in the next section.

7.3.2 *Two-staged IVDs*

The internal impedances of a single-decade IVD winding can be represented as shown in Figure 7.34.

The series impedances Z_q, Z'_q, etc. arise from the resistance and leakage inductance of the individual inter-tap windings (e.g. the individual strands in a rope construction). They cause an additional contribution $i_{\mathrm{mag}} \sum_{q=1}^{n} (Z_q + Z'_q)$ to the emf of the nth tap.

The input impedance of the device at the generator terminals is $Z = j\omega L + R$, where L and R represent the inductance and loss of the wound core. Z is much larger than the impedances Z_q and Z'_q so that I_{mag} is approximately U/Z, where U is the source emf. ωL and R are likely to be about the same size.

Typically, for 50×10 turns of 10^{-3} m diameter copper wire wound on a core of 10^{-4} m^2 cross-sectional area and 0.2-m mean magnetic path, ωL would be about $3 \times 10^4\ \Omega$, and the resistance and leakage reactance of the winding would be about $0.1 + j10^{-2}\ \Omega$ at $\omega = 10^4$ rad/s. The wire resistance will vary by 1% or more between strands, and the leakage inductance by the order of 100%. Several of the Z_q are involved in intermediate tap voltages, so that the output voltage could be in error in magnitude and phase by up to a part in a million of the input voltage.

The two-stage approach can effect an immediate improvement by providing a ratio winding with a very low value of I_{mag}. The circuit for a single decade is drawn in Figure 7.35.

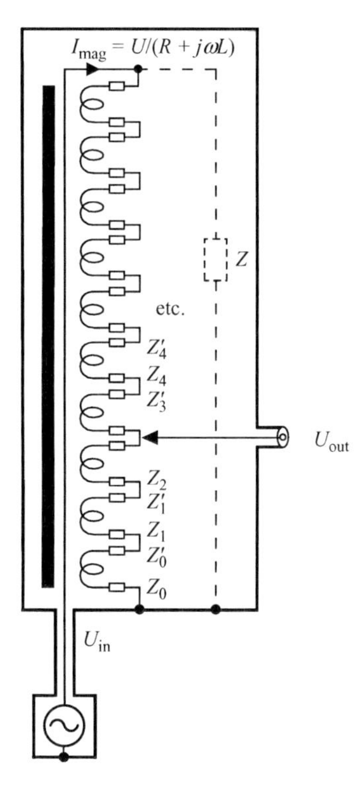

Figure 7.34 The internal impedances of a single-decade IVD

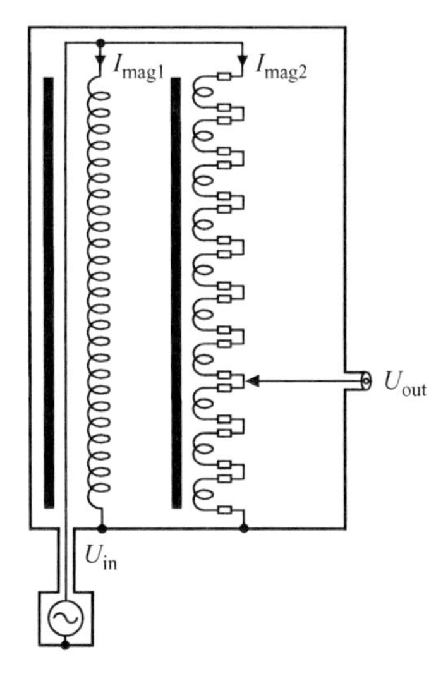

Figure 7.35 The two-stage principle applied to an IVD

The in-phase error in the output voltage at low or moderate frequencies (40–400 Hz) can be as low as a few parts in 10^9, but the quadrature voltage may be ten or more times larger.

Further decades of voltage division can be added by providing more windings and cores as before, with succeeding stages also having separate magnetising windings and two cores. Figure 7.36 is an example of a possible circuit.

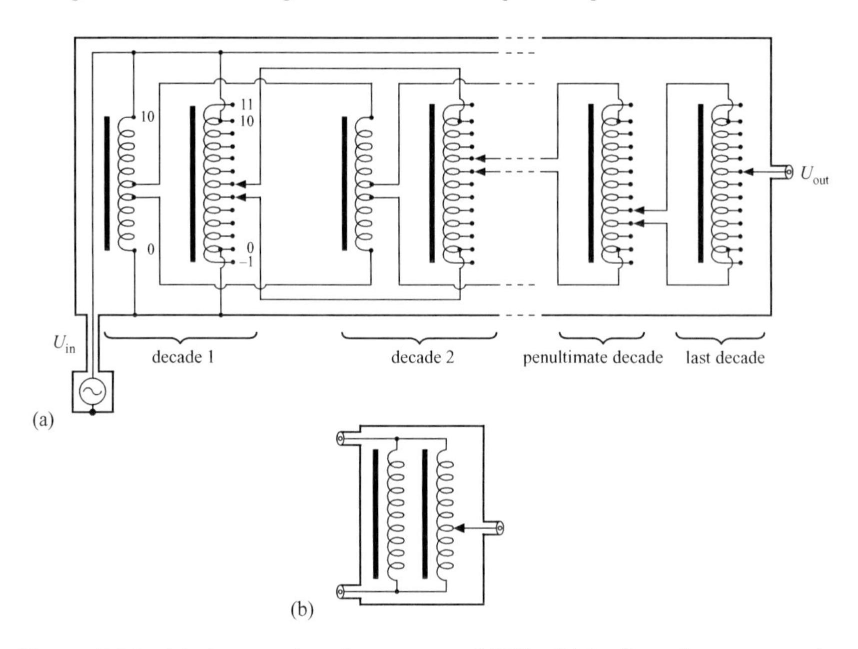

Figure 7.36 (a) A many-decade two-staged IVD. (b) Its formal representation

The less elaborate single-stage construction would be used for further decades after the first, second and possibly third, the relative loss in accuracy being by then insignificant. Again, with ratio windings having $10p$ turns and p turns, two successive decades can be combined on one core assembly needing only one magnetising winding.

Leakage inductance and inter-strand capacitance will be the same for a one- or two-staged transformer, and so a two-staged transformer will have similar high-frequency imperfections as a single-stage transformer.

7.3.3 Injection and detection transformers

In designing coaxial bridges, it is very useful to be able either to introduce a generator of a small additional voltage ΔU at a point along the inner of a coaxial cable (Figure 4.11a) or to detect the vanishing of a current I at a point along the inner (Figure 4.11b).

The duality principle suggests that both of these objectives can be met using the same device, which is either a coaxial injection or a detection transformer, depending on which of the roles it is fulfilling.

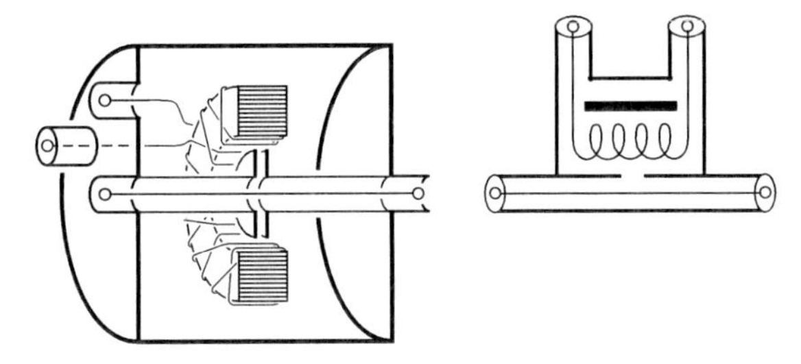

Figure 7.37 A pictorial section and a formal representation of an injection or detection transformer

A cutaway diagram (Figure 7.37) will make the construction clear.

A hollow cylindrical conducting can has coaxial connectors fitted axially to its ends. Inside the can, the connectors are joined by a shielded conductor. The shield is not continuous but has a short gap at the centre. By adding nearly touching metal discs to the shield on either side of the gap so that the gap is much deeper than its width, the capacitance between the inner conductor through this gap to objects in the annular interior of the can is made extremely small. Alternatively, the shield can have a short insulated overlapping section to serve the same purpose. A high-permeability toroid, bootlace wound with a large number of turns (typically 100), surrounds the coaxial line, and the ends of the winding are brought out to two coaxial connectors on the can. If only a single input or a detector connection is required, the other socket is shorted.

The gap at the centre of the shield locates the exact point along the inner conductor at which the current can be detected to be zero. Figure 7.38 shows this is so because the current through the distributed cable capacitances on either side of the gap flows through these capacitances and back through the outer conductor.

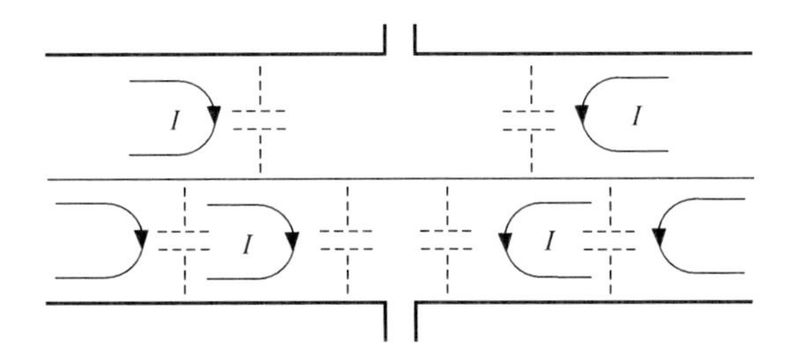

Figure 7.38 Capacitive currents on either side of the shield surrounding the inner conductor

This construction is a n:1 transformer whose primary is the n turns on the toroid and whose secondary is the single turn of the axial conductor. The gapped screen and the can form a complete electric screen, which isolates the primary from the secondary. The axial symmetry of the construction ensures that the device neither emits nor responds to external magnetic flux.

A construction in which the primary winding is isolated is also possible, as shown in Figure 7.39. Use of this device in a network will remove the need for one current equaliser.

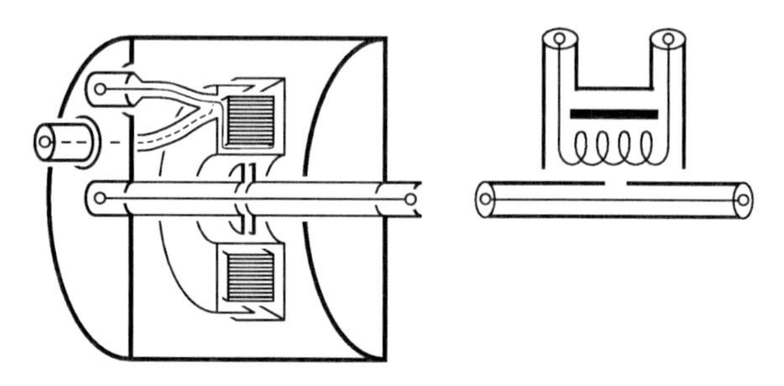

Figure 7.39 An injection/detection transformer having an isolated winding and its formal representation

A detector transformer, which responds to the total current in a coaxial cable, that is, to any imbalance in the supposedly equal and opposite currents in the inner and outer conductors, is also required. Figure 7.40 illustrates a suitable construction in which a high-permeability toroid is wound with a hundred turns, which begin and end on the inner and outer, respectively of a coaxial socket. The whole winding and connections to the socket are enclosed in a toroidal shield of high-permeability material, which is connected to the outer of the socket. The shield has an overlapped annular gap so that it does not constitute a shorting turn.

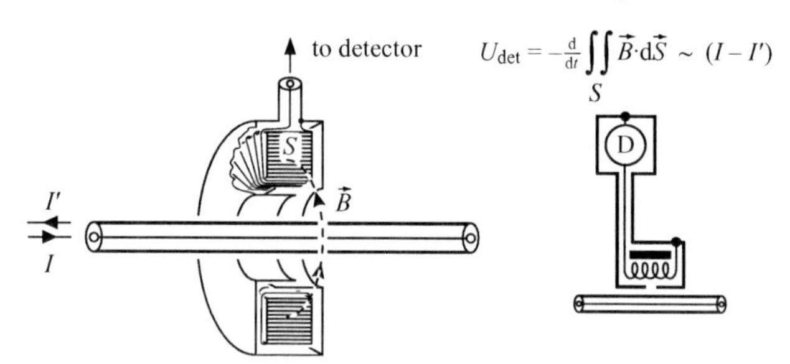

Figure 7.40 A cross section through a detection transformer, which detects or injects an unbalanced current in a coaxial cable and its formal representation

A cable to be tested for the presence of unbalanced current is threaded through the toroidal assembly, for which it is a single-turn primary winding, and any net current in the cable causes an emf to appear at the coaxial detector socket. The component can also fulfil the dual purpose of injecting a current into the low-impedance circuit of the outer conductors in a mesh.

7.3.4 *Use of an injection transformer as a small voltage source*

If a voltage generator of output U is applied to one input of the n-turn secondary (typically $n = 100$) and the other coaxial socket is shorted, a voltage of U/n is generated in the inner conductor at the centre of the device.

By virtue of the superposition theorem, the use of the device can be extended to generate independent in-phase and quadrature voltages in the inner conductor. The in-phase generated voltage U_1 and the output U_2 of the quadrature phase-shifting circuit are arranged to have low output impedances compared with the impedance of the 100-turn primary of the injection transformer which is typically $10^4\ \Omega$, and are applied one to each of the two ends of the primary winding. The schematic representation of the device is then as shown in Figure 7.41a. The practical circuit in Figure 7.41b could have $R_S = 10\ \Omega$ and $C_S = 0.1\ \mu F$ for an angular frequency $\omega = 10^4$ rad/s.

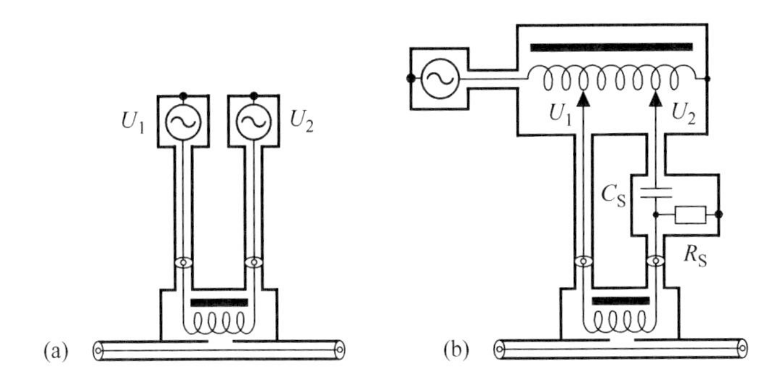

Figure 7.41 An injection transformer as a small voltage source

A typical figure for the output impedance of the generator and phase-shifting network is 10 Ω. The generator at either input to the primary is connected to the high impedance of the primary in series with the small output impedance of the other generator so that the phase of either generator is preserved and an emf equal to $10^{-2}(U_1 + jU_2)$ is generated in the coaxial inner secondary with an injection impedance of only $(10^{-2})^2 \times 10 = 10^{-3}\ \Omega$.

If the inductance of the primary is insufficiently large compared with the output impedance of the generator, the two-stage technique (see section 7.3.2) may be employed. The circuit is then as shown in Figure 7.42. A two-staged device contains two toroidal cores; the first is wound with a first stage or magnetising

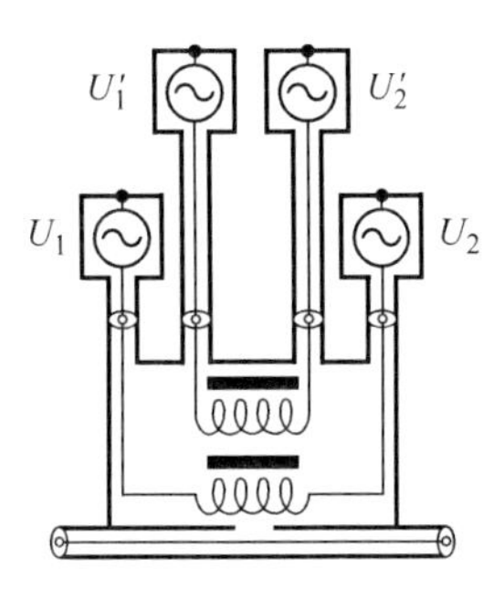

Figure 7.42 A two-staged voltage injection transformer

winding. The second toroid is placed alongside the first, and the second stage is wound threading both cores with the same number of turns as in the first stage. The pairs of coaxial inputs terminating these windings should be fed by two tracking in-phase and two tracking quadrature sources. The sources feeding the first-stage winding are corrected by the much smaller currents in the second-stage winding. The accuracy of the emf generated in the inner can be improved by this two-staged technique from the order of 1 in 10^3 to in-phase or quadrature ratio errors of 1 in 10^6 or better.

7.3.5 Use of an injection transformer as a detector of zero current

A current flowing down the central conductor threads the core and induces magnetic flux in it. This in turn induces an emf in the many-turned secondary, which can be sensed by a detector plugged into one of the output sockets, the other being shorted.

The precise point along the inner conductor at which the condition of zero current is sensed is defined by the position of the narrow gap in the outer, since up to this point some current can flow down the inner and across to the outer through the capacitance between them (Figure 7.38). In an equalised network, this current must return along the outer in the same direction as both sides of the gap are at the same potential, because they are shorted by the case of the device around the outside of the toroid, and no capacitive current will flow across the gap.

When this device is used merely to detect the presence or absence of a current, the accuracy of its ratio is unimportant, and a single-stage construction is all that is necessary. The factor of n^2 in the transformed impedance, where n is the number of turns on the secondary, can be very valuable when noise matching a detector having a high input impedance to a circuit of low impedance (see section 2.1.3).

7.3.6 Calibration of injection transformers and their associated phase change circuits

The in-phase calibration of an injection transformer A can be carried out with reference to an IVD of adequate accuracy. A circuit for a single-stage device is shown in Figure 7.43. A two-staged device would need a separate feed from a tracking IVD to supply its magnetising winding.

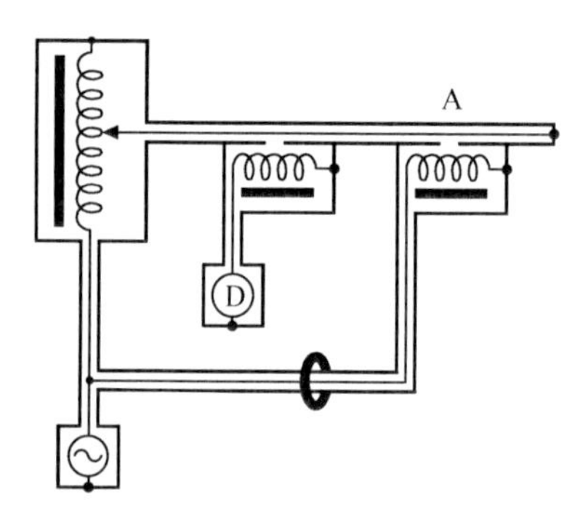

Figure 7.43 Calibrating the in-phase component of a voltage injection transformer

A null of detector D indicates zero current threading its associated detector transformer, which in turn indicates equality of potential between the IVD output and that of the injection transformer A so that the IVD reading for this condition calibrates the ratio of A.

A 90° phase change circuit and the injection transformer used with it are best thought of as a single entity to be calibrated together.

For this calibration, we need components of known value and phase angle, and the best components that are readily available are gas-filled or primary-standard quality fused-silica-dielectric capacitors, which have phase angles of the order of 10^{-6}, and small commercial electronic resistors of about 100 Ω value. The self-inductance of these resistors is approximately balanced by their self-capacitance as commonly constructed, and phase angles of less than 10^{-5} up to frequencies of 10^4 rad/s are usual (see section 6.4.11).

These components can be assembled into a bridge to calibrate a quadrature injection network as shown in Figure 7.44a.

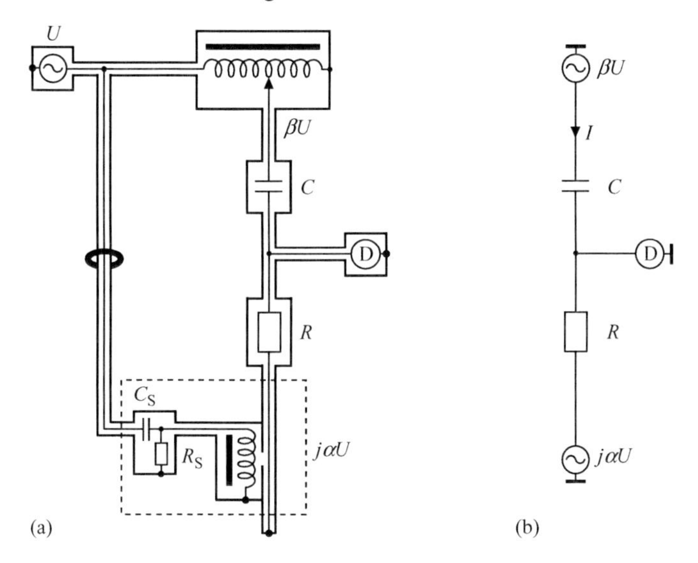

Figure 7.44 Calibrating a voltage injection network. (a) Coaxial circuit. (b) Principle of the circuit

As can be seen from Figure 7.44b, the bridge is balanced and the detector nulled if the current through the capacitor equals that through the resistor, that is, $\alpha = -\beta\omega CR$. The quadrature generator circuit is shown within broken lines. Component values of about $R = 100\ \Omega$ and $C = 100$ pF whose exact values are known are suitable for calibrating a phase change network of 10^{-2} attenuation for use with a 10^{-2} ratio injection transformer at a frequency of 10^4 rad/s.

In this discussion, only the source of a quadrature voltage and its injection have been considered. It is a straightforward matter to test the complete injection system with its two IVDs, one for the in-phase voltage and one for the quadrature. The former should be set to zero when checking the quadrature voltage and vice versa.

In most applications where a voltage is to be introduced into a network, the accuracy of the magnitude of the in-phase voltage is more important than that of the quadrature voltage. It is desirable, therefore, that any in-phase component introduced by the quadrature injection network should be trimmed to be as small as possible.

A very convenient, simple and stable trim component for the 90° phase change circuit associated with an injection transformer is made by putting a few turns of insulated copper wire wound up into a circular coil to form an inductor of value of the order of 100 μH in series with the resistor. The value of the resistor is so chosen that the resistance of the combination is just less than the nominal value required. The impedance of the combination must be much less than the input impedance of the injection transformer (about 10^4 Ω). R should therefore be 10 Ω or less. The phase of the injection thus generated is adjusted to 90° by decreasing the area of the inductor coil by squeezing it into an ellipse, and the magnitude is adjusted by shunting C with a trimming capacitor. The inductor is best connected to the end of the resistor at screen potential, as shown in the circuit of Figure 7.45.

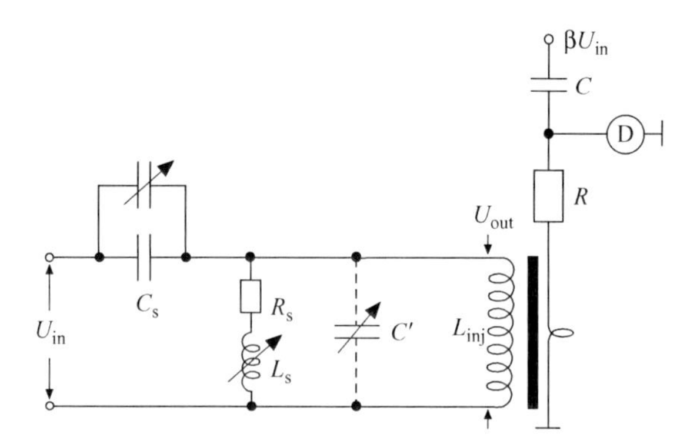

Figure 7.45 Trimming the accuracy of a quadrature voltage injection

This method of trimming has the advantage that if $(\omega CR)^2 \ll 1$, which is usually the case as ωCR is the attenuation introduced by the network, the circuit is trimmed correctly as to phase ($U_{out}/U_{in} = j\omega CR$) when $L/R = CR$. That is, the phase change is 90° and is frequency independent. The attenuation is proportional to frequency so that the same adjusted circuit can be used over a range of frequencies.

7.3.7 Voltage ratio transformers

These often form one of the major components of accurate impedance comparison bridges and are therefore very important devices. The ratio of their output voltages can equal the ratio of the number of turns to within 1 part in a million or so, and as we will see in sections 7.4.2–7.4.6, this ratio can be calibrated with even smaller uncertainty, to 1 in 10^9 if required. Attention to detail in construction is needed to ensure that this ratio is unaffected by external magnetic fluxes, that it is as independent as possible of the current drawn from the device by loading and is unaffected by changes in the mean potential of the exciting primary winding.

The core must have a large enough cross-sectional area so that the total flux in an unsaturated core is sufficient to provide the required output voltage from the secondary ratio windings. It is better if the secondary windings have comparatively few turns (of the order of a hundred), and so to attain the voltages of the order of a hundred volts used in many bridges to measure small admittance of, for example, 10 pF, cores of large cross-sectional area are required. An area of 10^{-3} m^2 is typical.

The magnetising or primary winding is best wound by an astatic technique (see section 7.2.1). This, together with a toroidal geometry and a toroidal magnetic screen as described in section 1.1.1, ensures a toroidal flux. The magnetic screen can also serve as one of two electrical screens between the primary and secondary windings. If these screens are maintained at the same potential by a single conductor having one end connected externally to the outer conductor of the coaxial lead connected to the primary winding and the other to a correct point on the outer conductors of the circuitry connected to the outputs of the secondary winding, all capacitive currents between the screens are reduced to a negligible value. An example of this situation is drawn in Figure 1.20. For a specific example of the application of this principle, see section 1.1.2.

It is sometimes better practice to separate the function of isolation from voltage ratio by having a separate isolating supply transformer. The ratio transformer then does not need internal electric screens, thereby simplifying its construction and enabling its properties to be more easily optimised. The supply transformer can have taps providing isolated voltages for other parts of the comparison network. Examples of this approach is discussed in chapter 9 when practical networks are described.

As pointed out in section 1.1.5, care must be taken because the screens are not equipotentials. If two screens have gaps at different places around the cross section, there will be a capacitive current between them, which must be provided via the networks connected to either screen. The effects of this current are nearly always undesirable, so it is best to put the gap in each screen at the same place around the cross section of the transformer. The correct topology is shown in Figure 1.27. It is highly desirable to have a magnetic screen for the whole transformer outside the secondary ratio windings so that external flux cannot enter and internal flux cannot escape. This screen should be connected to the secondary inter-winding screen.

A broken/dotted line represents a magnetic screen.

7.3.8 Two-stage construction

A two-staged ratio transformer can provide separate current and potential outputs to four-terminal-pair impedances as well as improving the definition and stability of the potential outputs.

Some of the bridge networks described by Kibble and Rayner [2] employed ratio transformers incorporating an isolated ratio winding to provide defined potentials and a 'second-core winding' to alter the overall potential difference across both ratio windings. For completeness, we include constructional details of this type of ratio transformer.

The construction is as shown in Figure 7.46. The winding connected to a current supply, which is independent and adjustable, is wound only around the second core.

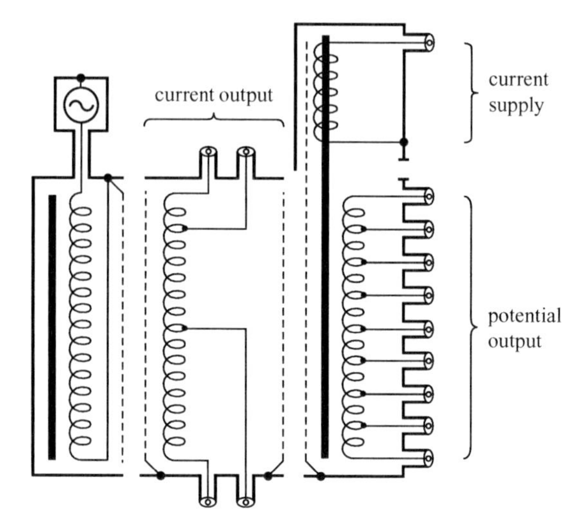

Figure 7.46 The construction of a two-staged voltage transformer

Winding and current supply together are termed a ‘second-core excitation’; their purpose is to alter the overall voltages generated in the potential windings without altering the ratio. For example, the inter-winding capacitance of the potential windings will cause capacitive currents to flow in these windings; these currents energise the second core and in turn cause the overall voltage of the potential windings to be altered, considerably increasing the in-phase component. A small second-core excitation is then required to restore these voltages to equality with those produced by the current windings. The second-core excitation supplies the current flowing in the inter-winding capacitances. Potential windings should provide no current to external components, but to minimise error arising from potential drop of any small current they do deliver, winding techniques that lead to low output impedances should be used. The method depicted in Figure 7.8 for reducing the output impedance of one section at the expense of another to provide a better compromise may well be also beneficially employed.

7.3.9 Matching transformers

These are transformers that have only one secondary winding. The impedance-matching ratio $(n_1/n_2)^2$ can match the output impedance of one sub-net to the input impedance of another for signal and/or signal-to-noise optimisation. This is further discussed in section 2.1.3. A common example is a transformer that matches the output impedance of a bridge to the input impedance of a null detector. Another, the dual of this application, is matching a source output impedance to a bridge input impedance. Matching transformers can often also conveniently provide isolation, and this is discussed in section 1.1.5.

7.3.10 Current ratio transformers

The above discussion in section 7.3.7 of voltage ratio transformers can also be applied to the operation of current ratio transformers by invoking the principle of duality. That is, all sources can be altered to detectors, and detectors to sources, and the roles of current and voltage transformers can be interchanged throughout a network. The resulting representation may help those who are more familiar with voltage transformers than current transformers in bridge networks.

The action of a single-stage current ratio transformer can be understood in the following way, with reference to Figure 7.47. The equality and opposing sense of the ampere turns, $n_1 I_1 = n_2 I_2$, is sufficient to ensure that they together induce zero flux in the core and therefore zero emf in the detector winding. The magnetic and electric screens connected to the low-potential ends of the toroidal windings ensure that there is no direct coupling between current and detector windings, that is, coupling other than the magnetic flux coupling in the core and the space within the toroidal detector winding. An accurate and stable ratio is obtained in this way, as in the case of a voltage transformer.

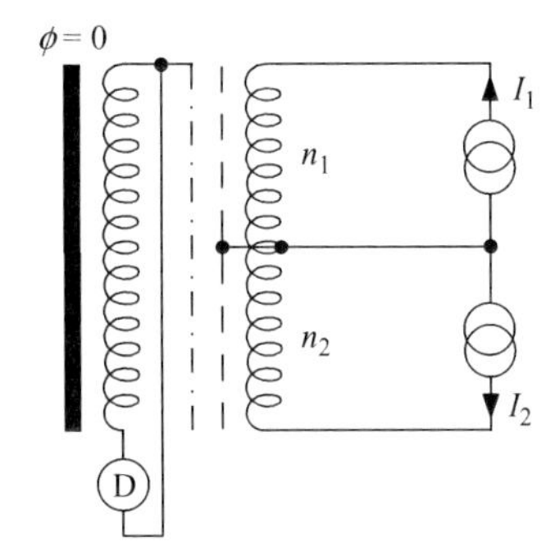

Figure 7.47 A current ratio transformer used to compare the ratio I_1/I_2 of two currents

The number of turns of the detector winding and the core cross section and reluctance can be selected to optimise the impedance to present to the detector for best signal-to-noise ratio.

The finite leakage inductance and resistance of the two ratio windings in conjunction with the finite parallel output admittance of the current sources and the internal, mostly capacitive, admittances of the windings cause ratio errors as illustrated in Figure 7.48. This is the dual of the situation in a voltage ratio transformer.

7.3.11 High-frequency construction

There are two key factors that need to be considered for high-frequency isolation, injection, detection, voltage or current ratio transformer construction, (i) the magnetic cores need to have high relative permeability and low losses over the desired frequency range, and (ii) inter-winding shunt capacitance needs to be minimised. These factors require the minimum number of turns needed to achieve a satisfactorily high input impedance of a transformer whilst enabling the self-resonant frequency to be as high as possible. Nanocrystalline high-frequency magnetic cores have a sufficiently high permeability for 1- and 10-MHz devices

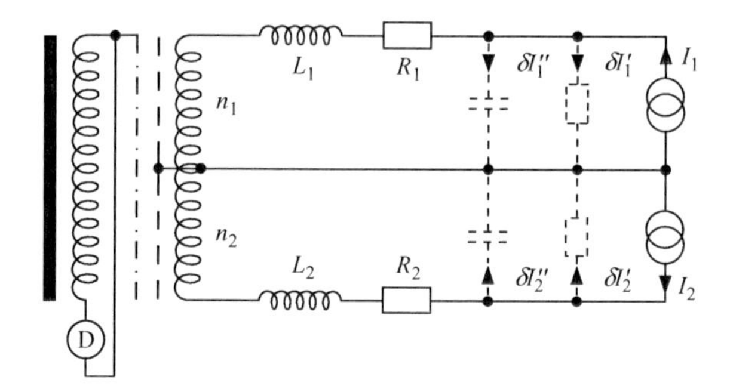

Figure 7.48 Winding impedances causing a ratio error in a current ratio transformer

(such as IVDs and transformers). Shunt capacitance in these devices is minimised by spacing the windings as far as possible from each other and from the toroidal core. For two-staged devices (such as two-stage injection/detection transformers, ratio transformers and IVDs), the same principles are applied as for low-frequency construction. Only electrical screens are necessary at high frequencies because generally magnetic fields can be confined by eddy currents induced in the shielding.

7.4 Calibration of transformers

Most of the types of transformers described in section 7.3 are exceedingly accurate ratio devices. The remaining small error in their ratios can be measured to obtain still higher accuracy, and this measurement constitutes a calibration of the transformer. The dependence of the ratio error on influences such as the voltage applied to the transformer, its temperature and previous magnetic history can also be measured.

7.4.1 Calibrating an IVD in terms of a fixed-ratio transformer

Figure 7.49 shows a simple circuit for finding the settings x of an IVD, which correspond to the 1/11, 2/11, ... 11/11 output ratios obtainable from the 11 sections of a 10:−1 voltage ratio transformer. These settings, which are nominally 0.0909..., 0.1818...., 0.2727...., etc., have the useful property of selecting every switch position of every decade of an IVD, thus revealing any wiring errors. The IVD is calibrated with a short in its zero tap, which is its usual defined condition, but the ratio transformer has nearly zero voltage at its −1 tap and therefore should be calibrated for this purpose, by one of the methods below, in this condition. Its other defining condition, of zero current through its output ports, is fulfilled to a sufficient approximation by the correct two-staged connection to the source and IVD shown. The ratio transformer is shown as having an in-built isolated primary winding, but a separate isolating transformer can be used instead.

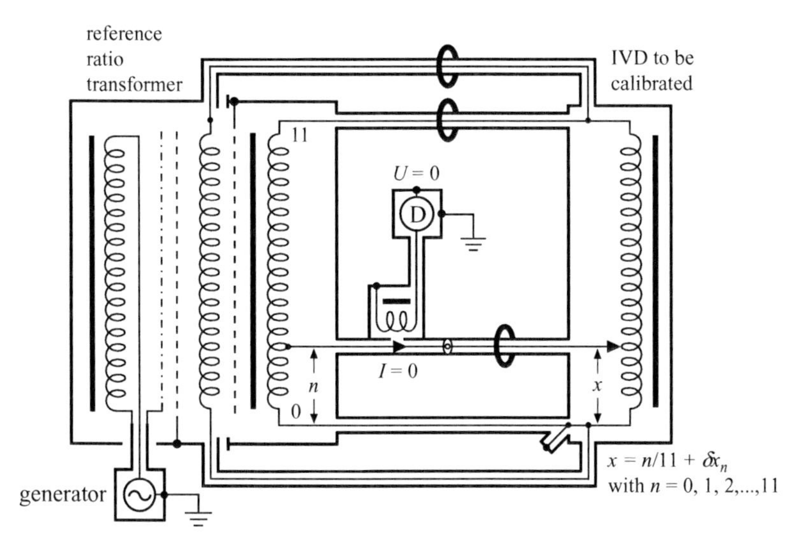

Figure 7.49 Calibrating an IVD with the output voltages of a fixed-ratio transformer

7.4.2 Calibrating voltage ratio transformers using a calibration transformer with a single output voltage

The most common method of calibration requires that other taps than the taps needed to establish the desired ratio (e.g. in the case of a $+10{:}-1$ ratio, the -1 tap, the 0 tap and the $+10$ tap). We will assume that all 12 taps $-1, 0, +1, +2, \ldots, +10$ are available. The principle is illustrated in Figure 7.50.

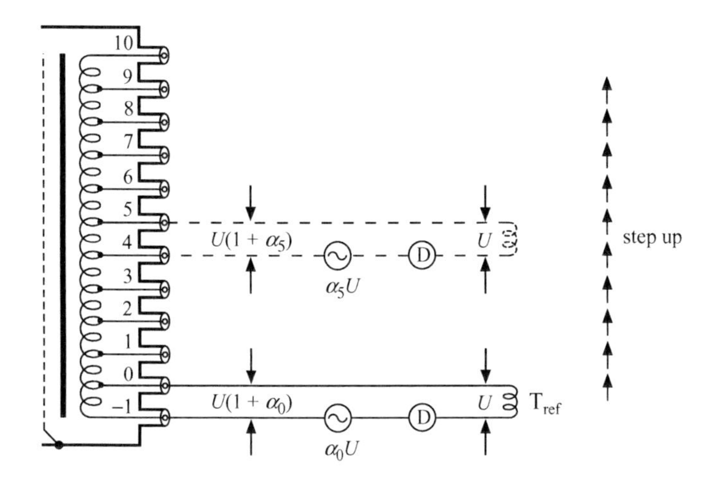

Figure 7.50 The principle of calibrating a transformer

The output voltage of the secondary winding of the calibrating transformer T_{ref} is adjusted, with the small auxiliary voltage source α_0 set to zero, until it matches that between the two adjacent taps 0 and -1 of the transformer to be calibrated. Then no current flows in the loop, as is shown by a null indication of the detector D.

Whilst keeping the output voltages of both transformers fixed, connections are made to successively higher pairs of output taps of the transformer to be calibrated and the detector again nulled by adjusting the auxiliary voltage source to the respective values $(\alpha_1, \alpha_2, \ldots, \alpha_{10}) \times U$. In general, α is a small complex quantity. The loop drawn in broken lines shows the connection made to determine α_5.

Thus, the voltages between successive taps of the transformer to be calibrated are $(1+\alpha_0)U, (1+\alpha_1)U, (1+\alpha_2)U, \ldots, (1+\alpha_{10})U$, and the voltage between the 0 and $+10U$ taps is the sum of these voltages.

$$U\left(10+\sum_{i=1}^{10}\alpha_i\right) = 10U\left(1+0.1\sum_{i=1}^{10}\alpha_i\right)$$

The $+10{:}-1$ voltage ratio of the transformer is

$$10U\left(1+0.1\sum_{i=1}^{10}\alpha_i\right) :-U$$

which departs from the nominal $+10{:}-1$ ratio by the fraction

$$0.1\sum_{i=1}^{10}\alpha_i \tag{7.5}$$

In the single-conductor practical circuit shown in Figure 7.51, the output voltage of the calibration transformer is assumed to be nominally equal to the inter-tap voltage of the transformer to be calibrated. The adjustable series voltage αU is supplied by an injection transformer, phase-shift network and dual output IVD as described in section 7.3.4. We have supposed for simplicity that the detector is an isolated, battery-powered type.

This simple 'bootstrap' approach works well enough to moderate accuracies at low frequencies but suffers from four serious drawbacks:

(i) If the overall potential of the surroundings is that of the zero tap, capacitive currents will flow from both transformers through all parts of the calibration loop to the surroundings, and some of these currents will flow through the detector and give incorrect balances.

(ii) As drawn, the circuit is that of a single-conductor, open-loop construction instead of a conductor-pair coaxial design. Consequently there will be many mutual inductive interactions between the loop and other parts of the circuit, again causing wrong balances.

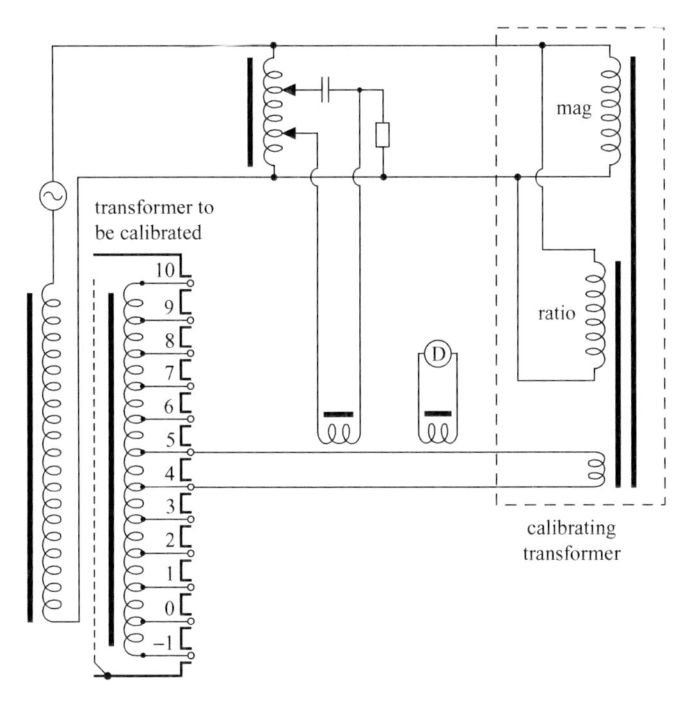

Figure 7.51 A schematic diagram of a practical circuit for calibrating a voltage ratio transformer

(iii) The use of an isolated detector may place a limit on the achievable sensitivity if its characteristics are not suited to the low-impedance circuit of the loop.

(iv) The relative constancy of the output voltages of the two transformers depends on the constancy of both of their primary-to-secondary voltage ratios. These cannot be expected to be as stable as the voltage ratio of, for example, two secondary windings of one transformer because of the inevitable lack of spatial coincidence of the primary and secondary windings owing to the necessary interposing of shields. The amount of stray flux in the intervening space, which therefore threads the secondary winding but not the primary winding, stays constant, but the flux in the core depends on the magnetic state of the core. This magnetic state is easily altered by temperature changes, mechanical shocks, changes in the source voltage for the whole network and the passage of time. If one of the interposed shields is a magnetic shield as well as an electric one, this situation is exacerbated. The effect may well limit the accuracy to 1 in 10^7.

The first three drawbacks can be overcome by appropriate arrangements of the circuit, albeit with some complications. Figure 7.52 is coaxial apart from the two tri-axial leads from the calibrating transformer to the two adjacent taps of the transformer to be calibrated. These leads should be routed closely together or

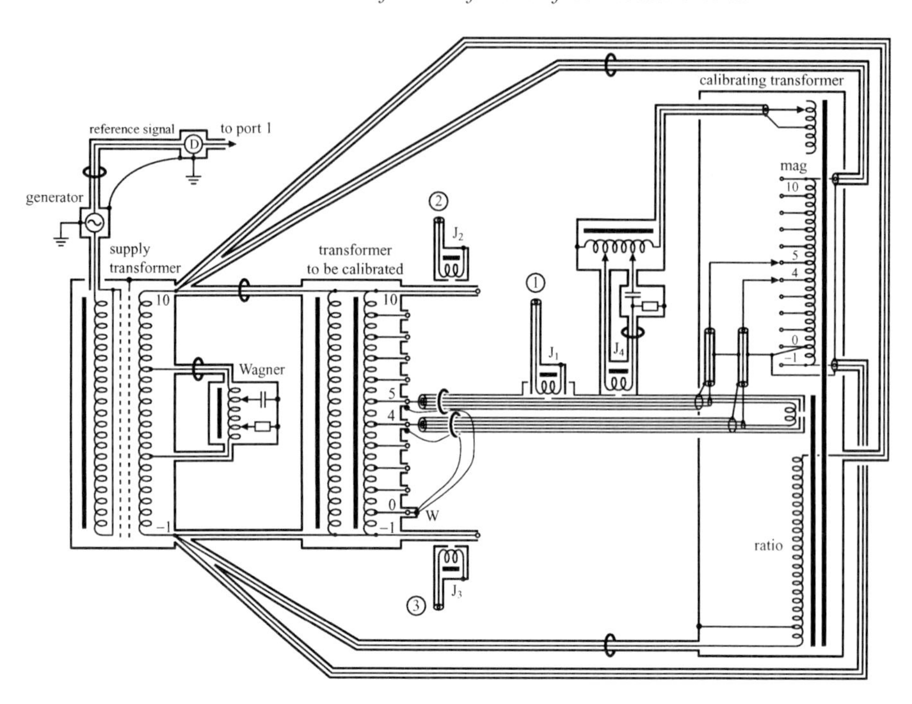

Figure 7.52 A coaxial screened circuit for calibrating a voltage ratio transformer

twisted to keep their loop area small. The intermediate shields of these leads are driven from taps on the magnetising winding of the calibrating transformer so that their potentials are equal to those of the inner conductors. The capacitive loads on these inner conductors are thereby reduced to an acceptably small value. The single conductors between the shorted tap W and the outer conductors of the output ports of the transformer being calibrated sense any small voltage between these outer conductors. These small voltages need to be eliminated from the measurement because the defined voltages of the transformer being calibrated are those found between the inner and its individual outer conductor of each port. These small voltages are transformed by the single conductors threaded through the toroidal cores into the calibrating leads so that they are properly accounted for.

Thompson [3] overcame the fourth drawback by eliminating the calibrating transformer and replacing it with a separate single winding section of the transformer to be calibrated, which is identical to the other ratio sections but not connected to them. The capacitances between this section and the other sections are accounted for in a particularly elegant way.

7.4.3 Calibration with a 1:−1 ratio transformer

Voltage ratio calibrations with uncertainties of the order of 1 in 10^9 are fairly readily achievable by the straddling method about to be described, but realising the full accuracy requires careful attention to the topology of the circuit and its electric screens.

The principle of the method is to replace the somewhat uncertain primary-to-secondary ratio of the calibrating transformer by the stable 1:−1 ratio of a bifilar secondary winding (see section 7.2.3). The calibration network now has two loops and two detectors, as shown in Figure 7.53, which shows only the secondary windings of the transformers. T_A is the transformer being calibrated; $n - m$, n and $n + m$ are three successive taps, which are nominally equally separated in voltage. T_C is the 1:−1 calibration transformer. The complex quantities $\delta_n U_n$, $\delta_C U_C$ and $a_n U$ represent small in-phase and quadrature voltages. The error $\delta_C U_C$ in the 1:−1 ratio can readily be eliminated by reversing the 1:−1 connections.

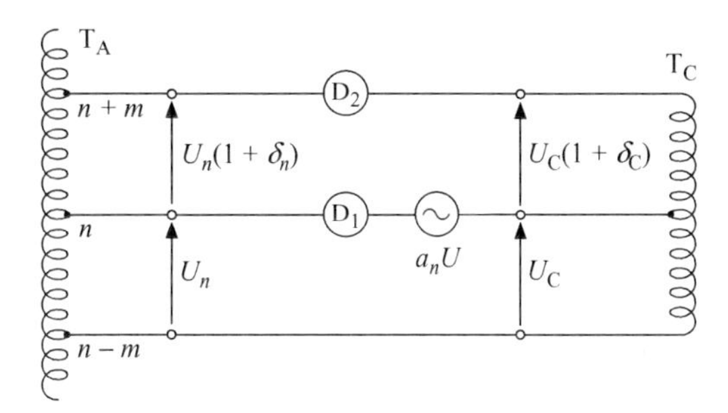

Figure 7.53 A 'straddling' method of calibrating a voltage ratio transformer

By altering its primary voltage, the total voltage $2U_C + \delta_C U_C$ developed by T_C is adjusted to be exactly equal to the voltage $U_n(2 + \delta_n)$ between the n and $n \pm m$ taps of T_A. Then the detector D_2 indicates zero. At the same time, the voltage difference between the middle taps of the two transformers is balanced and measured by adjusting the small in-phase and quadrature generator $a_n U$ to null the detector D_1.

These balances are achieved simultaneously by a quickly converging iteration. When observing D_2, the comparatively large impedance of D_1, which is arranged to have a greater impedance than the whole outer loop, ensures that D_2 responds primarily to the desired condition of equal total loop voltage. D_2 is shorted whilst bringing D_1 to a null.

It is important to realise that the impedance of the loops to the source, $a_n U$, is only the small leakage inductance and series resistance of the windings and loops. Hence, large currents will flow as a result of a small lack of balance between the voltages; this effect gives the method great sensitivity.

First, assume that $m = 1$, i.e. the taps are adjacent.

When both detectors are nulled,

$$U_n(2 + \delta_n) = U_C(2 + \delta_C)$$

This is the auxiliary balance condition that ensures that neither transformer is delivering current.

Also,

$$\frac{U_n(1 + \delta_n + a_n)}{U_n(1 - a_n)} = \frac{U_C(1 + \delta_C)}{U_C} \tag{7.6}$$

for the main balance condition of no current in D_1. In practice, $a_n U_n$ can be replaced by $a_n U$ to sufficient accuracy.

An equation similar to (7.6) is obtained by reversing the connections between T_A and T_C at the defining output connectors of T_A and rebalancing:

$$\frac{U_n(1+\delta_n+a'_n)}{U_n(1-a'_n)} = \frac{U_C}{U_C(1+\delta'_C)} \tag{7.7}$$

From (7.6) and (7.7), neglecting products of small quantities,

$$\delta_C + \delta'_C = 2(\alpha'_n - \alpha_n)$$

and

$$\delta_n + \alpha_n + \alpha'_n = \frac{\delta_C - \delta'_C}{2}$$

If the calibration transformer ratio is independent of the connections to the circuit, $\delta_C = \delta'_C$

and then

$$\delta_n = -a_n(\text{sum}) \tag{7.8}$$

where

$$a_n(\text{sum}) = a_n + a'_n$$

The recurrence relation ((7.8)) obtained by repeated observations a_n as T_C is stepped along the taps of T_A enables any ratio between the inter-tap voltages to be found. In particular, the departure of the transformer from a nominal 10:−1 ratio is

$$0.1\sum_{i=0}^{10}(10-i)a_n(\text{sum}) \tag{7.9}$$

If only the 10:−1 ratio is needed, then one can employ a scheme involving four, instead of ten, measurements where $m \neq 1$, with a consequent reduction of error. For example, one could use the three taps of T_C to connect successively to the −1, 0, +1 taps, then the −1, +1, +3 taps, then +1, +3, +5 taps and finally to the 0, +5, +10 taps, with the observed injected voltages $a_1(\text{sum})U_C$, $a_5(\text{sum})U_C$, $a_{10}(\text{sum})U_C$, respectively.

Figure 7.54 illustrates this scheme. The departure from a true 10:−1 ratio is

$$0.1[6a_1(\text{sum}) + 4a_3(\text{sum}) + 2a_5(\text{sum}) + a_{10}(\text{sum})] \tag{7.10}$$

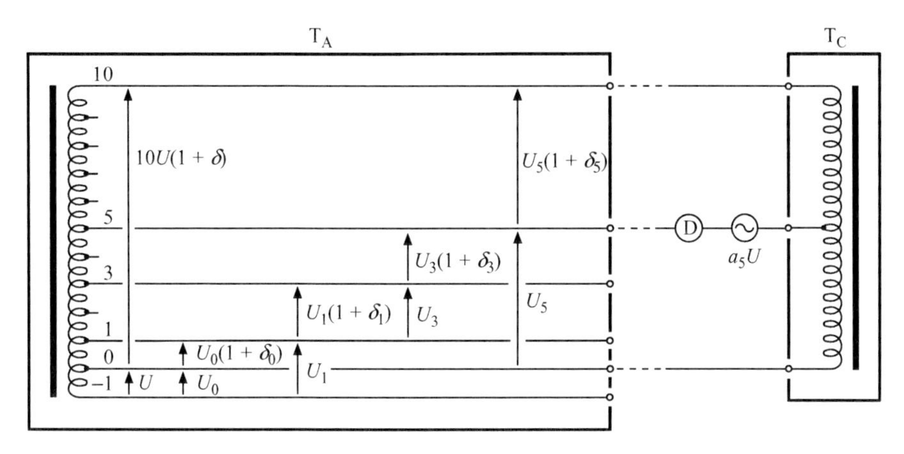

Figure 7.54 A scheme for calibrating a 10:−1 voltage ratio

Many other measurement schemes are possible, and a good way to look for systematic errors in the measurement is to measure the same total ratio by two schemes.

We now examine the various precautions necessary to realise this method accurately.

The property required of the calibration transformer T$_C$ is that the nominal 1:−1 ratio of its secondary windings shall be constant and independent of the overall potential of the ratio winding with respect to the rest of the circuit. Its electric screening and the screening associated with the detector D_1 are critical in meeting this requirement. The ratio must also be immune to the effect of any other external perturbations, for example, vibrations or external electromagnetic fields, especially at the calibration frequency. T$_C$ is not required to have an especially low output impedance. A possible design is shown in Figure 7.55.

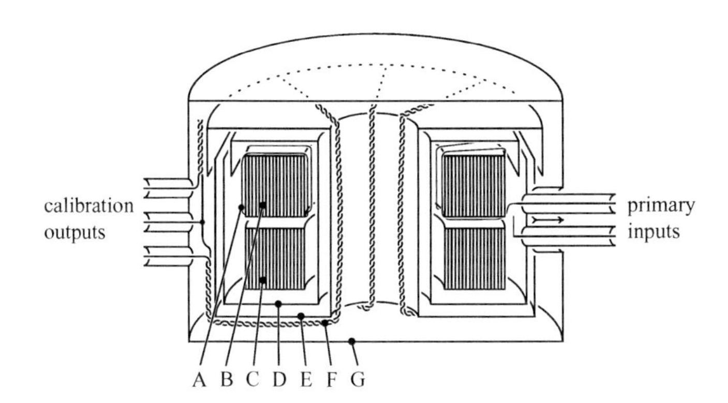

Figure 7.55 A section through a 1:−1 calibrating transformer

B is a high-permeability toroidal core wound with the primary winding A. The cross section of B and the number of turns of A are chosen to provide a sufficiently high input voltage rating and input impedance. C is a second core placed adjacent to A, and both are enclosed in a toroidal magnetic and electric primary shield D. The secondary electric shield E surrounds D and the twisted bifilar 1:−1 ratio winding F is wound on it, using sufficient turns to generate the maximum calibrating voltage needed. The ends are connected to form the three taps, which are then led out in separate coaxial screens through the magnetic and electric screen G, which surrounds the entire structure. Two adjacent tubular peripheral limbs (shown on the right) from this shield enclose the ends of the primary winding separately.

By always maintaining the screens E and G at the potential of the centre tap of the bifilar winding, the 1:−1 ratio of the bifilar windings is unaffected by being raised to various elevated potentials with respect to anything outside the space enclosed by E and G. That is, raising the potential of E and G and the enclosed bifilar winding with respect to the primary winding B and its shield D, or to external circuitry, will not affect the 1:−1 ratio.

7.4.4 The bridge circuit and details of the shielding

The bridge circuit and shielding are shown in Figure 7.56, configured for measuring the 0, +5, +10 ratio. Detectors (not shown) denoted by D_n are considered to be connected to the ports labelled *n*, although in practice, only a single detector plugged successively into these ports is necessary.

By looking at this figure, it can be seen that the secondary shield of T_C is extended from T_C to enclose totally the conductors from the taps of T_C up to the points A_1, A_2 and A_3 where the output terminals of the transformer to be calibrated, T_A, are defined as those between inner and outer. A small overlap of this shield with the outer conductors of the taps of T_A ensures complete electric shielding. This shield is elevated to the potential of the centre tap of T_C by being connected to the output of IVD I_4, a very small quadrature adjustment being provided by IVD I'_4, the phase change circuit and injection transformer J_4. The potential is correct when detector D_3 indicates zero, showing there is no current in the conductor connecting the shield to the centre tap of T_C.

As suggested in the diagram, the main detection transformer J_1 surrounds the inner conductor from the centre tap of T_C and its shield to A_2. Since this shield is (in general) at an elevated potential, capacitive currents will flow from it to the shield of J_1 and those from the left of the split in the screen of J_1 will thread J_1 and register on D_1. By shielding the left portion with a second screen connected by a wire, which passes around the outside of J_1, the capacitive currents going from the left and right sides to the right and left sides of the split screen of J_1 can be equalised until D_1 registers a null. Lack of complete equality leads to a finite detector indication. The extent of this effect can be measured, in order to make a correction for it, by temporarily connecting at the middle tap of the three taps of A_3 a multi-port connector in which the inners are brought together effectively at a point, as are the outer conductors. See Figure 7.57 for a suitable construction in

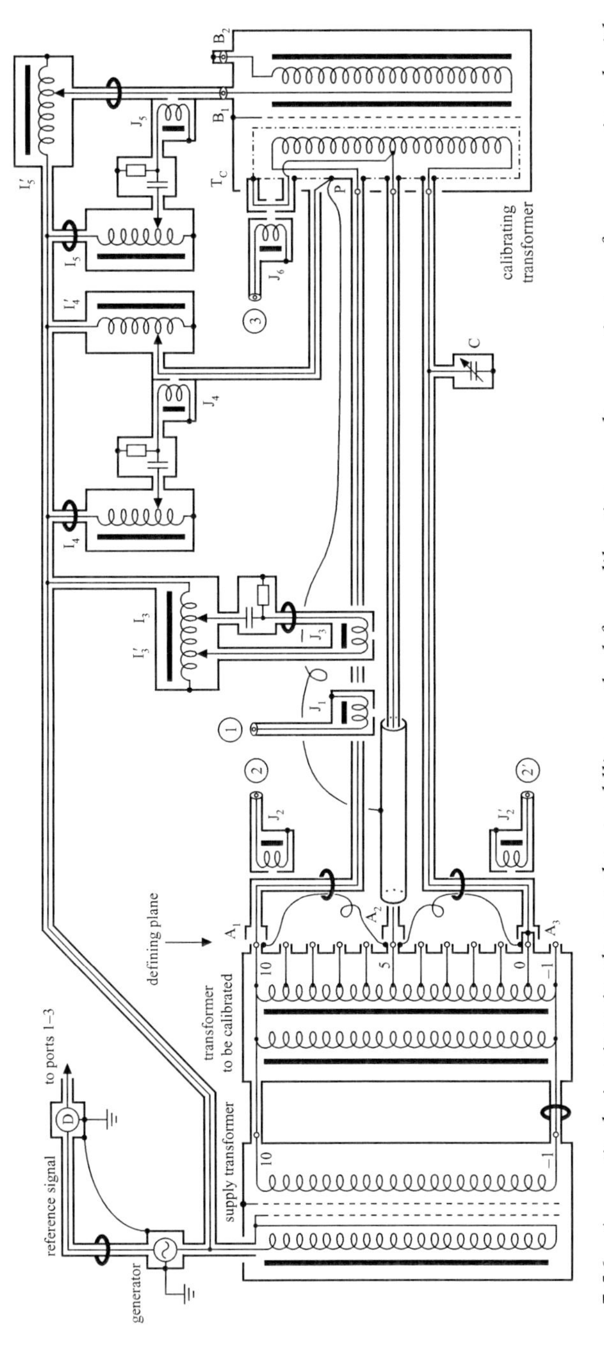

Figure 7.56 A practical circuit to implement the straddling method for calibrating a voltage ratio transformer equipped with defining transformers J_2 and J_2'

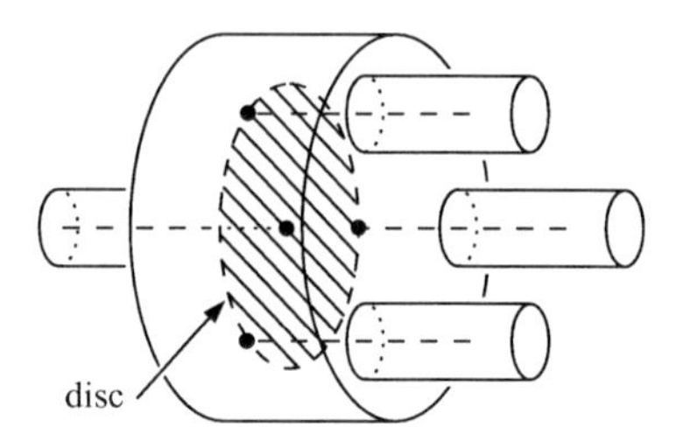

Figure 7.57 A multi-port connector where inner conductors are brought together at a point

which the inners are soldered symmetrically to a disc. This connector provides two further ports at the potential of A_2, and the other leads of the calibration transformer are removed from A_1 and A_3 and plugged into these. Also, the supply to the primary of T_C is reduced to zero without altering the network configuration. Any remaining small detector indication can be nulled with I_3 and I'_3, and the readings thus obtained are corrections for this zero error to the actual readings obtained when the bridge is reconnected as drawn.

Additionally, all three conductors to the right of J_1 are enclosed in a second all-embracing screen (not shown in Figure 7.56) also connected to P so that capacitive currents, which would have flowed to the individual screens, are intercepted and prevented from inducing error voltages in the inner conductors via the mutual inductance between inner and outer conductors.

The conductors from T_C are run closely together (to avoid emfs induced by external magnetic fields) until they separate at point A_2 to go to the taps of T_A at A_1, A_2 and A_3. An additional wire between the outers of the A_1 and A_2 sockets runs alongside the cable, and both wire and cable thread the high-permeability core E_1 once. There is the same arrangement between A_3 and A_2. By this artifice, any potential existing between the outer conductors at A_1 and A_3 is sensed and injected into the inner conductors by the cores so that the circuit correctly responds to the difference in potential between inner and outer of each of the various taps of T_A.

The defining conditions of zero current at the output connectors of T_A are fulfilled at the gap in the outer screens even though the total current detection transformers J_2 and J'_2 are located a short distance away from the connectors because they indicate zero total current in inner and screen. The capacitive current between inner and screen to the left of the defining plane of the transformer up to the gap in the outer must then have been supplied from the right along the inner conductor and does not constitute a load on T_A.

7.4.5 The balancing procedure

With reference to Figure 7.56, D_3 is first nulled by adjusting I_4 and I'_4. Then D_2 and D'_2 are nulled by adjusting I_5 and I'_5 to alter the overall inter-tap voltage of T_C to match that of T_A. Fine adjustment using continuously variable capacitors can help if operation of I_5 and I'_5 gives undesirable surges. Lack of a simultaneous null of D_2

and D_2' is due to unequal capacitances of the cables from T_C to the defining transformers at A_1 and A_3, and this can be eliminated by adjusting a small capacitance trimmer C when measuring the +1, 0, −1 tap combination. C may be required to be in the other outer straddling lead. D_2 and D_2' are then replaced by a temporary short for rapid convergence, and D_1 is nulled by adjusting I_3 and I_3'. Successive iterated balances of D_2, D_2' and D_1 rapidly lead to simultaneous nulls being registered.

As we indicated when outlining the principle of the method, it is necessary to invert the 1:−1 ratio and remeasure, to eliminate the small but finite departure of T_C from a perfect 1:−1 ratio. This inversion is accomplished by interchanging leads at A_1 and A_3 and reversing the current in the primary of T_C by interchanging the input and the shorting plug at B_1 and B_2. The connection of the primary's screen is unaltered by this reversal.

A check on the correctness of the method can be undertaken as follows. When A_2 is connected to the zero potential tap of T_A, it is also possible to connect two nominally equal capacitances by means of T-connectors at A_1 and A_3 to form components of a 1:1 bridge. The actual ratio of the capacitors can be eliminated by reversing the connections of this capacitance bridge at A_1 and A_3 and rebalancing. The mean of these observations should equal the departure from nominal of the ratio of the 1:−1 calibrating transformer. Then if the same value of δ_C is obtained from the transformer calibration network for other taps when A_2 is not at zero potential, we can be confident that the calibration network is giving accurate results.

The circuit of a suitable capacitance bridge is shown in Figure 7.58. The relative values of the capacitors must be stable over the time of observation to the accuracy required, which may be 1 in 10^9 or better. A pair of 1-nF gas-dielectric capacitors, which have nearly equal temperature coefficients and which are in the same temperature-controlled enclosure, would be suitable.

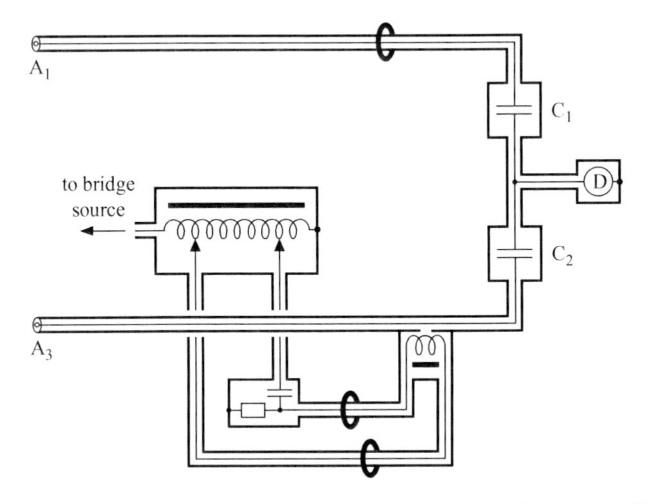

Figure 7.58 A capacitance bridge to check the accuracy of the straddling method

7.4.6 Calibrating voltage transformers by permuting capacitors in a bridge

Another method of deriving ratios of voltage transformers has been described by Cutkosky and Shields [4]. This method can measure a voltage ratio nominally equal to the ratio of two integers *m*/*n*. Figure 7.59 illustrates the principle. If all the admittances are nominally equal, the detector is near a null and can be nulled exactly by the small emf U_1 injected in series with, say, U_n. That is,

$$(U_n + U_1)(Y_1 + Y_2 + \cdots + Y_n) = U_m(Y_{n+1} + Y_{n+2} + \cdots + Y_{n+m}) \qquad (7.11)$$

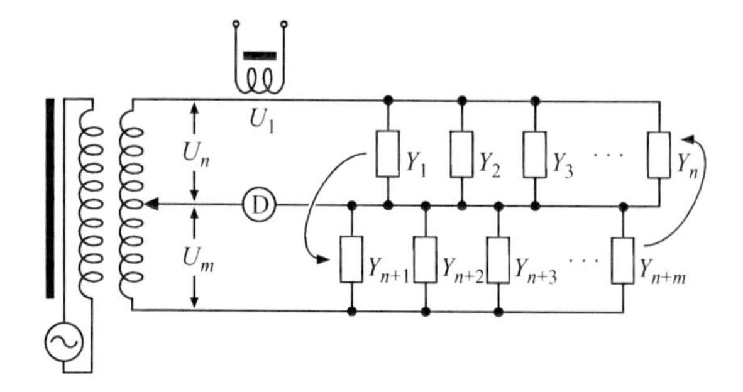

Figure 7.59 Principle of the 'permuting capacitors' method for calibrating a voltage ratio transformer

In Figure 7.59, by cyclic permutation, by transferring an admittance at each step to the other side of the ratio arms as indicated by the arrows, a series of $n + m$ similar relationships to (7.11) can be arrived at

$$(U_n + U_2)(Y_2 + Y_3 + \cdots + Y_{n+1}) = U_m(Y_{n+2} + Y_{n+3} + \cdots + Y_1)$$

etc.

On adding the $m + n$ relationships we have, since each individual admittance has voltage U_n across it n times and voltage U_m across it m times,

$$U_n n \sum_{i=1}^{n+m} Y_i + nY_{\text{mean}} \sum_{i=1}^{n+m} U_i = U_m m \sum_{i=1}^{n+m} Y_i$$

where we have assumed that $\Sigma \Delta Y_j U_j$ is a negligibly small sum of products. ΔY_j is the difference $Y_j - Y_{\text{mean}}$. Since

$$\begin{aligned} (n+m)Y_{\text{mean}} &= \sum_{i=1}^{n+m} Y_i \quad \text{and} \quad (n+m)U_{\text{mean}} = \sum_{i=1}^{n+m} U_i \\ \frac{U_m}{U_n} &= \left(1 + \frac{U_{\text{mean}}}{U_n}\right)\frac{n}{m} \end{aligned} \qquad (7.12)$$

The above analysis has assumed that U_n and U_m are not greatly perturbed by the loading effects of the Y_i and their associated connecting-cable shunt admittances. The complex arithmetic needed to account for this loading can be avoided by making the shunt admittances equal, for example, by T-connected shimming capacitors. There still remains the effect of the loading on the transformer ratio, and this has to be accounted for either by extrapolation to zero load (see section 5.3.9) or by using separate current and potential leads. A suitable circuit is shown in Figure 7.60 for the usual case of determining a 10:−1 ratio. Adjusting the second-core excitation and the adjustable source connected to the current winding nulls the detectors D_1 and D_2 associated with the defining transformers. Then no current is drawn from the potential winding. From the centres of these defining transformers through the symmetrical fan-out disc to the high port of the capacitors, the shunt admittances must be made equal, as above.

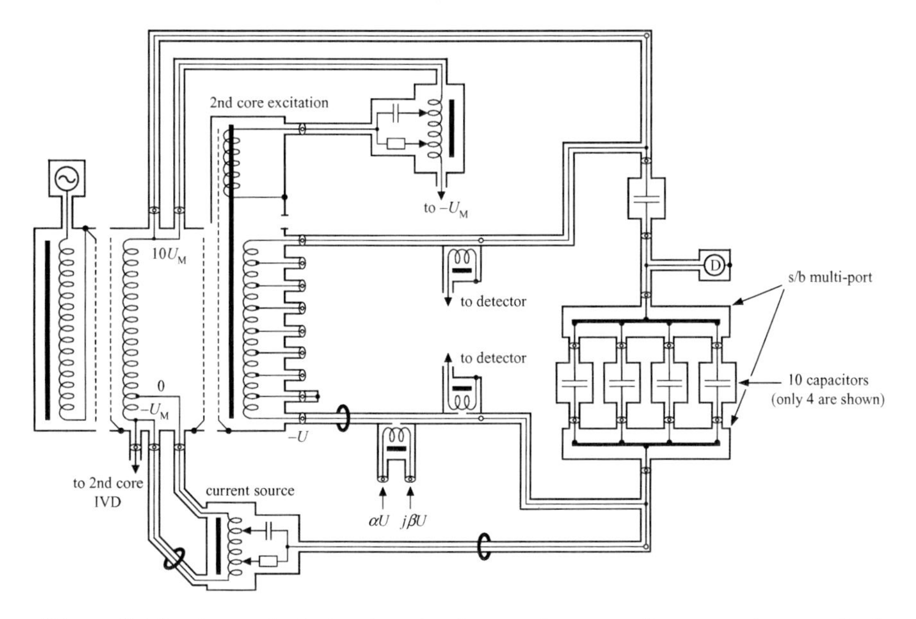

Figure 7.60 A practical circuit to implement the permuting capacitors method

The permuted admittances are usually capacitors as these components have least change of value with applied voltage. Resistors are not so suitable because self-heating alters their value with applied voltage more than capacitors, and they generate noise. Self-inductors are not precise standards for the reasons given in section 6.4.13.

The individual capacitors do not need to be treated as four-terminal-pair devices with respect to the detector connections unless the differences between the values of the capacitors, which this method also yields, are required because the detector cables and connections remain unchanged throughout permutation of the capacitors.

The voltage coefficients of the capacitors are needed to correct for the effect of changes in voltage they experience during permutation; these coefficients have to be separately measured by one of the methods described in section 6.4.8.

The permuting capacitors technique is particularly appropriate for determining the ratio of a voltage transformer at higher frequencies than a few kilohertz, and a suitable shielded housing for the capacitors has been described by Awan et al. [5] and Sedlacek et al. [6].

The arrangement described by Awan et al. [5] minimises the inductances in series with the capacitors. A cross section through the device is schematically illustrated in Figure 7.61. The 11 capacitors C_1–C_{11} are Negative-Positive-Zero (NPO) ceramic chip capacitors of nominally 50 pF value and are each two 100-pF capacitors in series selected from a large batch by combining small and large values in pairs so that the series capacitances are closely equal. The small size (3 mm $\times$ 1 mm $\times$ 0.5 mm), remarkable stability, low phase angle (7×10^{-5} rad), small temperature coefficient (0 ± 30 ppm/K) and negligible temperature hysteresis and voltage coefficient of these capacitors make them especially suitable for this application. For the higher frequencies, 1–10 MHz, 5-pF capacitors are more suitable to minimise the effect of residual inductances in the device.

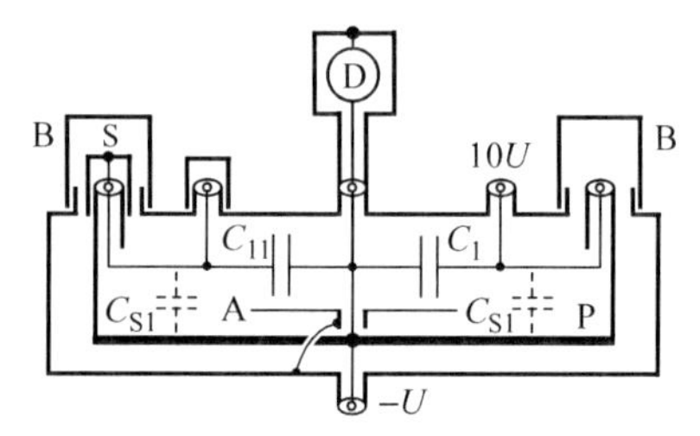

Figure 7.61 An arrangement of permuting capacitors suitable for frequencies up to 10 MHz [5]

C_1–C_{11} are connected to a central point (to which the detector D is connected) in a circular configuration similar to the spokes on a wheel to ensure the geometrical equivalence of each capacitor. The 'spokes' end on two concentric rings of SMB connectors. Ten of the capacitors can be connected in parallel to a circular copper plate P by inserting shorting plugs S. The centre of P is connected to a socket to which a bridge voltage of $-U$ can be applied. The remaining capacitor not so connected (C_1 in the figure) has the voltage $10U$ applied to it, as shown. The detector sides of C_1 to C_{11} are shielded from P by a plate A, which has vertical fins (not shown) to also shield the detector connections of each capacitor from one another. The equality of stray capacitances C_{S1} to C_{S11} ensured by their symmetry means that their only effect is to alter slightly and equally the effective values of C_1 to C_{11}. The exposed outer ring of connectors and the shorting plugs are shielded from the environment by caps B.

The small dimensions of the device and the proximity of go and return currents within it ensure that inductive effects should not affect the ratio by more than a few parts per million at frequencies of a few megahertz.

7.4.7 Calibration of current transformers

We do not discuss the calibration of these devices in general. A specific need will be addressed in section 9.1.6. We note, however, that as a consequence of reciprocity, a calibration method that is suitable for a voltage ratio transformer can also be applied to its dual device, a current ratio transformer, at least in principle, by interchanging the roles of detectors and sources.

7.4.8 Assessing the effectiveness of current equalisers

This aspect of impedance comparison networks has caused more difficulty than most others. It is important to realise that the equalisers in a network constitute a component of that network whose correct functioning together is as important for accurate results as any other component such as ratio transformers. Their function and the method of measuring their effectiveness have already been described in sections 3.2.1 and 3.2.2.

References

1. Cutkosky R.D. 'Active and passive direct reading ratio sets for the comparison of audio-frequency admittances'. *J. Res. NBS* 1964;**68c**:227–36
2. Kibble B.P., Rayner G.H. *Coaxial AC Bridges*. Bristol: Adam Hilger Ltd.; 1984. (Presently available from NPL, Teddington, TW11 0LW, U.K.)
3. Thompson A.M. 'Precise calibration of ratio transformers'. *IEEE Trans. Instrum. Meas.* 1983;**IM-32**:47–50
4. Cutkosky R.D., Shields J.Q. 'The precision measurement of transformer ratios'. *IRE Trans. Instrum.* 1960;**I-9**:243–50
5. Awan S.A., Kibble B.P., Robinson I.A. 'Calibration of IVD's at frequencies up to 1MHz by permuting capacitors'. *IEE Proc. Sci. Meas. Technol.* 2000; **147**(4):193–95
6. Sedláček R., Kučera J., Boháček J. 'A new design of permuting capacitors device for calibration of 10:1 high-frequency inductive voltage dividers'. *CPEM 2006 Digest* 2006;476–77

Chapter 8
General considerations about impedance comparison networks

In this chapter, we attempt to give some guidance on the design and use of impedance-comparing networks ('bridges') in general. Specific examples of various networks are described in chapter 9.

8.1 Designing bridge networks

Fulfilling the defining conditions of the impedances will often result in a network almost designing itself, but sometimes some subtlety is needed to obtain conditions that are provably equivalent to these defining conditions, as in the employment of combining networks.

Elegance, simplicity and a return to first principles usually obtain a much better result than resorting to ad hoc bolt-on solutions to remove the imperfections of the original concept. Then, not only should there be greater satisfaction with the final design and its operation but also attaining the desired accuracy will be more likely.

We find it better to separate the functions of isolation and generation of a precise voltage by having an isolation transformer separate from the ratio transformer, as shown in Figure 9.2 and subsequent examples. It also makes the principles of network design, in particular the Wagner circuit, clearer. See, for example, the use of a separate current supply to ensure that no current is drawn from the two potential ports of the ratio transformers in the networks described in sections 9.1.5, 9.2.3, 9.3.6, 9.3.7, 9.4.3 and 9.4.4. This design principle is a departure from the designs by Kibble and Rayner [1], which are based on a combined isolation/ratio transformer.

8.1.1 Applying coaxial techniques to classical single-conductor bridges

There is a considerable and valuable number of older bridge circuits [2] that are presented as single-conductor networks, often without much regard for interactions between their different parts, which can readily be converted to a coaxial network with its concomitant advantages. Single-conductor networks can conceptually be converted by taking one reference point, usually termed 'earth' or 'ground', and topologically stretching and extruding it as tubes over all the single conductors, components, detectors and sources. When current equalisers are added to each separately identifiable mesh, the result is a coaxial network.

Converting the familiar four-arm bridge to coaxial form, however, leads to the existence of large, but definite, shunt impedances from either end of every component to the coaxial outer conductor. The means of eliminating the effect of these shunt impedances from the bridge balance condition provides an example of the auxiliary balances needed in high-accuracy coaxial networks. These auxiliary balances have been described in section 5.1.

Figure 5.1 shows a single-conductor four-arm bridge, which is redrawn in Figure 5.2 as a completely screened network. Variable two-terminal impedances that add to the shunt impedances already present have been added to the four junction nodes between the components. To appreciate the purpose of these components, it is helpful to redraw the network of the outer conductors as a simplified line, as shown in Figure 5.3.

8.1.2 Placement of current equalisers

It might be supposed that provided each and every mesh of a network has one and only one current equaliser, there would be considerable freedom in choosing exactly which cables to put them in. It is usually found that where there is a choice, it is better to place equalisers in cables whose inner conductors carry the largest currents.

But there are some further constraints – for example, in impedance-measuring networks, placing a greater length of cable needed to make an equaliser in defining leads should be avoided as far as possible because of the greater and therefore more uncertain lead corrections needed (see section 5.4). Unfortunately, it will usually be found that one equaliser must be placed in a defining cable, and the additional length of cable and consequent cable correction accounted for.

Placement is assisted if the network diagram is drawn as single lines representing the cables that join at the nodes, which are usually the bridge components. If there is a T-connection into a cable, it can conceptually be slid along the cable until it coincides with a component. This does not alter the topology of the network.

The final placement can be checked by successively reducing the network by one mesh, containing one equaliser, at a time, until nothing remains, and double-checked by recourse to the theorem stated in section 3.2.2.

8.1.3 Wagner circuit (and when it is applicable)

The bridge described in section 9.1.4 is a four-arm bridge, which has a ratio transformer forming two of its arms. A required defining condition for reproducing the ratio of the transformer accurately is that there should be no current flowing through the short on the tap, as indicated in Figure 9.4. This condition is achieved by adjusting R_W and C_W, each of which might need to be on either side of the supply transformer. A property of a properly constructed ratio transformer is that its voltage ratio is hardly affected by shunt impedance loads at the output ports at the upper and lower bridge nodes, so the effect of these shunt impedances on the balance condition is often less than the required accuracy and no auxiliary detector network balance is required.

As a matter of good design, it is better to first eliminate all low and variable impedances to screen, if possible (e.g. inductive voltage dividers having a short to screen in one port) as adjustable auxiliary balance components because the Wagner balance would have to account for their lossy and variable input impedance. Ideally, there should be only one direct connection between the inner and the outer conductors of the coaxial bridge circuitry. Usually, this is a short in the zero port of a ratio transformer, and the Wagner circuit is adjusted so that there is no significant current through it.

8.1.4 Convergence

In designing bridge networks, adjustable sources of current or of voltage are often needed. To obtain satisfactory balance convergence, it is important to distinguish between the uses of these dual sources. In general, quasi-current sources of high internal impedance are required when a current is to be injected in parallel into a network node to augment or cancel an existing current. Quasi-voltage sources of low internal impedance are required when an adjustable voltage is to be injected in series.

A complete analysis of a bridge network to establish its convergence properties analytically is usually very complicated and, when completed, difficult to interpret. Fortunately, for the bridge networks described in this book, some general considerations are usually adequate. These networks usually consist of quasi-voltage sources such as IVDs, which are themselves connected across a voltage source, and voltage injection transformers and quasi-current sources where a comparatively high impedance is connected to the output of a quasi-voltage source with the objective of injecting a current into a network node.

Because quasi-voltage or quasi-current sources have only low or high rather than strictly zero or infinite output impedances, respectively, the various adjustments needed to make auxiliary and main balances in a network will never be entirely independent. Sufficiently rapid convergence can usually be achieved however, so that one, two or three iterations around these balances made in a sensible order will be sufficient to achieve simultaneous balances to the accuracy required. In general, it is better for current sources to have fixed-value impedances (capacitors or resistors) fed from variable voltage sources (IVDs).

The capacitances of a cable joining the C_L ports of two four-terminal-pair impedances are part of a Wagner circuit, and the adjustment of a combining network to eliminate the effect of the series impedance of this cable needs consideration. If the adjustment is made by eliminating the effect of a voltage from an injection transformer, all is well, but if the voltage is produced by inserting an impedance (a resistor or an open circuit) at a point along the cable, the exact balance of the Wagner circuit will depend on the position where the impedance is inserted. Therefore, to avoid an awkward convergence problem, the value of the inserted impedance should be as small as possible, consistent with obtaining the required sensitivity for the combining network adjustment. A value of a few tens of ohms is typical.

Examples

1. A Wagner balance is intended to balance the currents from inner to outer conductors on either side of a network through a designated zero-potential point (a short between inner and outer conductors at some point in a network, usually a tapping on a ratio transformer) (e.g. see Figure 9.2 and its accompanying description – a simple bridge with a Wagner arm). A current through an adjustable high impedance (usually an adjustable capacitance and an adjustable high resistance) between inner and outer conductors having high potentials between them will best perform this adjustment. Because high-valued resistors have a large phase angle and are not therefore good AC components, it is often better to use more moderate fixed values for components in series with the adjustable output of an IVD as in Figure 9.5. The convergence between the main balance and the Wagner balance in most bridges depends on the ratio of the output impedance of the potential ratio windings to the impedance of the Wagner adjusting components. Therefore, it is good practice to make the impedances that the Wagner components have to balance as large as practicable by minimising the capacitances of cables having significant voltage between inner and outer (i.e. minimise their length) and by avoiding having auxiliary IVDs with an output connected to screen.

   The potential windings of the ratio transformer should be designed to have minimal mutual output impedance by adopting one of the winding designs of sections 7.2.2–7.2.5. In particular, because the impedance of the 0:−1 winding is reflected as the square of the turns ratio into the apparent output impedance of the 0:10 winding by transformer action, the resistance of this section should be about 1/100 of that of the 0:10 sections. Since the length of its winding is 1/10 of that of the 0:10 sections, its cross-sectional area should therefore be about ten times that of the conductors forming the 0:10 windings. The ordered rope design in section 7.2.5 accomplishes this.
2. The voltages provided by the outputs of a ratio transformer need a small adjustment to balance a bridge network. A quasi-voltage source such as an injection transformer in series with an output will best perform this task, as described in section 7.3.3 and the accompanying text.

   If, however, both the outputs of a voltage ratio transformer must be altered without affecting their ratio, a current injected into an additional winding on a second core will do this simply and accurately. This current should come from a quasi-current source, as described in Figure 7.60 and the accompanying text.
3. If a 10:−1 voltage ratio transformer is constructed with an in-built isolated primary winding or if it is calibrated with a specified isolating transformer and interconnecting cables, no account need be taken of the defining condition of zero current in the short in its zero tap. Satisfying the defining conditions of zero current flow at the potential ratio taps is sufficient because the effect of non-zero current in the zero tap short is part of the calibrated ratio. But if the ratio transformer is calibrated independently of a separate isolating supply transformer, as is suggested in this book, the condition of zero current in this short should be satisfied by adjustable Wagner components.

The reader should now be able to recognise other examples of convergence considerations in bridge networks described in this book and be able to successfully design networks for other purposes.

8.1.5 Moving a detector to other ports in a bridge network

It is usual to employ only one detector in a coaxial bridge network. This detector is connected in turn to the various ports where a detector needs to be nulled, as, for example, shown in the various bridge diagrams in this book. The extra network mesh introduced by connecting the detector should always contain the current equaliser in the input detector lead so that no change in equalisation of the network occurs. Usually, only the detector sensitivity need to be optimised (see section 2.1.1) for the detector in the main balance position, as the other auxiliary balances are much less critical.

8.1.6 T-connecting shunt impedances for balance adjustment

In an ideal world, bridge network impedances would have extra ports connected directly to their internal defining points for the purpose of adding small shunt admittances to adjust their value slightly. Since existing impedance standards are not in general provided with these extra ports, it is necessary to consider how to connect a shunting admittance so that the shunt admittances and series impedances of its cables have minimum effect and how, for measurements of the highest accuracy, this effect can be calculated and accounted for. We assume, in this section, that the shunting admittances to be added can be defined to sufficient accuracy as two-terminal-pair components (see section 5.3.6) and that since the shunting admittances are usually small compared with the admittance of the shunted component, cable corrections to the shunting admittance itself are negligible.

If a two-terminal-pair component (corrected for the effect of its cables) is shunted, the result is just the parallel combination of the shunted and shunting impedances between the points where the T-connections are made. This is an example of the parallel connection of admittances described in section 5.7.

A four-terminal-pair component is best shunted by a high-impedance two-terminal-pair component by placing the T-connections in the C_H and P_L non-defining cables (Figure 8.1). The current drawn by the shunting connection in the supply (C_H) cable does not affect the definition of the shunted component, and

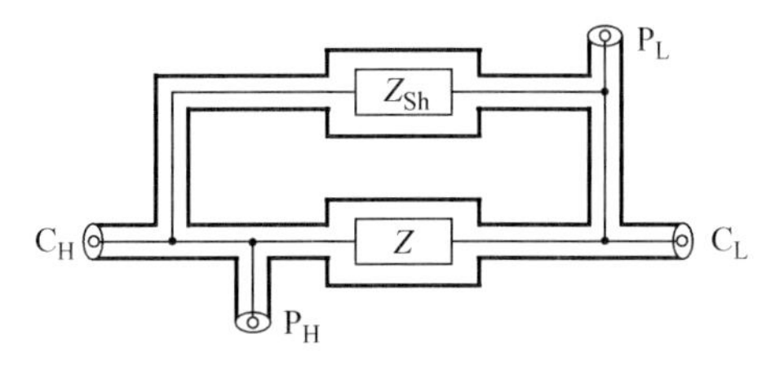

Figure 8.1 Shunting a four-terminal-pair impedance

although correction should be made for the small voltage drop down the P_L lead between the T-connection and the internal defining point of the shunted component, this correction will often be negligible.

If the T-connections are made into defining cables, it is necessary to consider the effect of the lumped shunt load introduced by the shunting impedance and its cables, and the situation is quite complicated.

8.1.7 Role of electronics in bridge design

In section 3.2.1, we have already described the use of amplifiers in feedback circuits to augment the efficiency of current equalisers. Feedback amplifiers have also been used to make some auxiliary balances automatically so that bridge balancing is simplified. In these and other applications of electronics in bridge networks, it is particularly important to consider the isolation of the electronic device and the effect of any extra meshes it might introduce into the network.

8.1.8 Automating bridge networks

It is tempting to consider removing the tedium of repetitive re-balancing of a bridge, by designing some degree of computer-controlled automation into the various bridge balances. This is very difficult, partly because remotely operated switches (relays) having the required degree of isolation between their magnetic energising coil and their contacts are rare and tend to be very expensive, and partly because writing a computer routine having the convergence properties conferred by a human operator is not a simple task. Nevertheless, at least one successful system has been achieved [3]. But, in general, it would seem that it is more efficient to employ a manual bridge network designed to be as ergonomic as possible, perhaps by employing analogue electronics to automate auxiliary balances (see section 5.1.1), and merely automate the recording and mathematical manipulation of the results. The latter approach requires computer-readable outputs of the adjustable components, and their isolation needs to be near perfect. An acceptable solution might be to extend switch shafts via an insulating section into a separate screening enclosure connected to the screen of the computer to operate within it switch wafers appropriately connected to the computer.

8.1.9 Higher-frequency networks

Higher-frequency networks are necessarily smaller and can be simpler than their low-frequency counterparts, principally because measurement uncertainties are greater and the consequent lower attainable accuracy is, nevertheless, sufficient for present needs. Simplifications are as follows: (i) current equalisers will often be unnecessary, (ii) extrapolation techniques can be used, which considerably simplify the measurement networks and (iii) symmetry of the bridge networks can be employed to substantially reduce offset errors.

One key difficulty that emerges at higher frequencies, particularly at frequencies above approximately 0.5 MHz, is that the resistive and capacitive components used in auxiliary balance devices (such as current sources, combining networks and Wagner balances) are no longer purely resistive or reactive but a combination of resistance, capacitance and inductance. Therefore, it is useful to measure the performance of these auxiliary components with a high-frequency LCR meter first, to determine whether they are suitable, before incorporating them in high-frequency bridge networks.

Sources in the megahertz frequency range normally have a 50-Ω output impedance. Sufficiently good impedance matching is needed when connecting such sources to high-frequency networks, but networks having input impedances ranging from few tens to few hundreds of ohms do not usually present a problem. High-frequency detectors, in particular lock-in amplifiers, need extra care. Certain lock-in amplifiers operating near the edge of their frequency range can have reduced performance in terms of gain, dynamic range and signal-to-noise ratio. Unexpected reference channel phase changes can occur because of the length of the connecting cable, leading to longer set-up times and increased bridge balance iterations because the in-phase and quadrature settings of the lock-in amplifier do not match those of the corresponding in-phase and quadrature voltages in the bridge network. However, this does not impact the accuracy of the bridge balance because attainment of zero detected voltage in both phases is independent of the phase of the reference signal.

As in any branch of metrology, measurement reproducibility is the key for debugging higher-frequency networks because then the effects of diagnostic changes deliberately introduced into the network can be unambiguously observed.

8.1.10 Tests of the accuracy of bridges

Given good quality standards, it is relatively easy to assemble and balance a bridge that, if proper attention has been paid to consideration of the signal-to-noise ratio, has a very great resolution. One can discern changes as small as 1 in 10^9, with a similar degree of stability of the balance. In this situation, it is rather easy to be misled into believing that the results obtained also have a similar accuracy. There are a number of general tests that should be applied to any untried network to uncover possible causes of systematic error. For each test, the bridge is supposed to be fully balanced in its normal state with the source emf and detector sensitivity at the highest level at which the bridge network will be operated, unless otherwise stated. We assume that the important injected voltages have been calibrated (see section 7.3.6) and the equalisers are evaluated (see section 3.2.2).

A two-terminal-pair component, which has zero admittance, is constructed as illustrated in Figure 8.2, where a break in the inner conductor of a coaxial cable has a complete screen across the break to intercept electric fields.

A two-terminal-pair component, which has zero impedance, requires that no voltage be generated across one port no matter how much current flows in or out of

Test	Desired result of test	Possible reasons why desired result is not obtained
(1) Main bridge source removed, replaced by a short.	Main detector reads zero (Johnson noise excepted).	Pick-up of electric or magnetic fields or a net current exists because of inadequate isolation by transformer screens or acoustic coupling.
(2) Detector gain decreased from its highest setting.	No change of zero detector reading.	Same causes as for (1). Intermodulation distortion.
(3) Source output varied by at least 2:1, keeping the relative harmonic content constant.	No change of zero detector reading.	Voltage-dependent transformer ratio (demagnetise the transformer?) or impedance having a voltage dependence. Intermodulation distortion.
(4) Source output constant, harmonic content varied.	No change of zero detector reading.	Frequency-dependent bridge, inadequate frequency selectivity of detector. Intermodulation distortion
(5) If a simple ratio bridge, measure two zero admittances.	Zero detector reading.	Inadequate number or placement of equalisers. Auxiliary balances insufficiently well understood.
(6) Measure two zero impedances (source matching may need altering).	Zero detector reading.	As for (5).
(7) A shorted turn around the core of each equaliser in turn.	A definite detector deflection.	(i) Currents in inner and outer conductors by chance balance in that mesh – retain the equaliser. (ii) More than one equaliser in the mesh. Remove all but one.
(8) Open coil (Figure 1.22) from bridge source passed near bridge cables and components. (A sensitive test; assess what is a tolerably small response. Very useful for low-impedance circuits.)	Zero detector reading.	Inadequate magnetic screening of components. Meshes without equalisers.
(9) Metal plate (Figure 1.23) connected to a high potential from the source, passed near bridge cables and components.	Zero detector reading.	Cable outer conductors or component cases are inadequate screens. Unused coaxial ports without a conducting cover.

the other. It is important not to include any mutual inductive coupling in the direct impedance. Figure 8.2b illustrates a suitable arrangement.

Four-terminal-pair versions of these two-terminal-pair components merely require the addition of extra ports by adding T-connectors at either end.

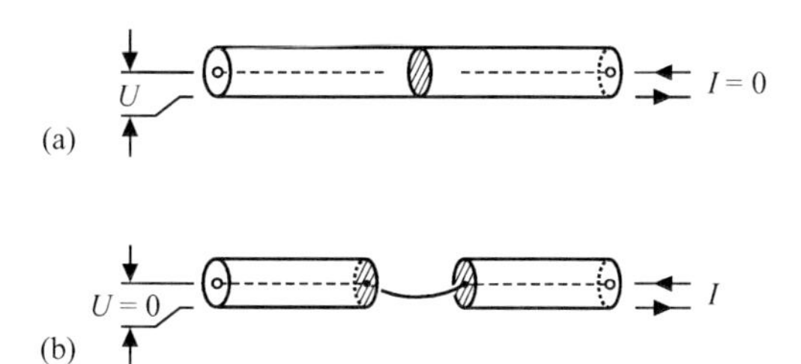

Figure 8.2 *(a) A two-terminal-pair zero admittance. (b) A two-terminal-pair zero impedance*

References

1. Kibble B.P., Rayner G.H. *Coaxial AC Bridges*. Bristol: Adam Hilger Ltd.; 1984. (Presently available from NPL, Teddington, TW11 0LW, U.K.)
2. Hague B. (revised by Foord T.R.) *Alternating Current Bridge Methods*. 6th edition. London: Pitman; 1971
3. Robinson I.A. *et al.* (Private communication)

Chapter 9

Bridges to measure impedance ratios

We now bring together all the concepts and devices described so far to show how they may be combined to make practical high-accuracy bridges. The number of networks that may be devised is endless, and therefore, we have chosen just those examples that either are of considerable practical importance to national standards laboratories or illustrate the use of some device or concept. In principle, we show how to relate like impedances of different values and then how to generate the impedance standards of R, L and M from the unit of capacitance. The reader should feel encouraged to devise other networks to serve other special needs.

It is certainly a major feature of the remarkable elegance of the subject that in principle, bridge comparison networks can be built from just five dual pairs of components. These we have introduced in an ad hoc manner, so that the significance of the pattern may well not have been apparent. The pairs are

1. impedances and admittances,
2. sources and detectors,
3. voltage and current ratio transformers,
4. injection and detection transformers and
5. combining networks and current sources.

We will often present a network as an admittance bridge; of course the distinction between this and an impedance bridge is purely one of algebraic convenience.

Although it is possible to construct any network, and in particular any of the classical four-arm bridges, in conductor-pair form, we do not give examples of this; instead, it seems more valuable to make use of the ready availability or ease of construction of transformers of variable or fixed ratio to look at impedance measurement problems afresh and, starting from the defining conditions for the components to be compared, design appropriate networks, which make use of the desirable properties of transformers.

Variable components, for example, decade resistance boxes, variable capacitors and switched tapped mutual inductors, were formerly used exclusively for adjustable components. While these are still useful when fitted with coaxial terminals, especially since many laboratories still have a good stock of them, we in general advocate using a good-quality admittance of fixed value supplied from an adjustable voltage source (an inductive voltage divider (IVD)) as the combination has a greater stability against time and environmental effects.

In bridge networks, which need a detector in more than one position, only one actual detector needs to be used. It can be connected into the various detector positions in turn to make the associated adjustments.

9.1 Bridges to measure the ratio of like impedances

9.1.1 *A two-terminal IVD bridge*

Figure 9.1 illustrates a two-terminal bridge based on an IVD as a voltage ratio device. This single-conductor network does not fall within the philosophy of this book; it is discussed here only to establish the principle of the coaxial bridge, which is described in the next section. $A_{1,2}(1+jD_{1,2})$ are the admittances to be compared. When the detector is nulled, that is, the current through it is zero, the points either side of it are at the same potential. With respect to the potential of the right-hand side of the detector, the potentials at the high terminals of A_1 and A_2 are then $(1-n)U$ and nU, respectively, being determined by the turns ratio of the IVD. We can then equate currents at their low terminals.

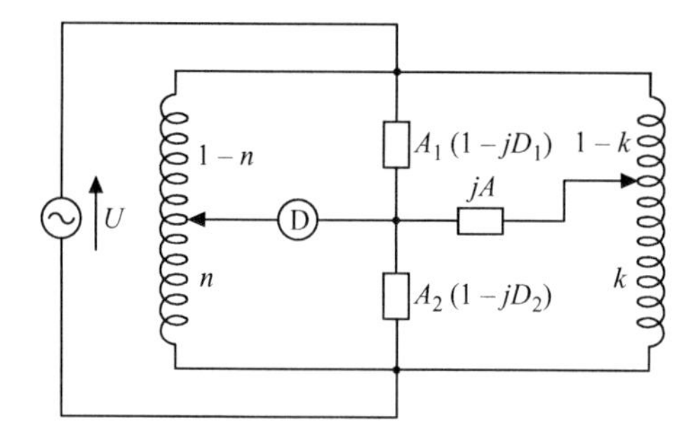

Figure 9.1 The principle of a two-terminal bridge based on an IVD

$$(U-nU)A_1(1+jD_1)+jA(k-n)U=nUA_2(1+jD_2)$$

Equating real and imaginary parts,

$$A_1=\frac{A_2 n}{1-n} \tag{9.1}$$

and

$$D_1=D_2+\left(\frac{A}{A_2}\right)\frac{(n-k)}{n} \tag{9.2}$$

Equations (9.1) and (9.2) express an unknown A_1 in terms of known standards A_2 and A.

For example, if components 1 and 2 are lossy capacitors having loss tangents D_1 and D_2, i.e.,

$$A_{1,2}(1+jD_{1,2})=j\omega C_{1,2}(1-jD_{1,2})$$

jA would then be a conductance $G = 1/R$. If a small conductance is required, it can be obtained from a T-network (see section 6.4.10).

The bridge equations relate the capacitance values

$$C_1 = \frac{C_2 n}{1 - n} \tag{9.3}$$

and the loss tangents

$$D_1 = D_2 - (\omega C_2 R)^{-1} \frac{(k - n)}{n} \tag{9.4}$$

The last equation represents a frequency-dependent quadrature balance, but as standard capacitors usually have very small loss tangents, this does not cause any difficulty because this balance is much less critical than that relating the capacitances.

Note the connection of the source and the IVDs to the high defining terminals (see section 5.3.3) of the two-terminal components A_1 and A_2. The current through the components is drawn directly from the source, and the IVDs, being high input impedance devices, draw little current. Hence, there is only a small potential drop along the connecting wires between the IVDs and these terminals.

9.1.2 A two-terminal-pair IVD bridge

This bridge, shown in Figure 9.2, is suitable for comparing the ratio of two like admittances. It is a coaxial version of the bridge described in the previous section. The direct connection at O between the detector and the IVD is replaced by the indirect connection via the outer conductors of the coaxial cables.

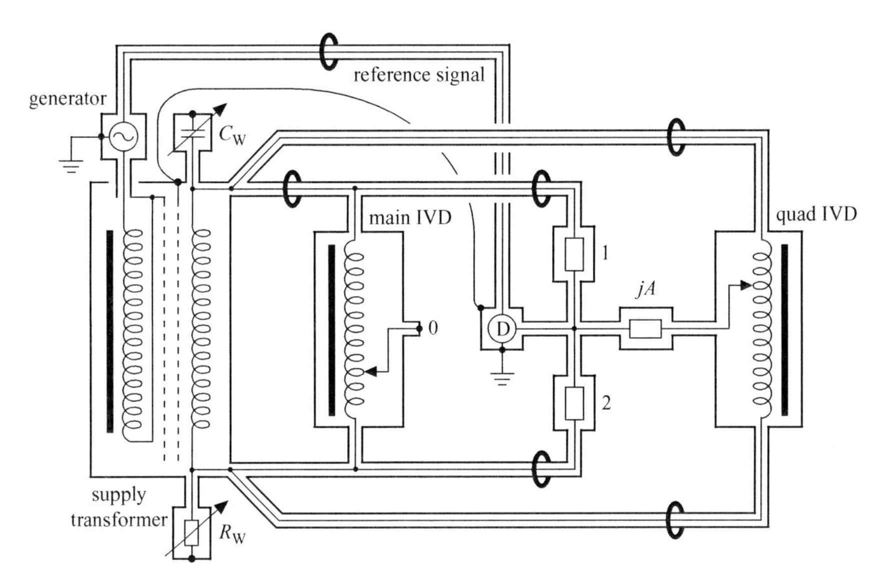

Figure 9.2 A two-terminal-pair bridge based on IVDs

Notice that the shorted variable output of the left-hand IVD is now labelled as the reference point 0. The small potential drops incurred by the currents returning along the outer conductors to 0 from the point where the isolated detector is connected become part of the measurement and satisfy the two-terminal-pair definitions of the components. Each mesh of the network has been provided with a current equaliser to ensure the good definition of these currents and potentials.

Wagner components R_W and C_W balance the capacitances and associated losses between the inner circuitry and outer conductors and shielding containers of the components. R_W and C_W, either of which may need to be connected in either position, are adjusted by replacing the short at port 0 by the detector and adjusting them to obtain a null. The defining condition of the IVD, which demands that there shall be no current through this short, is then fulfilled. This adjustment can be made more conveniently by leaving the detector in its usual position and simply temporarily removing the short, adjusting R_W and C_W to again obtain a detector null, replacing the short and iterating this procedure as necessary. (Once is usually enough.)

The shunt admittances of the cables from the upper and lower output ports of the main IVD to the components 1 and 2 are loads on the outputs of the IVD, which affect the output voltages because of its finite internal winding resistances and leakage inductances. To obtain greater accuracy, corrections must be applied for these loadings. The corrections can be experimentally determined by connecting a known shunt capacitance, which is an addition to the known cable capacitance, to each IVD output in turn and extrapolating the observed change in the bridge balance back to zero total capacitance. The Wagner balance will need readjustment to account for each of the two added capacitances.

When accurate and calibrated IVDs are used in this simple bridge, it is capable of a measurement uncertainty of less than 1 ppm for moderate values of $|Y_2|/|Y_1|$, say, $10{:}1 > |Y_2|/|Y_1| > 1{:}10$.

9.1.3 *A four-terminal-pair IVD bridge*

With but little extra elaboration, the network of the previous section can be adapted to compare two like impedances defined in a four-terminal-pair manner. This network is especially useful for comparing low impedances for which four-terminal-pair definition is necessary to obtain the required accuracy. By using the adaptor described in section 5.3.10, two- and four-terminal components can also be measured on this bridge.

The circuit is shown in Figure 9.3. Often, in view of the other approximations to be made, Wagner components will not be necessary. The theory of the Kelvin double bridge given in section 5.6.1 is applicable, and we make use of the combining network described in section 5.6.2 to connect in series the two components to be compared.

In this network, we also introduce a method of making the quadrature balance by injecting in series with the main IVD zero output voltage a quadrature voltage derived from a second IVD and phase-changing network as described in section 7.3.4.

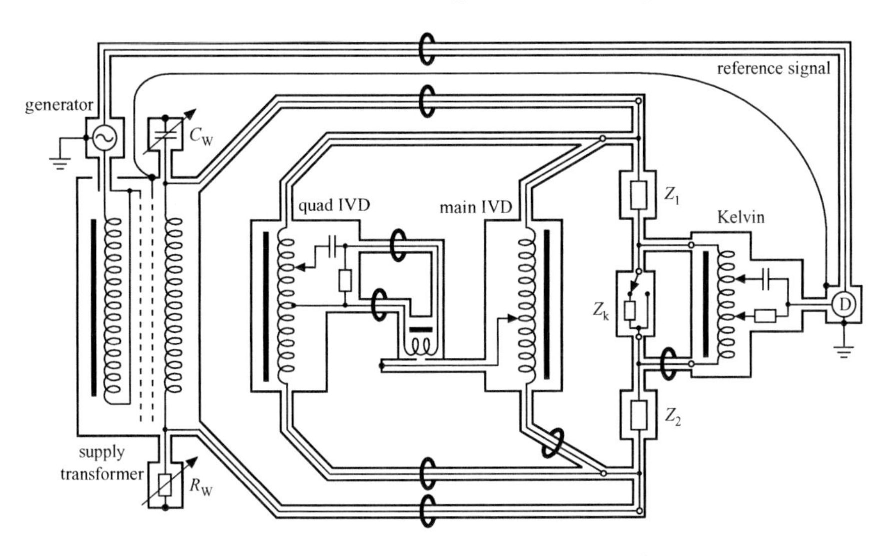

Figure 9.3 A four-terminal-pair bridge based on IVDs

The bridge is balanced by first changing the switch position to make the value of Z_k small and adjusting the main and quadrature IVDs in turn to null the detector. Then the switch is changed back to the position shown so that Z_k is increased to be of the same order of magnitude as Z_1 and Z_2, and the detector is again nulled by adjusting the settings of the combining network IVDs. This procedure is iterated until a null is obtained whether Z_k is inserted or not. The null obtained with Z_k small is the critical one, and the adequacy of this combining network auxiliary balance can be tested with Z_k small by changing the settings of the combining network by an amount comparable with, but larger than, the fineness of their adjustment and observing whether the main balance is significantly affected. The principle behind this testing procedure is very important and should always be invoked to test the satisfactory adjustment of the auxiliary balances of a bridge.

The chief approximations involved in this network arise from the voltage drops along the cables to the main IVD and the current the main and quadrature IVDs draw through the high-potential ports of Z_1 and Z_2 in violation of the defining condition. The principal voltage drop arises from the magnetising current of the single-stage main IVD flowing through the resistance and inductance of these cables; this source of error (which causes large quadrature errors at low frequencies) can be largely overcome by using a two-staged IVD (see section 7.3.2) with its magnetising winding driven directly from the source. There remains the inevitable voltage drop caused by the current flowing through the shunt capacitance of the cables and the series impedance $R/2 + j\omega L/2$ of the cables (see section 3.2.6), in addition to which the difference of the currents flowing through the shunt admittances of these cables causes voltage drops in the output resistance and leakage inductance of the IVD. The cables to the combining network have a very small potential difference between inner and outer and do not contribute significant error compared to the above causes.

These errors limit the accuracy of the network to perhaps about 1 in 10^7, but the network is nevertheless useful for comparing low impedances. The resolution obtainable is adequate as a detection transformer (section 2.1.4) can also provide impedance matching between a low bridge impedance and the comparatively high input impedance of a typical detector. The quadrature balance is provided by injecting a quadrature voltage $j\beta U$ in series with the IVD output, and the approximate bridge balance equations follow straightforwardly from equating the division of potential ratios across the components and across the IVD plus the injected voltage if $D_{1,2} \ll 1$. They are, in the notation of section 9.1,

$$A_1 = \frac{A_2 n}{1-n} \tag{9.5}$$

and

$$D_2 = \frac{D_1 + \beta}{n(1-n)} \tag{9.6}$$

9.1.4 *A two-terminal-pair bridge based on a 10:−1 voltage ratio transformer*

The comparison of two like impedances whose values are nominally in a 10:1 ratio is a frequently encountered problem. A 10:−1 voltage ratio transformer (see sections 7.3.7 and 7.3.8), which has been calibrated (see sections 7.4.1–7.4.6), is the ratio component in the bridge drawn in Figure 9.4, and the small additional

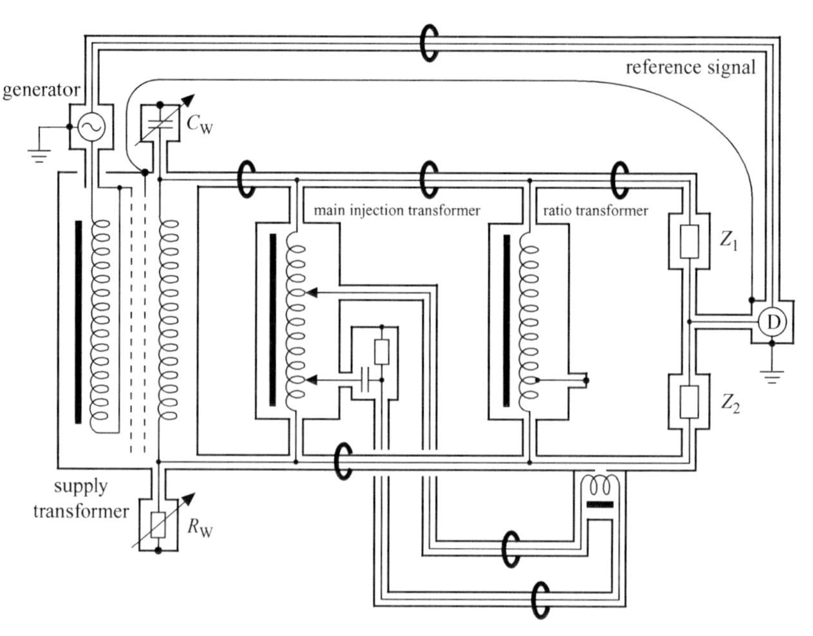

Figure 9.4 A two-terminal-pair bridge based on a 10:−1 voltage ratio transformer

variable voltage needed to balance the bridge can be obtained from an injection transformer (see section 7.3.3 and 7.3.4). The whole bridge network is driven by an isolating transformer (see section 1.1.5) to preserve the isolation of the bridge.

The nominal 10:−1 transformer ratio is defined as the ratio U_1/U_2 where U_1 and U_2 are the voltages appearing across the upper and lower exit ports of the transformer under conditions of zero current flow through these ports. For moderate impedances Z_1 and Z_2 and cable shunt admittances, this ratio will not be greatly affected by the finite current drawn from the coupled low output impedances of the transformer. If the inaccuracy this introduces is unacceptable, a calculated correction can be made for the loading effects, or the four-terminal-pair values of Z_1 and Z_2 can be obtained by the extrapolation technique (see section 5.3.9) at the expense of performing extra balances.

The voltage presented to Z_2 is $U_2(1 + \alpha + j\beta)$, where αU_2 and $j\beta U_2$ are the in-phase and quadrature injected voltages, respectively, so that the equation

$$\frac{U_2(1 + \alpha + j\beta)}{Z_2} = \frac{U_1}{Z_1} \tag{9.7}$$

holds when the detector is nulled.

Notice that the incorporation of a well-constructed source-isolating transformer (see section 1.1.5) makes the connection of a phase-sensitive detector easier (see section 1.1.2).

9.1.5 A four-terminal-pair bridge based on a two-stage 10:−1 voltage ratio transformer

We now describe a network in which both the defining conditions for the transformer output voltages and the four-terminal-pair definition of the components to be compared are satisfied. The complexity of the network is greater than that described in the last section, but nevertheless, achieving the detector balances at the various ports is straightforward.

The bridge is shown in Figure 9.5. The like impedances Z_1 and Z_2 are to be compared: their nominal values are such that $Z_1 \approx 10Z_2$. They are defined by the four-terminal-pair convention of section 5.3.8.

The all-important ratio of the voltages across the impedances is measured under conditions of zero current at the defining points of the impedances. The voltage ratio is varied to balance the bridge by injecting a small, known voltage ΔU from the injection IVD and associated injection transformer. Simultaneous zero current at the output of the two potential windings as sensed by the detection transformer is achieved by iterative adjustment of current sources 1 and 2.

At the low-potential end of each impedance, the defining current leads are joined with provision for injecting a voltage with an injection transformer connected to the generator and adjusting the Kelvin network to null the detector. After making this auxiliary balance, the injected voltage is set to zero by removing the

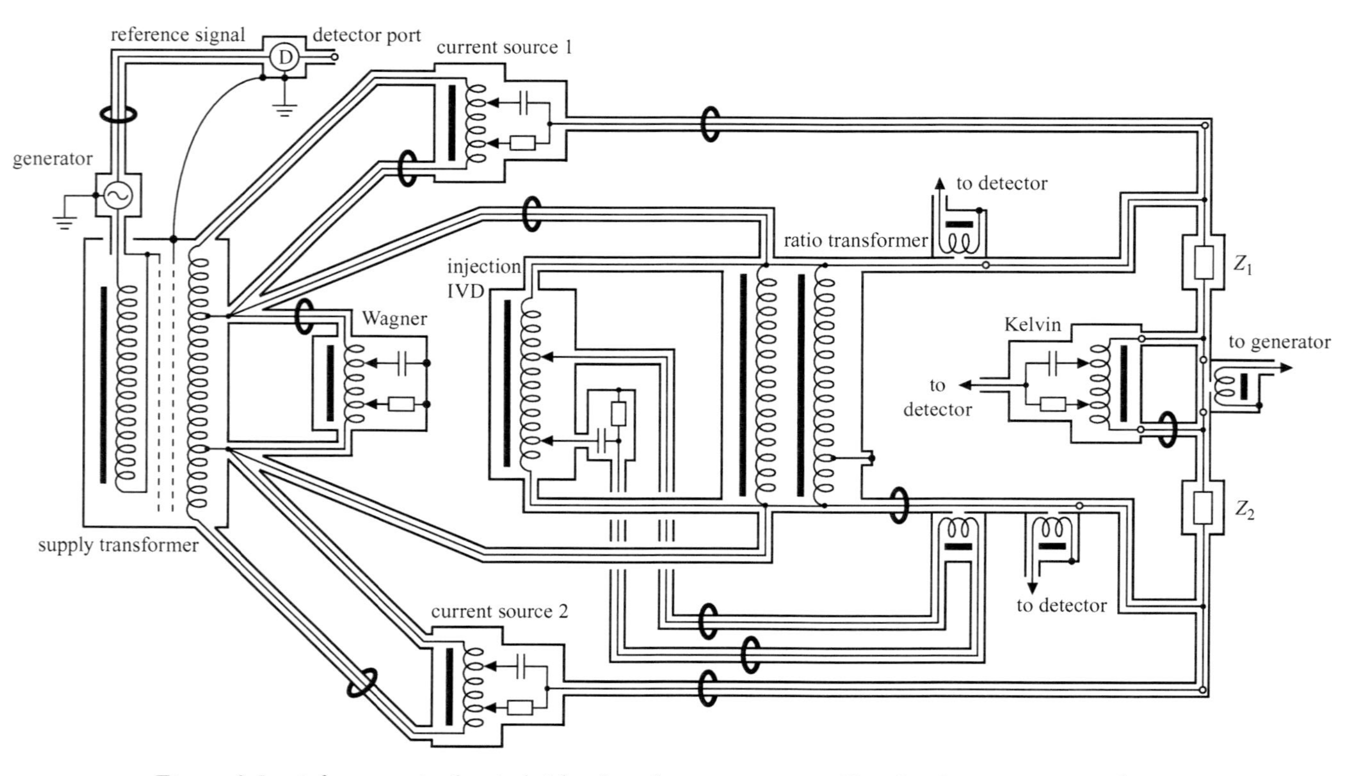

Figure 9.5 A four-terminal-pair bridge based on a two-stage 10:−1 voltage ratio transformer

connection to the generator and replacing it with a shorting plug. Then, if necessary, ΔU is readjusted. Convergence is very rapid, and only a few iterations of these balances are necessary to achieve a simultaneous null.

The detection transformers and associated detectors when nulled ensure that the condition of zero current at the outputs of the potential windings is satisfied. The injection IVD, phase-change circuit and injection transformer provide adjustable in-phase and quadrature voltages, respectively. By two-staging these (see section 7.3.4), the necessary injection accuracy of better than 10^{-5} for comparatively large values of $\Delta U/U$ is achievable.

The adjustable voltage sources are all driven from the first-stage winding of the ratio transformer to avoid loading the potential ratio winding of this transformer.

9.1.6 Equal-power resistance bridge

In this elegant network, devised by Cutkosky [1], the same power is dissipated in each of the resistors being compared. This avoids corrections for the load coefficients of the resistors and optimises the signal-to-noise ratio when the most significant noise contribution is their Johnson noise at room temperature. It is a combination of a voltage ratio bridge in which the higher-valued resistor contributes most noise with a current ratio bridge in which the noise of the lower-valued resistor dominates. It is fully described by Kibble and Rayner [2].

9.2 Bridges to measure the ratio of unlike impedances

9.2.1 R–C: the quadrature bridge

The purpose of a quadrature bridge is to relate resistance R and capacitance C by means of the basic circuit of Figure 9.6.

$$\frac{U}{jU} = \frac{1/j\omega C}{R} = \frac{1}{j\omega CR} \tag{9.8}$$

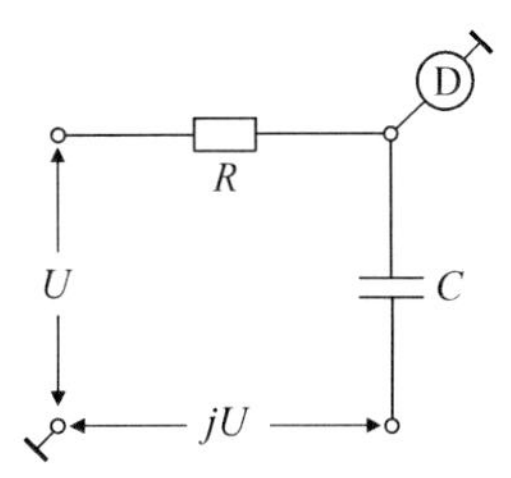

Figure 9.6 The principle of a quadrature bridge to relate resistance and capacitance

It is not practically possible to generate jU from U with anything like the required accuracy, and the remainder of this section is concerned with the ingenious solution to overcome this.

A second bridge, whose balance condition is

$$\frac{jU}{-U} = \frac{1/j\omega C_2}{R_2} = \frac{1}{j\omega C_2 R_2} \tag{9.9}$$

is combined with the first bridge, as shown in Figure 9.7.

$-U$ can be generated from U with a 1:−1 ratio transformer, and if the departure from exactness of the transformer is eliminated by interchange of its outputs, 1 in 10^9 accuracy for this ratio is achievable.

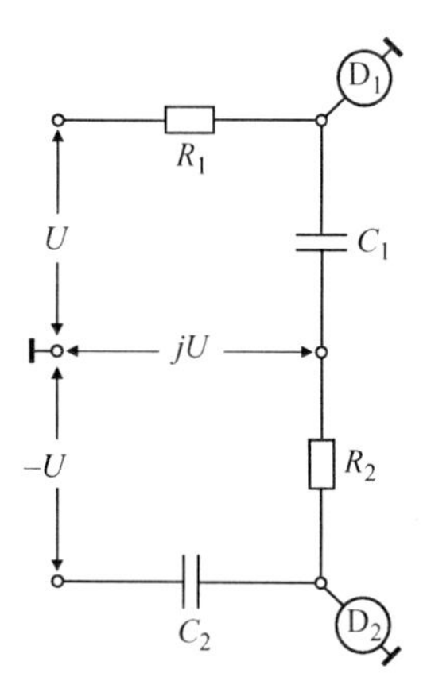

Figure 9.7 Combining two quadrature bridges

Now a way must be found so that a condition equivalent to a simultaneous null of D_1 and D_2 can be obtained that does not depend on the exact value of jU. This can be done if the potentials at A and B are combined via auxiliary impedances R_a, R_b, C_a and C_b to be sensed by a single detector D. Applying linear network theory, the response of D to jU by itself can be found if the sources $+U$ and $-U$ are set to zero. When this is done, the *whole* network can be redrawn as a twin-T bridge (Figure 2.2). Remember that the short thick lines to which 0, R_a, C_b and jU are connected represent network points that are connected together (by the outer conductors of coaxial cables in the eventual design). If all the capacitances and all the resistances are nearly equal, the twin-T balance condition $\omega RC = 1$ can be achieved by adjusting R_a and C_b. The network of R_a, R_b, C_a and C_b is called the detector combining network.

Finally, jU can be generated from voltage sources $+2U$ and $-2U$ and the adjustable components C_J and R_J, which again act in conjunction with the rest of the network, as shown in Figure 9.9.

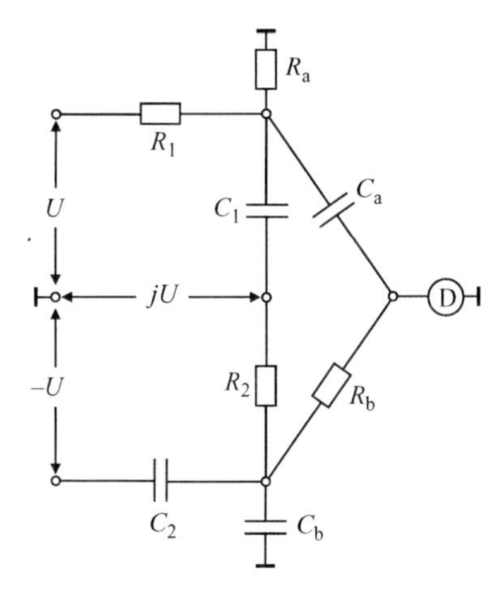

Figure 9.8 The principle of a quadrature bridge coupled to a detector combining network

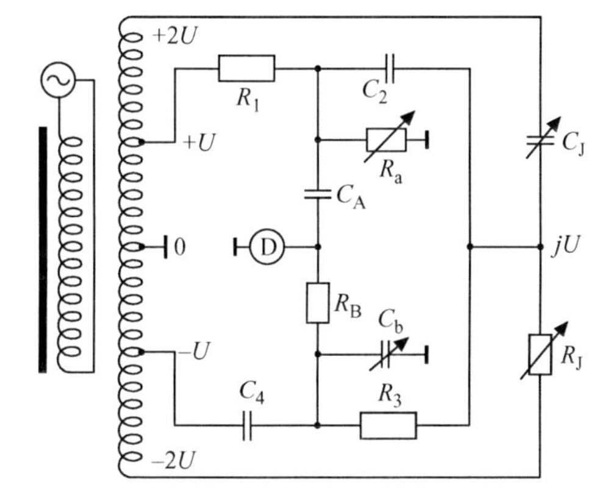

Figure 9.9 The principle of a complete quadrature bridge

The main balance condition

$$\omega^2 C_1 C_2 R_1 R_2 = 1 \tag{9.10}$$

can be achieved approximately by altering the value of one of these components to null the detector. Then the detector combining network can be radically unbalanced by temporarily altering the value of one of its components and C_J and R_J adjusted to again null the detector. This is termed the j-source balance. Then C_J or R_J can be temporarily altered, so that their value can be reset after the detector combining network is rebalanced to null the detector. This procedure, which has achieved the approximate adjustment of the main balance and the product of two auxiliary balances, should be iterated until all three balances are simultaneously achieved, remembering that the accuracy depends only on the product of the two auxiliary balances. These, therefore, need each only be made to an accuracy of, say, 1 in 10^4 if 1 in 10^8 accuracy is to be achieved for the main balance.

For simplicity in the above explanation, we have assumed that C_1, C_2, R_1 and R_2 are ideal, that is, they have no small quadrature component. In practice, they will each have a small phase angle φ and can be expressed as $C_1(1 + j\varphi_{C1})$, etc. On inserting these corrected values into (9.8) and (9.9) and equating separately the real terms, (9.10) is again arrived at if second and higher products of the φ terms are neglected. The imaginary terms yield the condition

$$\varphi_{C1} + \varphi_{C2} + \varphi_{R1} + \varphi_{R2} = 0 \tag{9.11}$$

where the small phase defect of the $+U/-U$ transformer ratio has been eliminated by taking the mean of forward and reverse transformer connections.

This network was devised principally for deriving the unit of resistance, the ohm, from that of capacitance, the farad, as realised by a calculable capacitor. More recently, this derivation has also been operated in reverse to derive the farad from the ohm, as realised from the quantum Hall effect [3].

A quadrature bridge can also set up a standard of phase angle, starting from two capacitors, known to have very low loss factors to derive the phase angle of resistors, which is much more difficult to do otherwise. The principle is to take three similar resistors, each of which is provided with an adjustable trim to alter their phase angle as described in section 6.4.11 and connect them, two at a time, into a quadrature bridge. The phase angle imbalance of the bridge reflects the sum of the phase angles of the two resistors and the two capacitors (9.11). Let the known or estimated sum of the phase angles of the two capacitors be ϕ_C, the phase angles of the resistors be ϕ_1, ϕ_2 and ϕ_3 and the magnitude of the quadrature adjustment required to balance the bridge be $\delta\phi$. Then

$$\phi_C + \delta\phi_a + \phi_2 + \phi_3 = 0$$

$$\phi_C + \delta\phi_b + \phi_3 + \phi_4 = 0$$

$$\phi_C + \delta\phi_c + \phi_4 + \phi_2 = 0$$

Adding the first two equations and subtracting the third,

$$\phi_3 = -\frac{\phi_C + \delta\phi_a + \delta\phi_b - \delta\phi_c}{2} \tag{9.12}$$

The quadrature adjustment of the bridge can be altered by $-\phi_3$, and the phase angle trim on resistor 3 altered to balance the bridge. This procedure can be repeated using similar equations corresponding to the other two resistors, and the final adjustment of all three checked, with two at a time in the bridge.

9.2.2 The quadrature bridge – a two-terminal-pair design

This simple and elegant design was originally conceived by Thompson. The principal components C_1, C_2, R_1 and R_2 were originally regarded as being two-terminal-pair components defined at the ends of their cables, which are therefore a part of these components. The accuracy of the bridge is improved by treating the resistors as four-terminal-pair components, as shown in Figure 9.10. The limitation of the definition of the resistors to a few parts in 10^8 by the lack of reproducibility of the contact resistances of their connectors is removed. The capacitors cause no problems; the lead and contact resistances of their terminating connectors only affect slightly their phase angle. The cables of the two-terminal-pair capacitors C_1 and C_2 stay with them when they are transferred to other bridges, but account has to be taken of the added effect on their values in this network of the cables internal to the resistors up to the internal defining points of the resistors. The loading effect on the $1:-1$ ratio transformer of the auxiliary 100-pF capacitors and their cables on the high-potential ports of C_2 and R_1, if these cables are similar, is symmetrical and does not affect the result. Use of these capacitors to determine the $1:-1$ voltage ratio applied to R_1 and C_2 avoids considering the change in loading effects on the ratio transformer if it were to be reversed instead.

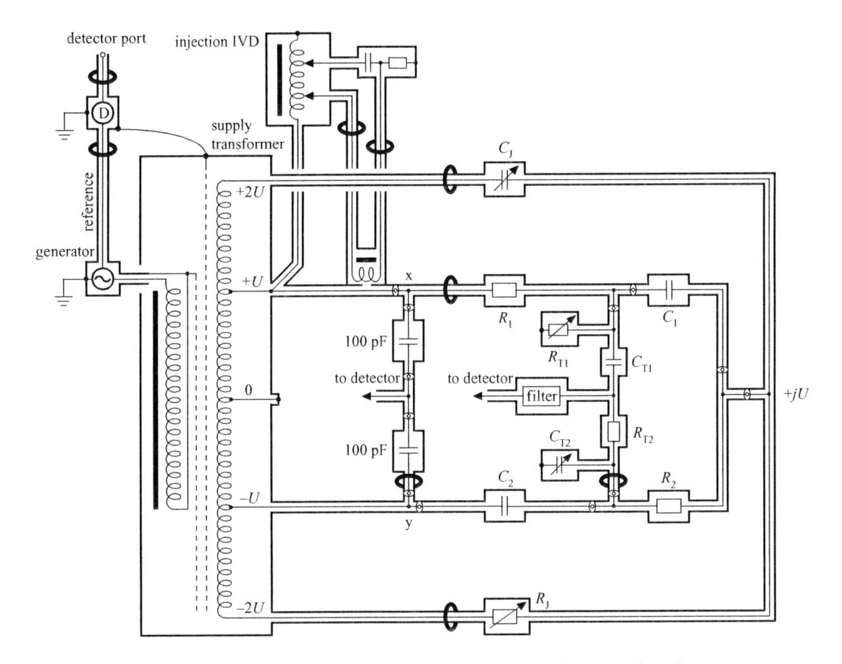

Figure 9.10 A two-terminal-pair quadrature bridge

The potential jU at the junction of R_2 and C_1 is set up by the auxiliary adjustable capacitor C_J connected to the $+2U$ tap and the adjustable resistor R_J

connected to the $-2U$ tap. The detector combining network is C_{T1}, C_{T2}, R_{T1} and R_{T2}. α_1 is the mean of the small in-phase injected voltages needed to balance the auxiliary 100-pF capacitance bridge and with the capacitors together with their connecting leads interchanged, and α_2 is the injected in-phase voltage needed to balance the quadrature bridge.

The balance procedure is to first break the connections at X and Y and connect the bridge side of them together into the shorted centre tap of the supply/ratio transformer, thus setting $\pm U$ to zero, and altering the components from the $\pm 2U$ taps to null the detector to properly adjust the nominal $-jU$ voltage. Then, after restoring the connections at X and Y, the $-jU$ adjustment is temporarily changed in a restorable manner by, for example, shorting the variable capacitor C_J, and the detector network is balanced by adjusting C_{T2} and R_{T1}. Finally, the short across the variable capacitor is removed, and the injected voltage is adjusted to again balance the detector. The whole balance procedure is iterated enough times so that no readjustment of α_2 is required to within the target accuracy. Whether the $-jU$ and detector combining network balances have been made with sufficient accuracy can be tested by altering in turn, for example, C_J and C_{T2} by a fraction, say, 1% of their value. Since the bridge balance is affected only by the product of the setting accuracy of these components, the bridge balance should not be affected by more than 1% of the setting accuracy of the other.

The relationship

$$1 + \alpha_1 + \alpha_2 = \omega^2 C_1 C_3 R_2 R_4 \tag{9.13}$$

holds when the bridge is balanced.

9.2.3 The quadrature bridge – a four-terminal-pair design

Improved definition of the capacitances C_1 and C_2 can also be obtained by adopting a four-terminal-pair configuration for them at the expense of increased complexity in the bridge design. A circuit has been devised and published by Cutkosky [1], and the version that the authors have used is drawn in Figure 9.11.

The four-terminal-pair components C_1, C_2, R_1 and R_2 are all joined in series by combining networks Kelvin 1, 2 and 3 (see section 5.6.2) in a way that is equivalent to realising their defining conditions. Networks Kelvin 1, 2 and 3 are adjusted so that voltages ΔU_1, ΔU_2 and ΔU_3 injected by temporarily connecting the generator in turn to their corresponding injection transformers have no effect on the main detector, thus producing in turn a condition equivalent to ΔU_1, ΔU_2 and ΔU_3 being zero. As in the previous section, C_{T1}, C_{T2}, R_{T1} and R_{T2} form a detector combining network, which, together with the rest of the bridge, is a twin-T bridge, which can be adjusted by altering the values of the detector combining network so that the main detector is immune to departures of the auxiliary source created by the adjustable current source networks, R_J and C_J (see section 5.6.2) from the required value jU. Only one source and one detector are required. They are plugged in turn to the ports concerned to make the main and the auxiliary balances, and all ports except the ones they are currently moved to are closed with short-circuiting plugs.

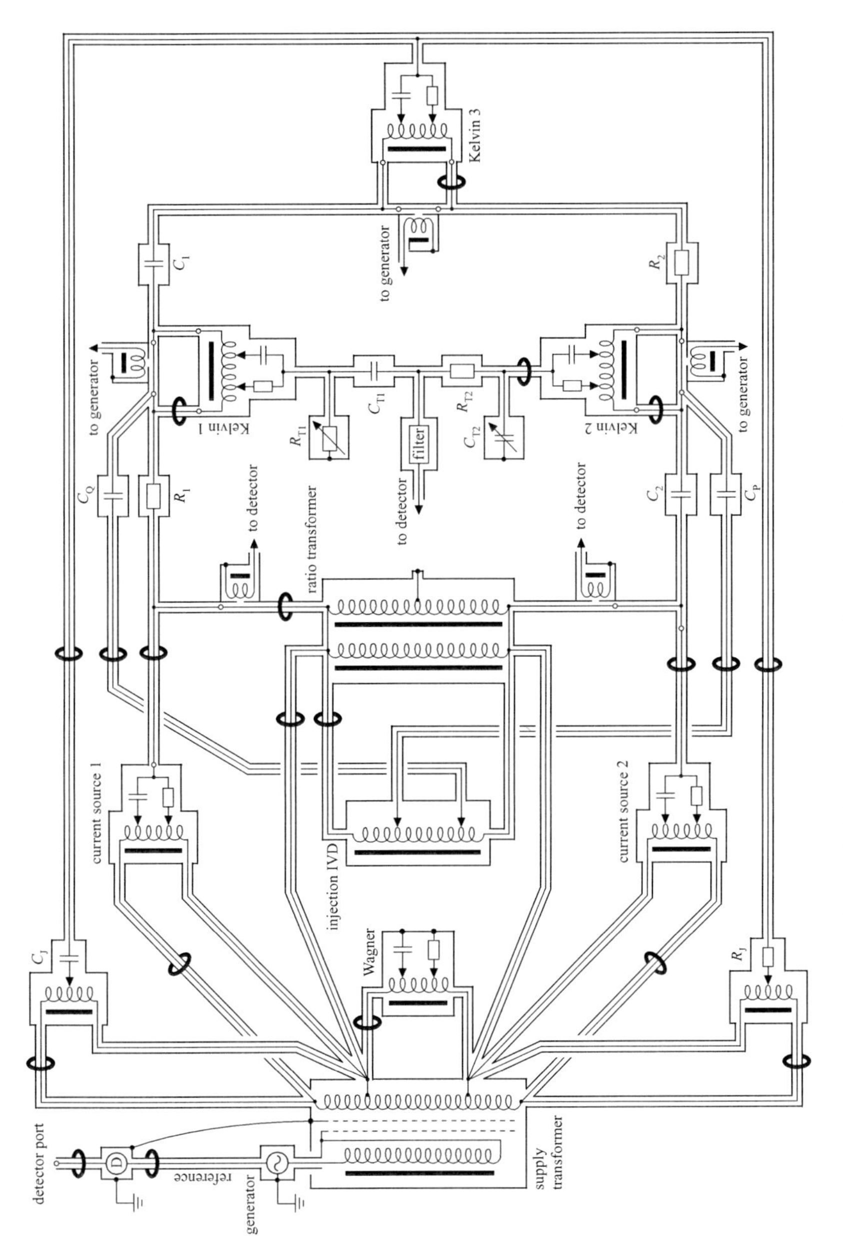

Figure 9.11 A four-terminal-pair quadrature bridge

The defined potentials $\pm U$ are applied to the upper potential ports of C_2 and R_1 via defining transformers by the 1:−1 ratio transformer, the condition $I = 0$ being achieved by adjusting the current sources 1 and 2 to null the detector when it is connected to the corresponding defining transformer.

The detector combining network is adjusted by altering C_{T2} and R_{T1} to null the main detector when either C_J or R_J is temporarily maladjusted. C_J and R_J are adjusted by nulling the main detector when either C_{T2} or R_{T1} is temporarily maladjusted.

The main balance adjustment is then made with the injection IVD, which, in conjunction with C_P and C_Q, have the effect of putting small variable capacitances in parallel with C_2 and R_1.

The departure of the ratio transformer from a perfect 1:−1 ratio is eliminated by reversing the connections at the output ports of its associated detection transformers and at the inputs to it from the supply transformer. This also reverses the sense of the injection IVD, so that their readings corresponding to zero voltage applied to C_P and C_Q need not be determined. The defining condition of zero volts at the centre port of the ratio transformer is obtained by adjusting the Wagner components, so that the main detector is again nulled when the short there is temporarily removed.

All these adjustments are iterated as necessary, but convergence is very rapid. The main detector is coupled into the bridge via a harmonic filter (see section 2.1.2).

There are 20 meshes in the network shown, and each is provided with an equaliser whose effect on the total network is evaluated by the method described in section 3.2.2.

The voltage applied to C_P is $(0.5 - k_1)/2U = (1 - 2k_1)/U$, where k_1 is the setting of the associated IVD output, and hence, the admittance of C_2 is augmented by $C_P(1 - 2k_1)$, so that the bridge balance condition, (9.10), becomes

$$\omega^2[C_1 + C_P(1 - 2k_1)]R_1C_2R_2 = 1 \tag{9.14}$$

On reversing the ratio transformer,

$$\omega^2[C_1 + C_P(1 - 2k_1')]R_1C_2R_2 = 1 \tag{9.15}$$

and taking the mean of these two results, which eliminates any small departure from a nominal 1:−1 ratio of T_2, we finally have for the in-phase balance condition

$$\omega^2[C_1 + C_P(k_1' - k_1)]R_1C_2R_2 = 1 \tag{9.16}$$

For the quadrature adjustment, the phase defect represented by the capacitance C_Q shunting R_1 is similarly modified by $2(k_2 - k_2')C_Q$, so that the relation of (9.11) between the phase defects becomes

$$\varphi_{C1} + \varphi_{C2} + \varphi_{R1} + \varphi_{R2} + \omega R_1 C_Q (k_2 - k_1') = 0 \tag{9.17}$$

This bridge has been constructed with components of nominal value $C_1 = C_2 = 1$ nF, $R_1 = R_2 = 10^5$ Ω and $C_P = C_Q = 1$ pF. The nominal frequency was $\omega = 10^4$ rad/s.

With $U = 10$ V, giving 1 mW power dissipation in the resistors and a detector time constant of 3 s, a resolution and accuracy approaching 1 in 10^9 can be achieved.

9.2.4 Bridges for measuring inductance

Most national standards laboratories calibrate inductors in terms of the traceable SI values of capacitors by means of the four-arm Maxwell-Wien bridge illustrated in Figure 9.12. The balance equations are

$$L_x = C_3 R_1 R_2$$

and

$$R_x = \frac{R_1 R_2}{R_3}$$

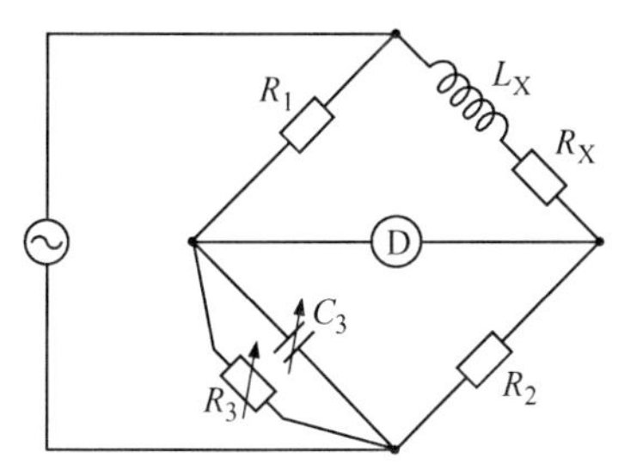

Figure 9.12 Principle of the Maxwell-Wien bridge

By choosing appropriate values of C between 10 pF and 10 nF and R_1 and R_2 between 10 Ω and 1 MΩ, Hanke *et al.* [4] calibrated their laboratory standard inductors, which had values between 100 μH and 100 H at frequencies between 50 Hz and 1 MHz.

The basic network requires the addition of a shield and a Wagner network (see section 8.1.3) to adjust the shield potential to that of the detector. Despite its simplicity, this bridge can attain a relative accuracy of about 10 ppm at lower frequencies, reducing to about 1% at higher frequencies. These accuracies are entirely adequate for calibrating the standards of commercial laboratories and, in any case, are limited to the extent to which inductance standards can meet their defining conditions.

Côté [5] has constructed a coaxial four-terminal-pair version of the Maxwell-Wien bridge, which attains only a similar accuracy despite greater complexity. He also reports that the convergence of his network is poor, making the bridge somewhat tedious to balance.

Sedlacek and Bohacek [6] have described a network, which calibrates inductance in terms of resistance, which is very closely related to the quadrature bridge described in section 9.2.1 for relating capacitance to resistance.

Kibble and Rayner [2] describe a very simple resonance technique to measure a high-Q special inductor in terms of capacitance. Other values of inductance are measured in terms of this inductor by an inductance ratio bridge.

9.3 AC measurement of quantum Hall resistance

By using the multiple-series connection scheme of a quantum Hall resistance device (see section 6.7.1), AC quantum Hall resistance measurements can be made as for any other two-terminal-pair impedance standard. Also, because of the special properties of the quantum Hall effect (see section 6.6.2), it can provide self-contained combining networks (see section 6.7.1). This simplifies the measuring bridges, which compare quantum Hall resistance devices with each other or with conventional impedance (resistance or capacitance) standards. These modified bridge networks as well as some special bridge networks for the measurement of the contact and longitudinal resistances to characterise AC quantum Hall resistance devices are described in the following sections.

9.3.1 AC contact resistance

Contact to the two-dimensional electron gas can be achieved by bonding wires to lithographic AuGeNi contacts or by soldering thin wires to tin ball contacts. Non-ideal or poor contacts can cause two effects. First, the small residual currents flowing through the potential contacts (as a consequence of the multiple-series connection scheme) produce a small and calculable voltage drop at the contact resistances, which must be taken into account. Second, poor contacts can cause a systematic deviation of the Hall resistance from the quantised value, and this is often accompanied by excess noise and irregular structures in the plateau. It is therefore necessary to measure the contact resistances.

The AC contact resistance can be measured, for example, by a three-terminal-pair measurement, which is simple and fast and yields a resolution of about 0.01 Ω. The scheme is shown in Figure 9.13. It yields the sum of the contact impedance and the series impedance of the particular lead. This is just the connecting impedance, which is relevant to the multiple-series connection scheme (see section 6.7.1). The connecting impedance is mainly real and within a good approximation equal to the sum of the contact resistance and the series resistance of the particular lead. As measuring the series resistance of the individual quantum Hall resistance cables is

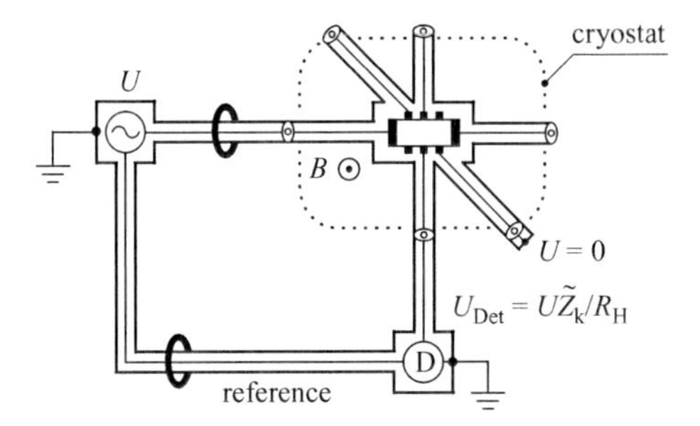

Figure 9.13 Three-terminal-pair measurement of the connecting impedance of an individual quantum Hall resistance lead

part of the measurement of the cable parameters (see section 6.7.2), this can be subtracted to find the actual contact resistance. See [30,31] of chapter 6.

If desired, it is also possible to determine the small imaginary part of the contact impedance. This requires a phase adjustment of the detector and can be achieved, for example, if a resistance of a few ohms, chosen to have a negligible phase angle, is temporarily inserted into the short-circuited lead. The phase of the detector is adjusted so that the change of detector signal is in phase. The small quadrature component of the connecting impedance can then be measured as the quadrature reading of the detector. This quadrature component is usually mainly caused by the series inductance of the particular lead.

The impedance of all contacts can be measured successively by reconnecting the source, the detector and the short-circuit plug to the appropriate terminals of the quantum Hall resistance device. The applied voltage and the gain and sensitivity of the detector can be measured by interchanging the source and detector. For lithographic AuGeNi contacts, the current applied to the potential contacts should be considerably smaller than the current applied to the current contacts because the potential contacts have smaller dimensions and tolerate less current and because this is closer to the actual situation of the multiple-series connection scheme.

A contact is called ohmic if the contact resistance shows no current dependence. Good ohmic contact resistances are usually of the order of 1 Ω or less and show no frequency dependence, so that the AC and DC contact resistances are believed to be practically the same (on the evidence from only a few individual typical AC quantum Hall resistance devices).

The AC contact resistances can also be measured as a function of frequency. If resonances occur, this may indicate mechanical vibrations of excessive long bond wires (see section 6.7.2).

9.3.2 AC longitudinal resistance

The AC longitudinal resistance is a key parameter for investigating the quality of a quantum Hall resistance device and for understanding its AC characteristics. For ungated and split-gated devices, all the frequency- and current-dependent structures in the plateaux of R_{xx} apply in a very similar way to R_H. The measurement of R_{xx}

yields information on the relevant capacitances as a function of the magnetic field as well as on the frequency and current dependence of the associated AC losses. Also, the homogeneity of the capacitive effects along the device edges can be tested, and the effect of gates and shields can be investigated. Furthermore, the individual measurements of the Hall and longitudinal voltages along a mesh with three (or more) nodes can be tested for consistency. Invoking Kirchhoff's law, they should add up to zero. Finally, the frequency dependence of the longitudinal resistance can be extrapolated to zero as a test of whether or not the quantum Hall resistance device meets the DC guidelines, without the need for a separate DC R_{xx} measurement. Altogether, the AC measurement of R_{xx} yields a lot of valuable information. The measuring circuits for the low- and high-potential side of a quantum Hall resistance device are different and are therefore described in separate sections 9.3.3 and 9.3.4, respectively.

9.3.3 Measuring R_{xxLo}

Figure 9.14 shows the equivalent circuit of a R_{xx} measurement at the low-potential side of a quantum Hall resistance device. The high-potential side of the equivalent circuit is the same as shown in Figure 6.42 and thus not explicitly shown in Figure 9.14. The terminals of the quantum Hall resistance device selected for the R_{xxLo} measurement are labelled A and B.

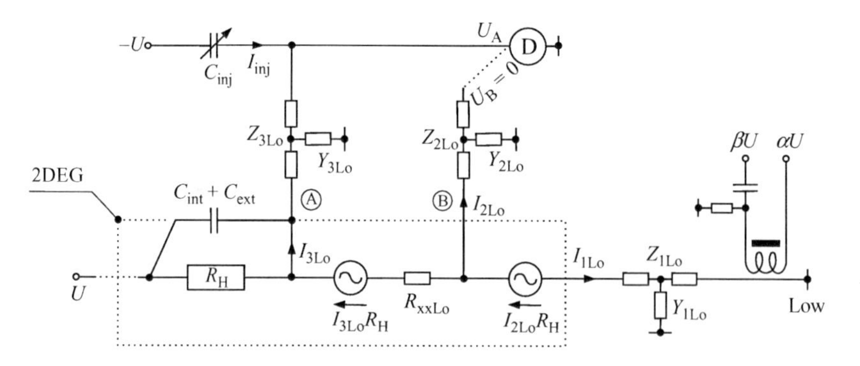

Figure 9.14 Equivalent circuit of a R_{xx} measurement at the low-potential side of a quadruple-series quantum Hall device

The admittance of the lead going to A in the presence of a non-zero potential at the A-terminal drives an unwanted AC current through the A-terminal, and the resulting Hall voltage falsifies the R_{xxLo} measurement. To eliminate this unwanted current, small in-phase and quadrature voltages are injected into the low junction of the quantum Hall resistance until the voltage at the port B, U_B, becomes zero. As a result, the capacitive current in lead B is zero; it is not exactly zero in lead A but so small that no significant in-phase voltage drop occurs along lead A. Therefore, the contact and lead resistances of leads A and B have no significant effect on the R_{xxLo} measurement.

Also, the capacitance between the A-terminal and the high-potential side of the quantum Hall device drives a current through the A-terminal. This capacitance consists of an internal contribution, C_{int}, through the two-dimensional electron gas and an external stray capacitance to high-potential conductors, C_{ext}. In contrast, the capacitance between the A-terminal and conductors connected to the shield, which are close to the device, have no significant effect because the voltage across these capacitances is practically zero. As a first step, a capacitive current I_{inj} is injected into the A-terminal by means of a low-loss capacitance standard C_{inj} (Figure 9.14); 1 pF is a suitable capacitance value. Then the voltage at the port A, U_{A}, is measured. The injected quadrature current is adjusted until the detector signal is minimal (i.e. until the quadrature component due to the parallel capacitances is compensated). The phase of the lock-in amplifier is then adjusted so that an imbalance of the injection capacitance C_{inj} is precisely in quadrature. R_{xxLo} is now calculated from

$$\frac{U_{\text{A}}}{I} = R_{\text{xxLo}} + j\omega R_{\text{H}}^2 \times \left[C_{\text{int}}(1 - j\tan\delta_{\text{int}}) + C_{\text{ext}}(1 - j\tan\delta_{\text{ext}}) - C_{\text{inj}}\left(1 - j\tan\delta_{\text{inj}}\right)\right] \tag{9.18}$$

The first three terms including the internal and external capacitances and associated loss factors $\tan\delta_{\text{inj}}$ and $\tan\delta_{\text{ext}}$, respectively, are the required AC longitudinal resistance. The term $\tan\delta_{\text{inj}}$ describes the phase error of the injected quadrature current arising from the loss factor of the injection capacitor or a phase error of the voltage applied. The measurement is sensitive to phase errors because the pre-factor ωR_{H}^2 in (9.18) is large, even if the internal and external capacitances are very small. The quadrature component of the longitudinal voltage is easily observed, being about three orders of magnitude greater than the resolution of the measurement. If the loss factors are well below 10^{-3}, the associated in-phase components are negligible.

If a quantum Hall device is bonded for a quadruple-series connection scheme, the longitudinal resistance can be measured with AC at both pairs of low-potential terminals. The potential of the unused low connection is extremely close to zero, and so it makes no significant difference if a cable is connected to the low junction of the device or it simply remains open.

Figure 9.15 shows the circuit of a practical R_{xxLo} bridge. The phase-sensitive detector follows a single-ended preamplifier. Instead of the capacitor C_{inj} being adjustable, a capacitor with a fixed value is supplied from the adjustable output voltage of an IVD. This IVD is not directly connected to the sine generator but to the high junction of the quantum Hall resistance to avoid phase errors, as mentioned above. The supply transformer provides isolation and drives gates or shields, which are not explicitly shown.

The potential at port B may change when a field sweep is carried out (without rebalancing the injection system), or it might be non-zero due to an imperfect balance or due to drift effects. To greatly reduce the resulting error, lead B can be

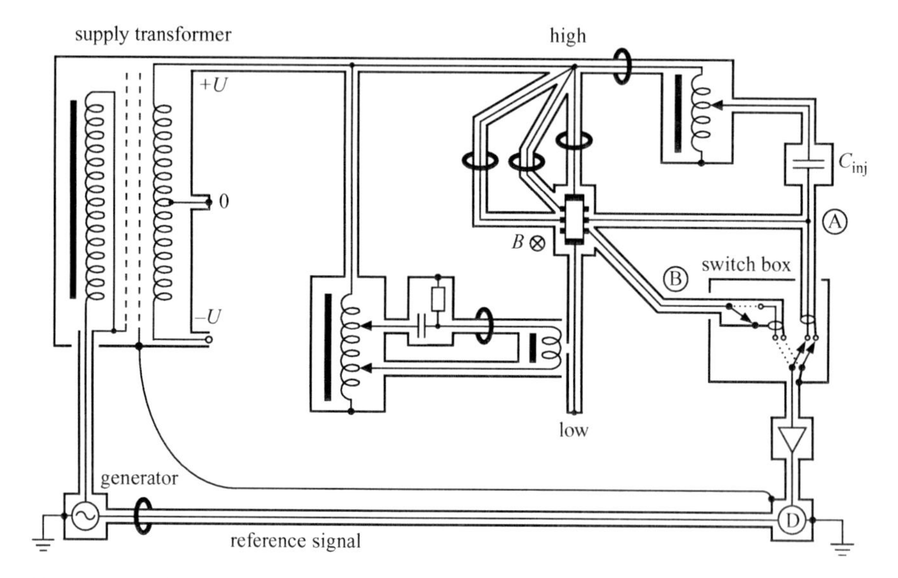

Figure 9.15 R_{xx} *measurement at the low-potential side of a quantum Hall device*

short-circuited after nulling the potential at port B. A switch box can be used for this purpose and for changing the detector between ports A and B.

In general, when a non-zero voltage, for example, the longitudinal voltage, is measured at the output port of a bridge and the detector or the detector lead has an input impedance that is comparable with the output impedance of the bridge, the bridge is loaded and the displayed voltage is smaller than the voltage of the unloaded bridge. The loading factor L is the ratio of the observed voltage to that which would be measured if there were no loading. It can be calculated and is given by

$$L = \frac{1}{\sqrt{[1 + (R_{\text{out}}/R_{\text{det}})]^2 + (\omega R_{\text{H}} C_{\text{det}})^2}} \tag{9.19}$$

R_{out} ($= R_{\text{H}}$) and R_{det} are the output impedance of the bridge and the input impedance of the detector, respectively. C_{det} is the parallel capacitance of the detector lead. The loading function L decreases with frequency and can be either calculated or directly measured when the bridge is imbalanced by a known amount. For small longitudinal voltages, this loading correction is usually negligible and thus not explicitly quoted in (9.18).

9.3.4 Measuring R_{xxHi}

The equivalent circuit for the AC R_{xx} measurement at the high-potential side of a quantum Hall device is shown in Figure 9.16. The low-potential side of the equivalent circuit is the same as shown in Figure 6.42 and thus not explicitly shown in Figure 9.16.

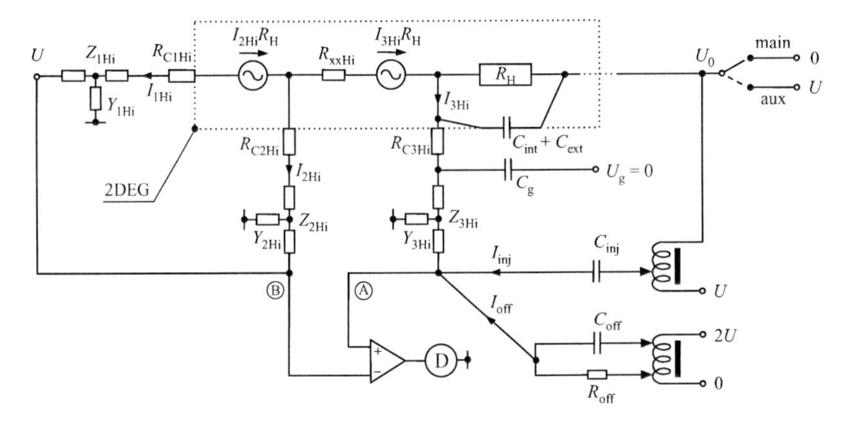

Figure 9.16 Equivalent circuit of a R_{xx} measurement at the high-potential side of a quadruple-series quantum Hall device

The main difference with respect to the R_{xxLo} measurement is that the A- and B-terminals are now at high potential and cannot be balanced to zero as can be done at the low-potential side. Therefore, the admittance $Y_{4\mathrm{Hi}}$ of lead A draws a large current $I_{4\mathrm{Hi}} = Y_{4\mathrm{Hi}}U$ through the A-contact of the quantum Hall device. The corresponding Hall voltage $I_{4\mathrm{Hi}}R_{\mathrm{H}}$ of that current amounts to about $0.1U$ and completely swamps the longitudinal voltage IR_{xxHi}. It is practically impossible to have a current detector sufficiently close to the A-contact, so $I_{4\mathrm{Hi}}$ cannot be directly measured and nulled by any current injection system. Several other but smaller voltages are also superposed on the longitudinal voltage (e.g. the common-mode rejection signal of the differential preamplifier). All these unwanted voltages have the common property that they do not directly depend on the Hall current I and can be combined as one offset voltage, U_{off}. The details of the individual contributions will turn out not to be relevant, and they are thus not listed here in detail. Analysing the equivalent circuit of Figure 9.16 yields for the differential voltage detector

$$\begin{aligned} U_{\mathrm{Det}} &= IR_{\mathrm{xxHi}} + U_{\mathrm{off}} + IR_{\mathrm{H}}\Pi \\ &\quad + jI\omega R_{\mathrm{H}}^{2}\left[C_{\mathrm{int}}(1 - j\tan\delta_{\mathrm{int}}) + C_{\mathrm{ext}}(1 - j\tan\delta_{\mathrm{ext}}) - C_{\mathrm{inj}}\left(1 - j\tan\delta_{\mathrm{inj}}\right)\right] \end{aligned} \tag{9.20}$$

with $U_{\mathrm{off}} = UY_{4\mathrm{Hi}}R_{H} - I_{\mathrm{off}}R_{\mathrm{H}} + \text{higher-order terms}$

$$\text{and} \quad \Pi = \prod_{k=1}^{n-1}\left(\frac{\tilde{Z}_{k\mathrm{Hi}}}{R_{\mathrm{H}}}\right) \tag{9.21}$$

The compensation of the offset voltage can be achieved by injecting a current I_{off} through a capacitor C_{off} and a resistor R_{off} into lead A. To distinguish between the longitudinal voltage IR_{xxHi} and the offset voltage U_{off}, the following procedure is employed: the low junction of the multiple-series connection is temporarily switched from low to high potential. This configuration is called the 'auxiliary mode'. As a result, the Hall current I flowing through the quantum Hall device (and thereby the corresponding longitudinal voltage) is switched off, while the offset voltage remains the same. In this configuration, the offset compensation system is balanced until the detector is nulled.

$$I = 0 \qquad \Rightarrow \qquad U_{\text{Det}} = U_{\text{off}} = 0 \tag{9.22}$$

When the low junction of the multiple-series connection is now reconnected to low potential ('main mode'), the offset voltage U_{off} remains zero and the detector voltage divided by the main Hall current I gives the AC longitudinal resistance.

$$\begin{aligned} \frac{U_{\text{Det}}}{I - R_{\text{H}}\Pi} &= R_{\text{xxHi}} + j\omega R_{\text{H}}^{2} \\ &\times \left[C_{\text{int}}(1 - j\tan\delta_{\text{int}}) + C_{\text{ext}}(1 - j\tan\delta_{\text{ext}}) - C_{\text{inj}}\left(1 - j\tan\delta_{\text{inj}}\right)\right] \end{aligned} \tag{9.23}$$

To compensate for the internal and external capacitances of the quantum Hall device, a capacitive current I_{inj} is injected into terminal A via a high-quality capacitance C_{inj} (Figure 9.16). This injection system is set up in such a way that the capacitive current is injected only in the main mode (and not in the auxiliary mode) and is adjusted until the detector signal is minimal. The phase of the lock-in

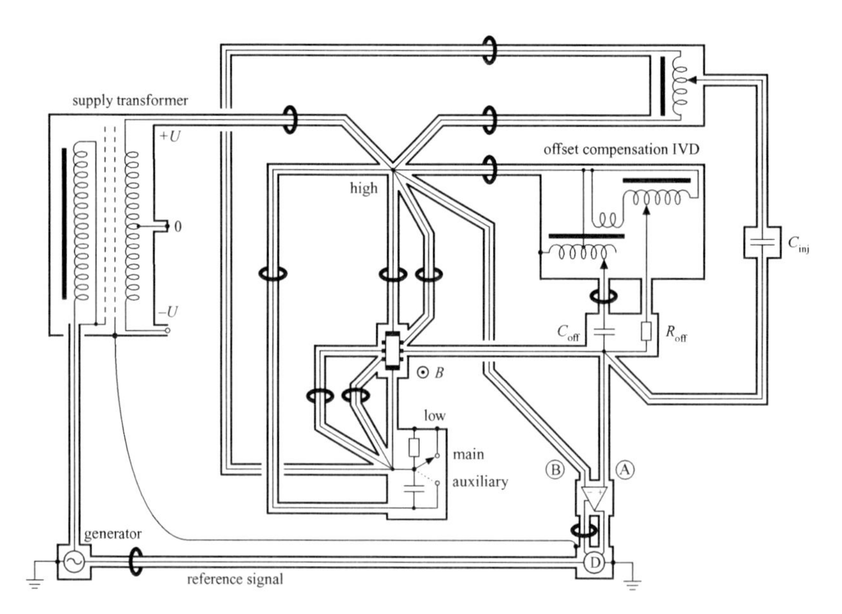

Figure 9.17 Circuit diagram of the AC R_{xxHi} bridge

amplifier is adjusted so that an imbalance of C_{inj} occurs precisely in quadrature. Then only the in-phase component remains, while the quadrature component, if desired, can be determined from the value of C_{inj} and the setting of the particular IVD.

Equation (9.23) is formally identical with (9.18) for the low-potential side of the AC quantum Hall resistance, apart from the term $R_{\text{H}}\Pi$ on the left-hand side (defined in (9.21); see also (6.48)). This correction term occurs because lead B is not directly connected to the respective contact of the quantum Hall device but to the high junction of the multiple-series connection scheme. The potentials at both points are slightly different because the small residual current $I_{3\text{Hi}} = \Pi I$ causes a small voltage drop along the penultimate high-potential lead. In the case of a quadruple-series connection ($n = 4$), Π is typically smaller than 10^{-10} and therefore negligible. In the case of a triple-series connection ($n = 3$), Π has a typical magnitude of a few parts in 10^8. It is therefore necessary to apply a small correction, which can be calculated with sufficient accuracy from the measured connecting impedances $\tilde{Z}_k$ (see section 6.7.2).

Figure 9.17 shows the scheme of a practical R_{xxHi} bridge for a quantum Hall device in triple-series connection scheme. The supply transformer provides isolation and drives gates or shields, which are not explicitly shown. To ensure that the voltage driving the offset compensation system does not change when altering between the two configurations, the input terminal of the corresponding IVD is connected to the high junction of the multiple-series connection and not directly to the generator or to the supply transformer. Direct connection would cause significant loading effects.

Re-plugging the relevant leads to set the low junction of the multiple-series connection temporarily to a high potential would give rise to a significant systematic error. Instead, a switch is used so that the impedance of the outer conductors and of the whole R_{xxHi} bridge remains the same for both configurations. However, the bridge output impedance is not exactly the same in the two modes. In the auxiliary mode, the in-phase current through the quantum Hall resistance is switched off, while at the same time, an extra capacitive current is switched on in those particular low-potential leads, which are temporarily at a high potential. This loads the common-mode rejection of the detector and may cause a significant error. It can be avoided by adding a matched capacitor and a 12.9-kΩ resistor into the switch box, as shown in Figure 9.17, to ensure that the bridge impedance remains the same for both configurations.

The value of the required offset compensation capacitance C_{off} is of the order of 10 nF. Its impedance at a frequency of 1 kHz is comparable to R_{H} and further decreases with frequency. This would seriously load the R_{xxHi} bridge and falsely reduce the detector signal (see (9.19) in section 9.3.3). To avoid this loading effect, the IVD driving the offset capacitor can be centre tapped so that the voltage across the offset capacitor is larger and a correspondingly smaller capacitance can be used, for example, $C_{\text{off}} = 400$ pF. Also, the capacitance of the detector lead loads the R_{xxHi} bridge, and thus, it should be kept small. By these measures, the loading effect can be kept reasonably small and, if necessary, a small correction can be applied ((9.19) in section 9.3.3). Because the resolution of the R_{xxHi} bridge is about seven orders of magnitude greater than the compensated capacitance of lead A, the bridge is sensitive to very small fluctuations and drift of the combined capacitance of lead A and the offset compensation capacitor. Therefore, an offset capacitor with a good short-term stability is required, and lead A should not be disturbed during the measurement.

The required value of the offset compensation resistor R_{off} is of the order of 20 MΩ. Such a resistor is found to show unwanted fluctuations and generates a high thermal noise power, and its parallel capacitance shows fluctuations. A better stability and less thermal noise have been achieved by replacing the 20-MΩ resistor by a 200-kΩ resistor. Then more dials of the corresponding IVD have to be balanced (six instead of four), and so the improvement in stability comes along with a loss of sensitivity. If the IVD driving the offset resistor is provided with an internal step-up transformer (with a ratio of e.g. 50), this compensates for the loss of sensitivity and improves the performance of the R_{xxHi} bridge.

The previous paragraphs describe how to measure R_{xxHi} at the last pair of high-potential terminals. If a quantum Hall device is bonded in quadruple series, R_{xxHi} can also be measured at the penultimate pair of high-potential terminals. The last, unused high-potential lead can either be connected to the high junction of the multiple-series connection or simply remain open.

9.3.5 A simple coaxial bridge for measuring non-decade capacitances

Commercial capacitance bridges may yield inaccurate results when trying to measure non-coaxially three-terminal capacitances through long leads to the quantum Hall device in a cryostat. The simple coaxial bridge suggested as an alternative below is a very useful tool that can carry out an accurate, low-noise measurement even at a very low voltage level.

The tool shown in Figure 9.18, a simple coaxial two-terminal-pair ratio bridge for measuring non-decade capacitances and their associated loss factors, is a very useful tool for investigating quantum Hall devices. An example of an application is the measurement of the capacitance between the two-dimensional electron gas of a device and a back or side gate. This capacitance can be measured, for example, as a function of the magnetic field or as a function of frequency and current. Another application is the measurement of the loss factor of different materials, which are selected for mounting a device.

The unknown capacitance is measured in terms of the capacitance of an air-dielectric low-loss rotary vane capacitor whose maximum value can be chosen to be, for example, either 1 pF or 10 pF and whose minimum value is 50 fF. A ratio transformer, which can be configured for different ratios like 1:1, 10:1 or 1:10, allows a capacitance range from 5 fF to 100 pF to be continuously covered. The phase of the detector is adjusted such that an imbalance of the rotary vane capacitor causes a completely in-phase response. The loss factor of the unknown capacitance is then calculated from the quadrature component of the detector signal. Assuming

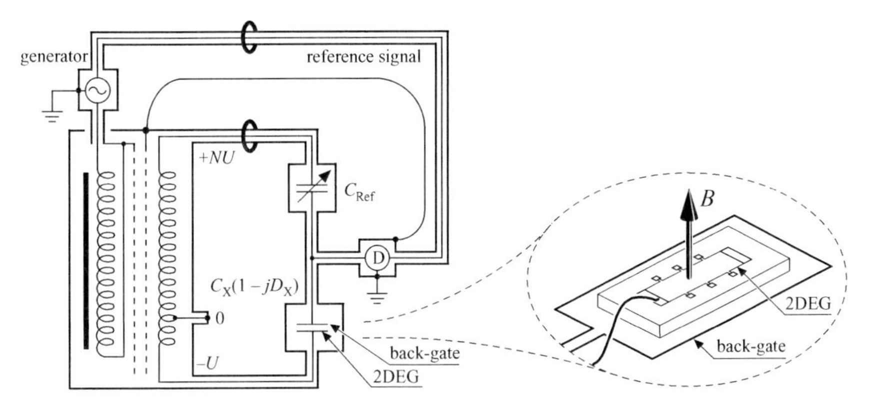

Figure 9.18 A coaxial two-terminal-pair ratio bridge for the comparison of non-decade capacitances

that the rotary vane capacitor has a sufficiently low loss factor and the ratio transformer has a sufficiently small phase error, this simple bridge yields an uncertainty of less than 1×10^{-5} for the loss factor, even at a low voltage level of 0.1 V. If necessary, a Wagner balance can be added.

9.3.6 Coaxial resistance ratio bridges involving quantum Hall devices

Quantum Hall resistance has the advantage that its quantised value depends only on unchangeable fundamental constants and is precisely independent of time, temperature, barometric pressure and all the other side effects that conventional resistance artefacts suffer from. Another advantage is the very low thermal noise power of the cryogenic quantum Hall device compared to room temperature resistance standards. Furthermore, the currents flowing through the defining Hall potential contacts are automatically minimised by the multiple-series connection scheme, and they are even smaller than can be manually adjusted in a conventional four-terminal-pair bridge. Therefore, a quantum Hall device in multiple-series connection can automatically take over the function of both a current source and a Kelvin arm. This property simplifies conventional four-terminal-pair bridges (see sections 9.1 and 9.2) as shown in the following examples.

Figure 9.19 shows a ratio bridge for comparing a triple-series connected quantum Hall device with a conventional four-terminal-pair resistance standard. While a ratio bridge for comparing two conventional four-terminal-pair standards requires a Kelvin network and two current sources with their associated current detection transformers, the very low self-adjusted residual current in the defining potential leads of the quantum Hall device is so small that the Kelvin network and one current source can be omitted.

If a ratio bridge is to compare two quantum Hall resistances directly with each other, the remaining current source and the associated current detection transformer can be omitted (Figure 9.20). Not only this further simplifies the ratio bridge but

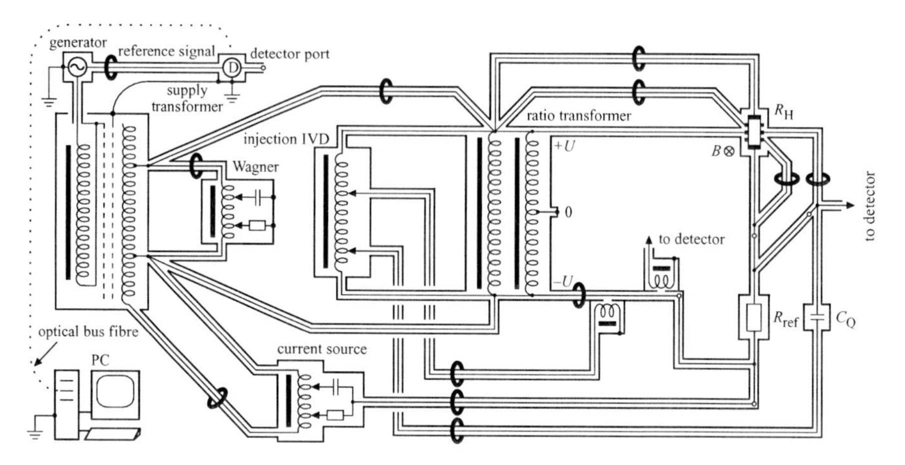

Figure 9.19 Comparing a triple-series connected quantum Hall device, R_{H1}, with a four-terminal-pair resistor, R_{ref}

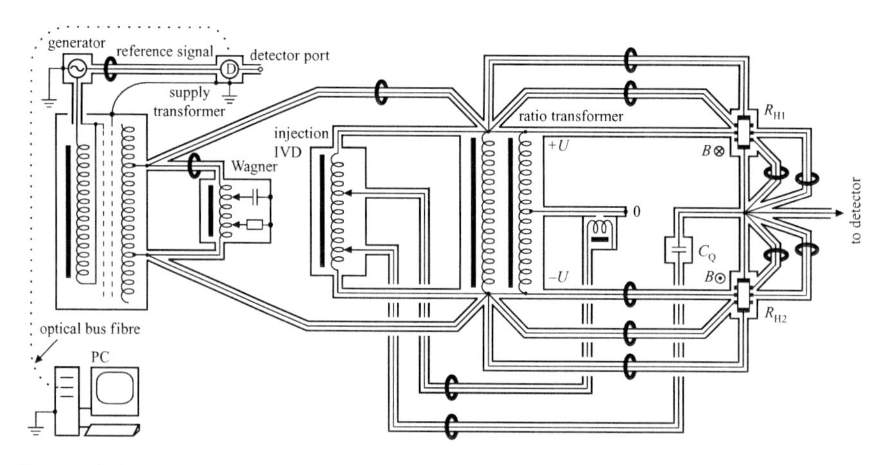

Figure 9.20 Comparing two triple-series connected quantum Hall devices, R_{H1} and R_{H2}

also, because both quantum Hall devices are operated at low temperature and no other room temperature resistors are required in this bridge network, the noise level is very low and a very high resolution is possible. To simplify the drawing, Figure 9.20 shows the two quantum Hall devices as if they would experience a magnetic field of different polarity.

9.3.7 A quadrature bridge with two quantum Hall devices

Multiple-series connected quantum Hall devices can also be directly incorporated in a quadrature bridge ([26] of chapter 6). This simplifies the bridge network in the same way as for the ratio bridge described in the previous section. Whereas a conventional four-terminal-pair quadrature bridge is the most complicated coaxial bridge network usually encountered (Figure 9.9 and section 9.2.3), requires a somewhat tedious balancing procedure and suffers from a poor signal-to-noise ratio, the remarkable properties of quantum Hall resistance can greatly improve all these aspects. Three combining networks and four injection/detection transformers can be omitted, as shown in Figure 9.21. As a result, the balancing procedure and convergence are also improved. Finally, cryogenic quantum Hall resistances permit a very low overall thermal noise level if the resistors of the detector combining network are also at a low temperature and if a low-noise harmonic filter (see Figure 2.4 and section 2.1.2) and a low-noise preamplifier (see section 2.3.1) are used. Altogether, the operation of such a quadrature bridge is, compared to a full four-terminal-pair quadrature bridge with room temperature resistance standards, amazingly simple.

9.4 High-frequency networks

Callegaro [8] has published a useful review, containing an exhaustive list of references, of this subject.

The networks in this section are modelled closely on those designed for kilohertz frequencies described in sections 9.1 and 9.2. The main variation is that the components are smaller and the lengths of interconnecting cables shorter to maintain the dimensions of the network much less than a quarter wavelength at the operating frequency. Because only more modest accuracies are needed at higher frequencies, some simplifications are possible to help fulfil this need. In particular, current equalisers where a cable is threaded only once through a core are usually adequate for frequencies up to 1 MHz. At frequencies greater than about 1 MHz, they can be omitted entirely because the go-and-return inductance of a cable is about a factor of 10 less than the loop inductance and resistance of its outer conductor even if the loop inductance is not enhanced by threading a core.

Because of their small size, Sub-Miniature type B (SMB) connectors are suitable for cable terminations and the ports of components, but they are somewhat delicate. To conserve their working life, a suitable tool should be constructed and used so that only axial force is applied to plug and unplug them.

Small magnetic cores wound from nanocrystalline metal ribbon of micrometre thickness have sufficient permeability for frequencies up to 1 MHz to construct transformers described in this section. High-permeability ferrite cores are needed for higher frequencies. In general, transformers can be kept below self-resonance with sufficient input impedance by using far fewer turns than at kilohertz frequencies. This will also minimise the parasitic inductances, resistances and interwinding capacitances, which cause departure from their nominal voltage ratio.

9.4.1 An IVD-based bridge for comparing 10:1 ratios of impedance from 10 kHz to 1 MHz

Figure 9.22 is a diagram of this bridge. The high-potential ports of the four-terminal-pair impedances Z_A and Z_B are connected with identical cables to the ratio windings of the two-staged in-phase ratio IVD. The high-current ports of Z_A and Z_B

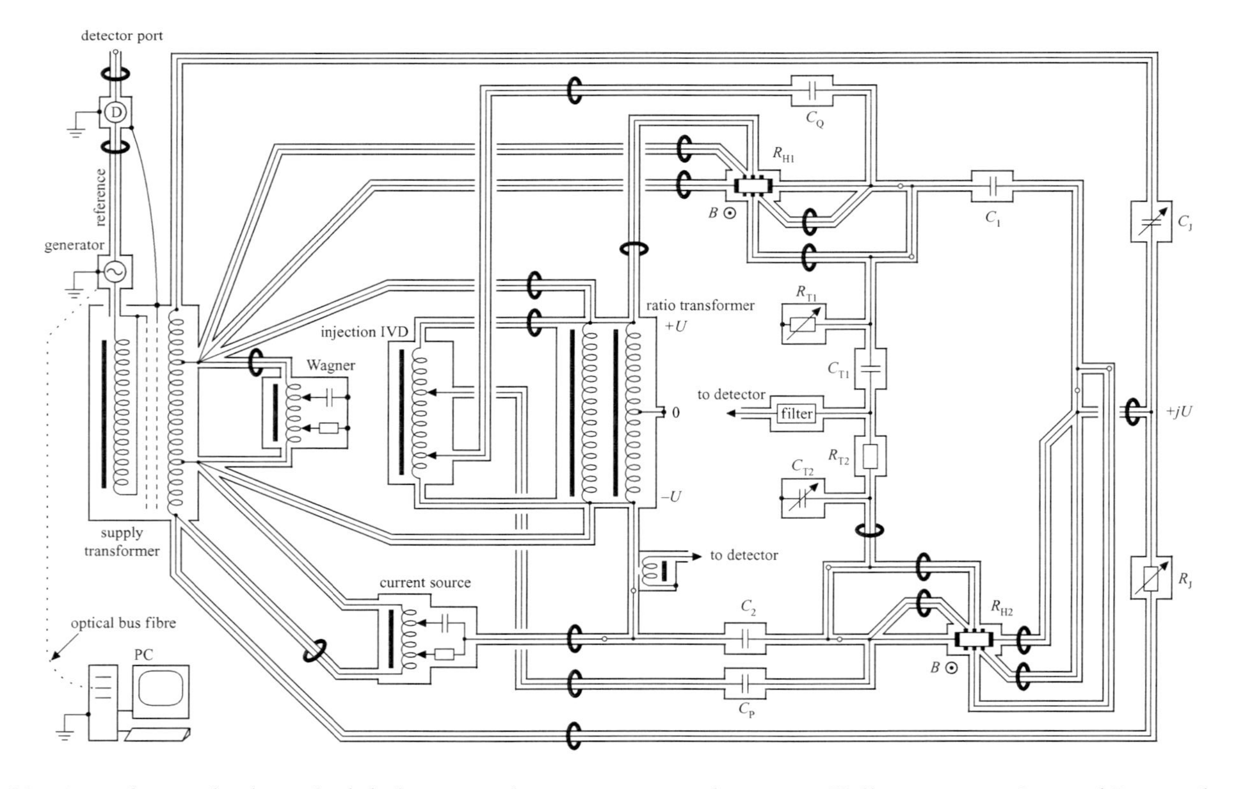

Figure 9.21 A quadrature bridge, which links two triple-series connected quantum Hall resistances, R_{H1} and R_{H2}, and two capacitances, C_1 and C_2, according to $\omega^2 R_{H1} R_{H2} C_1 C_2 = 1$

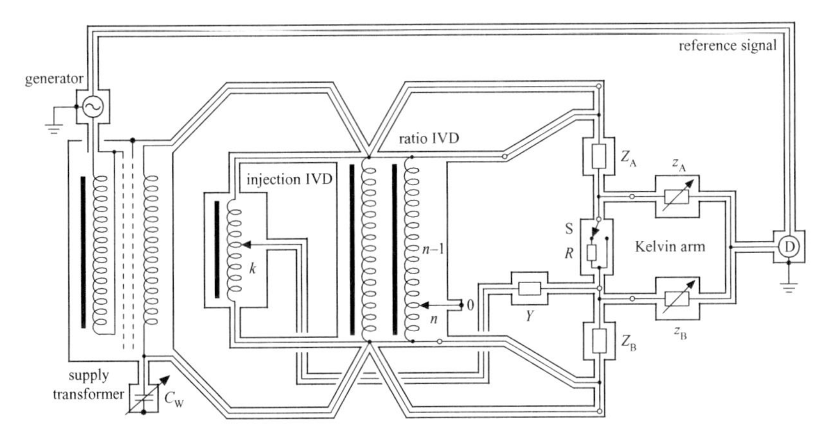

Figure 9.22 A quasi-four-terminal-pair IVD-based bridge for frequencies up to 1 MHz

are connected to its magnetic winding to minimise the current in the potential connections. The defining condition of zero current at the IVD ports is determined by extrapolation of the results obtained from altering the lengths of these cables. The adjustable impedances z_a and z_b at the low-potential ports of Z_A and Z_B are small 5–10 pF trimmer capacitors, which form combining network. The low-current ports of the standards are connected via a 500-Ω resistance R, which can be shorted by a switch S. S is opened for balancing the combining network and closed to balance the bridge. These balances are iterated as necessary. The condition of zero current through the potential reference point O of the bridge is obtained by adjusting the Wagner capacitance C_W so that the detector is nulled when the shorting plug at O is removed.

If Z_A and Z_B are capacitive, $1/Y$ should be a resistance, and the bridge balance equations are

$$C_A = \frac{C_B n}{1 - n} \tag{9.24}$$

where n is the ratio IVD setting

and

$$D_A = D_B - Y(\omega C_B)^{-1} \frac{k - n}{n} \tag{9.25}$$

where k is the injection IVD setting.

An accuracy of 0.01% has been achieved when measuring the ratio of 100-pF and 1-nF gas-dielectric capacitors.

A general method for calibrating the IVD voltage ratios similar to sections 7.4.1 to 7.4.5 has yet to be developed, but particular ratios including 10:−1 can be calibrated using the permuting capacitors device [9], described in section 7.4.6 in the simple bridge drawn in Figure 9.23. Alternatively, a ratio transformer in a

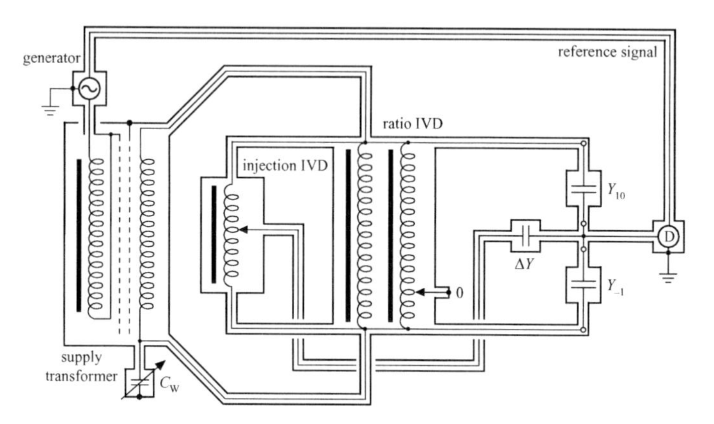

Figure 9.23 Calibrating a 10:−1 ratio transformer or IVD with a permuting capacitors device

bridge can be calibrated by balancing the bridge with the 0.1-pF and 1-pF or the 1-pF and 10-pF adjustable capacitors described in section 6.5.8. The transformer ratio at any frequency can then be inferred from their values obtained by measuring their 1-kHz values and assuming their frequency dependence as calculated from their self-resonant frequencies. This method will, in general, be less accurate than the permuting capacitors method, but it may be adequate and is quicker to implement.

9.4.2 A bridge for measuring impedance from 10 kHz to 1 MHz based on a 10:−1 voltage ratio transformer

This network (Figure 9.24) is similar in construction to that of the previous section. The ratio IVD is replaced by a 10:−1 fixed ratio voltage transformer, and current detection transformers have been added to the voltage ratio ports of the transformer. Detectors connected to the current detection transformers are nulled to ensure zero current through these ratio ports by adjusting the binary current sources 1 and 2 connected to the high-current connections of Z_1 and Z_2. The two-staged high-frequency injection IVD provides the main bridge balance via appropriate admittances Y_A and Y_B (in-phase and quadrature injection components). This IVD as well as the main 10:1 voltage ratio transformer are calibrated at frequencies up to 1 MHz by the high-frequency permuting capacitors technique (see section 7.4.6). As the accuracy of this complete 4TP bridge for capacitance or resistance standards is only 10 ppm at 1 MHz, current equalisers are not necessary, and the small frequency dependences of the injection admittances Y_A and Y_B have negligible effect.

9.4.3 Quasi-four-terminal-pair 1:1 and 10:1 ratio bridges for comparing similar impedances from 0.5 to 10 MHz

Figure 9.25 is the circuit diagram of a network that can be configured with an appropriate ratio transformer as a 1:1 or 10:1 bridge. The combining network is two coaxial cables whose lengths are also in the same ratio. This bridge is very

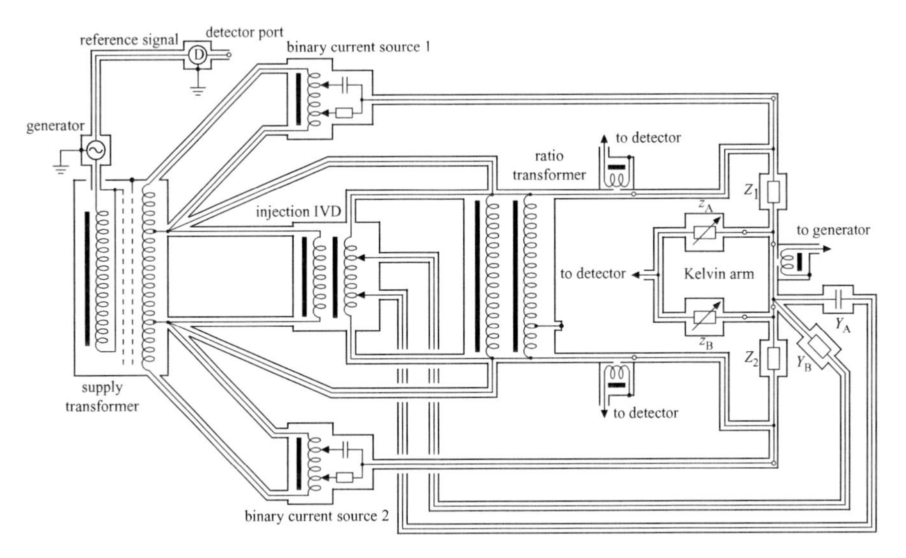

Figure 9.24 A complete 4TP 1-MHz impedance bridge based on a fixed 10:1 voltage ratio transformer and binary IVD current sources

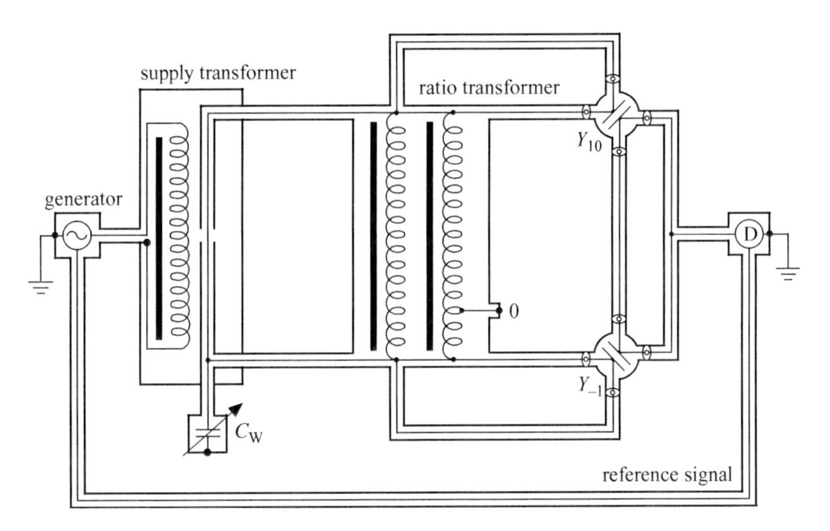

Figure 9.25 A bridge for comparing similar impedances from 0.5 to 10 MHZ

convenient for calibrating commercial high-frequency capacitance standards, particularly of the type illustrated in the lower half of Figure 9.29.

1:1 and 10:−1 two-staged ratio transformers can be constructed using high-frequency ferrite cores, which exhibit low reluctance and small loss over the 0.5–10 MHz frequency range. They remain below self-resonance if both stages are wound with about 10 spaced turns of 0.3-mm-diameter poly-tetraflouride-ethelene

(PTFE) insulated wire, and their impedance does not unduly load a 50-Ω source. For the 1:1 transformer, the first core has ten turns, the second core is placed axially next to it and the combination wound with ten turns, interleaved with the exposed portions of the first-stage winding so that turns of similar potential are adjacent, and tapped after five turns. The ends and tap of the second-stage winding terminated on three SMB connectors soldered into one end cap of the 22-mm-diameter, 25-mm-long copper cylindrical enclosure. The first stage terminates on two SMB connectors soldered into a small sub-plate and bolted to but insulated from the other end cap.

The 10:1 ratio transformer is broadly similar, except that its first-stage winding has 11 turns. The second-stage winding, a single axial conductor between connectors soldered into the centre of each end cap constitutes the -1 winding when a shorting plug is inserted into the connector having the appropriate end of the ten-turn winding connected to its inner pin. The shorting plug can be removed to adjust C_W to make the Wagner balance.

9.4.4 A four-terminal-pair 10-MHz 1:1 resistance ratio bridge

The circuit diagram is shown in Figure 9.26. The circuit again follows low-frequency coaxial bridge principles. The isolation transformer and the provision of the 1:1 voltage ratio are combined into a single transformer. This reduces the complexity of the bridge significantly, and this is important for accurate high-frequency measurements in order to keep the circuit simple and therefore the interconnecting cables short. The capacitance between the primary and secondary windings of the isolation transformer is 0.8 fF. Z_1 and Z_{-1} are the main four-terminal-pair resistance standards, and their high-current terminals are driven from the $+2U$ and $-2U$ potential terminals of the isolation/ratio transformer via adjustable binary parallel capacitive and resistive components having 5-bit coarse resolution. The high-potential terminals of the standards are connected to the $+U$ and $-U$ terminals of the isolation/ratio transformer via detection transformers D_1 and D_2. The '0' tap of the transformer is the bridge zero defining point, being the only point where its inner and outer conductors are shorted together. The low-current and low-potential terminals of the standards are connected to the switch S and a combining network consisting of a 100-Ω miniature variable resistance z_A and a fixed 50-Ω resistance z_B. The bridge balance procedure is the same as that of a similar low-frequency bridge except that the bridge in-phase and quadrature balance condition is calculated by extrapolation from the small change produced in the detector reading by shunting a known impedance ΔZ across one of the standards and readjusting the auxiliary balances. The linearity of the detector is verified separately, and the error from the frequency dependence of ΔZ can be sufficiently small. Typically, if Z_1 and Z_2 are nominally 1 kΩ, ΔZ could be between 10 and 100 kΩ. The error of the 1:1 transformer ratio is eliminated by interchanging the connections to it and taking the mean of the results.

The complete bridge system is contained within a 0.25-m radius. This is sufficiently smaller than the 7.5-m quarter wavelength at 10 MHz that all the bridge

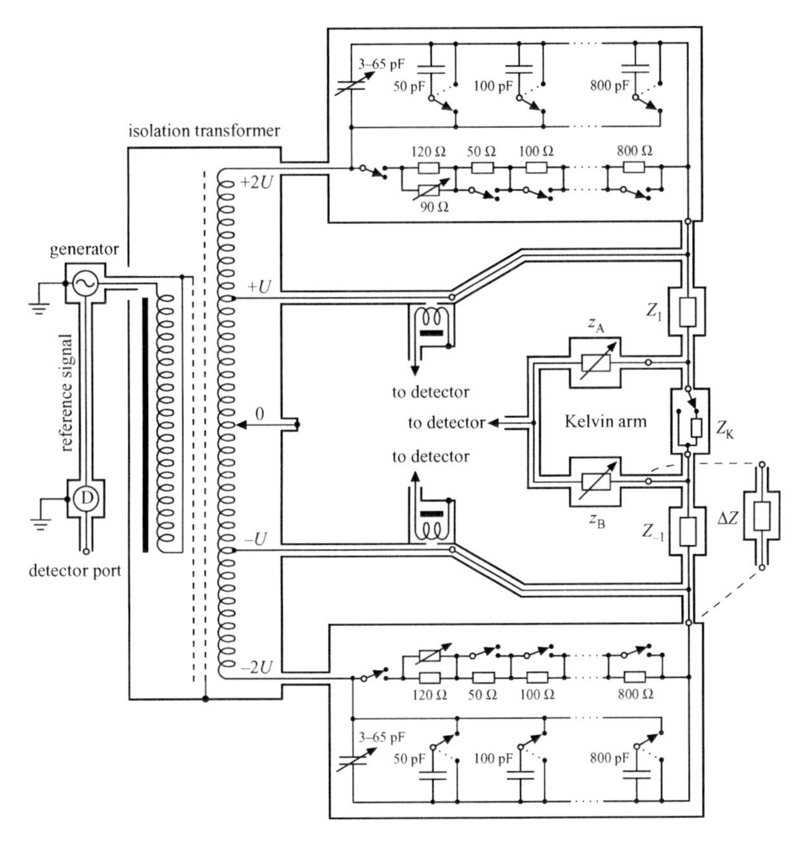

Figure 9.26 A 1:1 four-terminal-pair 10 MHz resistance ratio bridge

components can be modelled by lumped parameters. Also, reflections caused by differing characteristic impedances between the cables, connectors and standards are not significant. The correct functioning of the bridge can be tested by comparing two identical high-frequency calculable coaxial resistance standards (see section 6.5.3).

9.4.5 A 1.6- and 16-MHz quadrature bridge

A quasi-four-terminal-pair quadrature bridge system operating at 1.592 MHz is shown in Figure 9.27. The bridge operates in a similar manner to the conventional low-frequency quadrature bridge system, discussed in more detail in sections 9.2.2 and 9.2.3. Measurements confirming correct operation have been made by directly comparing the calculated and measured frequency dependence of two 100-pF capacitance and two 1-kΩ resistance standards between 1.592 kHz and 1.592 MHz (10^4 rad/s and 10^7 rad/s, respectively). Operation at 15.92 MHz (10^8 rad/s) is also possible using the same bridge system as well as the same capacitance standards but with two 100-Ω resistance standards. The network in Figure 9.27 compares the mean value of capacitors C_1 and C_2 to that of resistors R_1 and R_2. On the right is a

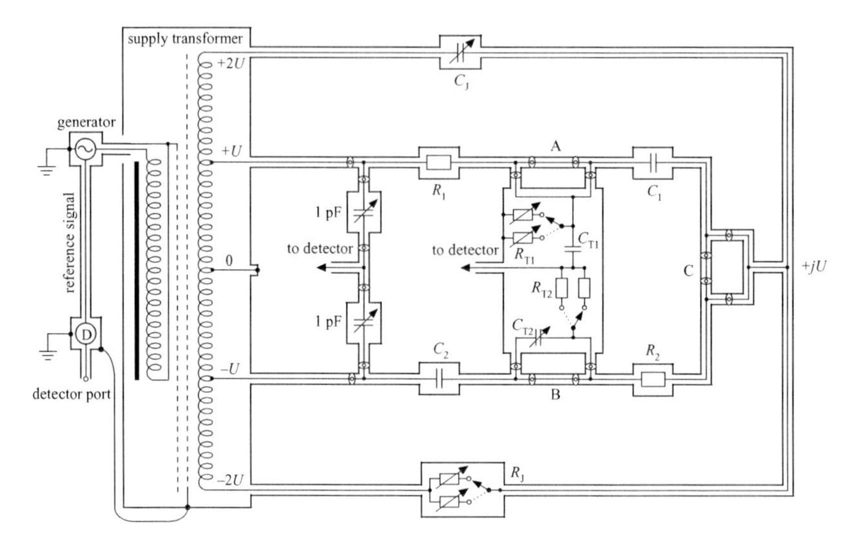

Figure 9.27 A quasi four-terminal-pair quadrature bridge operating at 1.6 or 16 MHz

high-frequency supply transformer having output voltages of $\pm U$ and $\pm 2U$. The ratio of the potentials at the P_H ports of R_1 and C_2 is measured by the sub-bridge consisting of 1-pF capacitors. These parallel-plate air-dielectric capacitors are designed for minimum change of value with frequency and are of as identical construction as possible, so the residual change with frequency is the same for each capacitor. They are provided with conducting screws, which impinge on their fringing electric fields, thus enabling their value to be finely adjusted (see section 6.5.8). Although provided with four-terminal-pair terminations, here only their H_P and L_C ports are connected so that they become two-terminal-pair components. A high-frequency phase-sensitive detector is temporarily connected to them and balanced by adjusting one or the other capacitor slightly, and the ratio of their values at 1 kHz is measured on a separate high-resolution bridge by carefully disconnecting them temporarily from the network. When reconnected, this ratio is also the required ratio of the potentials at the P_H ports of C_1 and R_4. The small and nearly equal input impedances of these capacitors (~3×10^4 Ω and ~3×10^3 Ω at 1.6 and 16 MHz, respectively), together with that of any identical cables, ensure that the departure from $I = 0$ defining condition at these ports is violated only to a small and nearly equal extent.

The j-network is connected to the $\pm 2U$ ports of the supply transformer by two short cables. It consists of C_J and R_J. The capacitor C_J is ~68 pF in parallel with a 5–65 pF miniature vane trimmer capacitor, housed in a small copper box. The resistor R_J is either a 560-Ω resistor in series with a 0–1 kΩ miniature trimmer resistor or a 56-Ω resistor in series with a 0–100 Ω miniature trimmer resistor, housed in another copper box. These resistor combinations can be switched into the circuit with a miniature toggle switch for bridge operation at 1.6 or 16 MHz, respectively. These components, and those in the detector combining network described below, have been verified on an impedance analyser to have adequate high-frequency performance, and since they form an auxiliary network adjusted to a balanced state, their exact values are immaterial. An insulated trimmer tool can be inserted through a small hole in their housing boxes to adjust their values. The components are mounted within the box sufficiently far from the aperture to ensure that the effects of emanating electric fields on the rest of the bridge network are negligible.

The detector combining network is similarly housed in a copper shielding box. Either a 1-kΩ or a fixed 100-Ω resistor R_{T2} is connected in the circuit by a switch for 1.6- or 16-MHz operation, respectively, and the 1-kΩ or 100-Ω adjustable resistors R_{T1}, similarly by another switch. The detector combining network is adjusted by altering the appropriate variable resistor and the variable capacitor to null the detector. The bridge network also contains three non-adjustable combining networks, A, B and C made from low-impedance coupling connectors joining the defining ports P_H or C_L and other connectors joining the P_L or C_H ports adjacent main bridge components in such a way that the potential or current pick-off points are in their centres. There are no current equalisers as the go-and-return impedance of the coupling connectors and cables is more than a factor of 10 lower than the loop impedances of their outer conductors, thus ensuring that current balancing in the interconnecting cables is adequate for the required level of accuracy of about 0.1% at 1.6 MHz.

On comparing the known values of R_1, R_2, C_1 and C_2 with those measured on the quasi-four-terminal-pair HF quadrature bridge, Awan and Kibble [10] found a difference of only 0.17% at 1.6 MHz.

9.4.6 Four-terminal-pair resonance frequency measurement of capacitors

The four-terminal-pair resonance frequency f_0 of capacitance standards can be determined by measuring with the circuit shown in Figure 9.28a a set of four parameters over a range of frequencies near f_0. The capacitances C_i and C_d are chosen to be small enough (typically 0.1 pF) not to perturb the measurement parameters significantly. These parameters are the two-terminal-pair impedances appearing in the impedance matrix (Z) defined in (9.26).

$$\begin{bmatrix} V_1 \\ V_2 \\ V_3 \\ V_4 \end{bmatrix} = \begin{bmatrix} Z_{11} & Z_{12} & Z_{13} & Z_{14} \\ Z_{21} & Z_{22} & Z_{23} & Z_{24} \\ Z_{31} & Z_{32} & Z_{33} & Z_{34} \\ Z_{41} & Z_{42} & Z_{43} & Z_{44} \end{bmatrix} \cdot \begin{bmatrix} I_1 \\ I_2 \\ I_3 \\ I_4 \end{bmatrix} \tag{9.26}$$

where $Z_{ij} = V_i/I_j (i,j = 1,2,3,4)$. For a linear network obeying reciprocity, $Z_{ij} = Z_{ji}$.

The four-terminal-pair impedance is, in the notation of Figure 9.28b,

$$Z_{4TP} = \left. \frac{V_2}{I_4} \right|_{I_2=I_3=V_3=V_4=0} \tag{9.27}$$

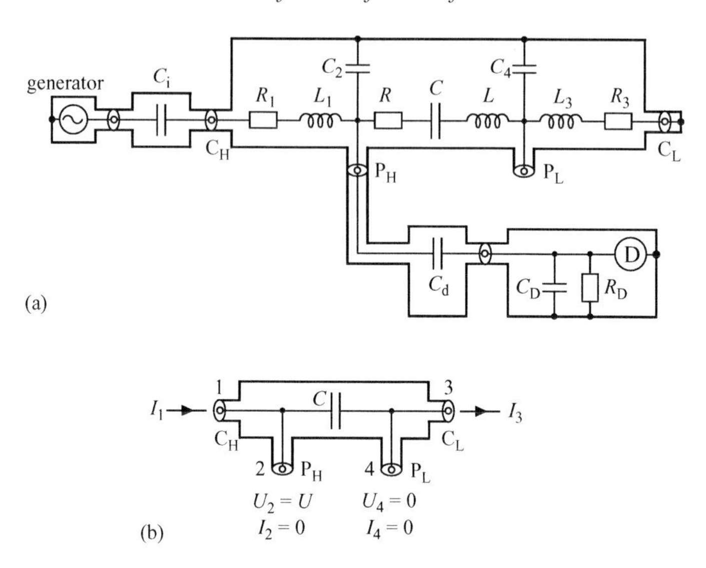

Figure 9.28 (a) A circuit for measuring the four-terminal-pair resonance frequency of a gas-dielectric capacitor C. (b) The four-terminal-pair definition of the capacitor

From knowledge of f_0, the frequency dependence of the capacitor for frequencies $f \ll f_0$ can be calculated from (6.43). It is also possible to calculate the dissipation factor of gas-dielectric capacitors from the -3 dB width of the measured resonance curve, that is, the ratio of the difference of the frequencies f_1 and f_2 to f_0, which are 3 dB down from the peak at f_0. Capacitors whose frequency dependence has been determined in this way can calibrate LCR meters and impedance analysers at frequencies that are sufficiently lower than f_0.

Equation (9.27) can be rearranged using the single-port definitions given in (9.26) to give [11]

$$Z_{4TP} = \frac{Z_{24}Z_{31} - Z_{21}Z_{34}}{Z_{31}} \tag{9.28}$$

9.4.7 Scattering parameter measurements and the link to microwave measurements

The instruments commonly used for impedance measurements between 10 MHz and about 300 GHz are two- or four-port vector network analysers. These instruments can also make four-terminal-pair measurements, Z_{4TP}, as detailed below.

In the so-called 'indirect *Z*-matrix method' [12,13], Z_{4TP} is estimated by measuring the individual elements Z_{ij} of the 4×4 *Z*-matrix of (9.26) with a one-port network analyser. The diagonal elements Z_{ii} are measured directly with the other three ports left open. The off-diagonal elements Z_{ij} are measured using the relation

$$Z_{ij} = Z_{ji} = \sqrt{Z_{jj}(Z_{ii} - Z_{ij}^{sj})} \tag{9.29}$$

where Z_{ij}^{sj} is the impedance measured at port i when port j is shorted and the remaining ports are left open. By combining (9.28) and (9.29), we obtain the Suzuki relation for a four-terminal-pair admittance in terms of single-port definitions:

$$Y_{4TP} = \sqrt{\frac{Z_{11} - Z_{11}^{s3}}{Z_{22}}}\left[\sqrt{(Z_{11} - Z_{11}^{s2}) \cdot (Z_{44} - Z_{44}^{s3})} - \sqrt{(Z_{11} - Z_{11}^{s3}) \cdot (Z_{44} - Z_{44}^{s2})}\right]^{-1} \tag{9.30}$$

and hence, $Z_{4TP} = (Y_{4TP})^{-1}$. The limitation of this method is that one-port measurements made at a given frequency with network analysers are not sufficiently accurate for an evaluation of Z_{4TP}, due to, for example, contact impedances, cable loading and matching, and instrument drift. These difficulties have been overcome by making measurements over a much larger bandwidth than the frequency range of interest to identify the magnitude of the residual and parasitic impedances. To date, this method has only been applied to a limited range of gas-dielectric capacitance standards.

Callegaro and Durbiano [14] showed that it is also possible to measure four-terminal-pair impedance standards using scattering parameters measured with a vector network analyser. Since an impedance matrix $[Z]$ and its scattering parameter matrix $[S]$ are related by

$$[S] = ([Z] - Z_0[I])([Z] + Z_0[I])^{-1} \tag{9.31}$$

where $[I]$ is the identity matrix of order n, they showed that a four-terminal-pair impedance can be written as

$$Z_{4TP} = 2Z_0[s_{21}s_{34} - s_{31}s_{24}] \cdot \left[s_{31} + \begin{pmatrix} s_{21}s_{32} - s_{31}s_{44} - s_{31}s_{22} + s_{41}s_{34} - s_{21}s_{32}s_{44} + s_{21}s_{34}s_{42} \\ +s_{31}s_{22}s_{44} - s_{31}s_{42}s_{24} - s_{41}s_{34}s_{22} + s_{41}s_{24}s_{32} \end{pmatrix}\right]^{-1} \tag{9.32}$$

where Z_0 is the characteristic impedance of the network. If the network is terminated with a cable of characteristic impedance $Z_0 = \sqrt{(Z/Y)}$, then the cable appears to the source to be infinitely long and other sections of the same characteristic impedance can be added without affecting the source. In contrast, with guided waves, a transverse electromagnetic wave in free space has a characteristic impedance of $\sqrt{(\varepsilon_0/\mu_0)} = 377\ \Omega$.

Although there are some advantages of performing s-parameter measurements to determine a four-terminal-pair impedance, the method is generally only accurate for impedance values close to Z_0. In this case, since $|s_{ij}|^2 \leq 1$, most of the significant terms in (9.32) are transmission *s*-parameters (i.e. s_{21}, s_{24}, s_{31} and s_{34}). Reference plane definitions and the effect of impedance-matched connector adaptors are less critical than in reflection measurements. Furthermore, Z_{4TP} can be directly computed from (9.32) instead of relying on an equivalent electrical model of the device being measured. A four-port vector network analyser can make the measurements in a single run, but with a two-port vector network analyser, the matrix elements of [*S*] have to be measured sequentially by connecting two terminal pairs at a time while terminating the remaining two terminals with matched loads, as illustrated in Figure 9.29.

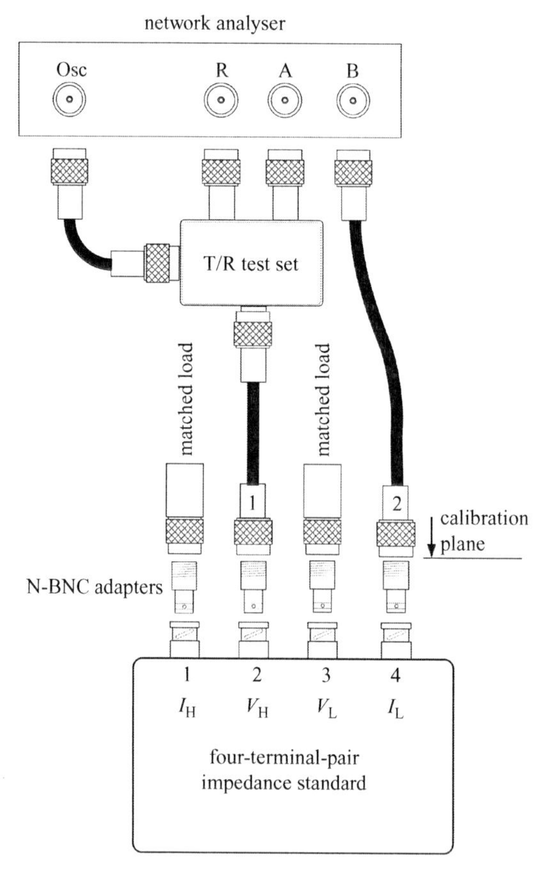

Figure 9.29 An example of a connection between a four-terminal-pair impedance to a two-port vector network analyser. The particular connection scheme shown is to measure s_{22} and s_{42}

For example, scattering parameter measurements of a General Radio 1-nF capacitance standard calculated from the above expressions shows that the apparent capacitance of the standard is about 30% higher at 10 MHz compared to its 1 kHz value (Figure 9.30b).

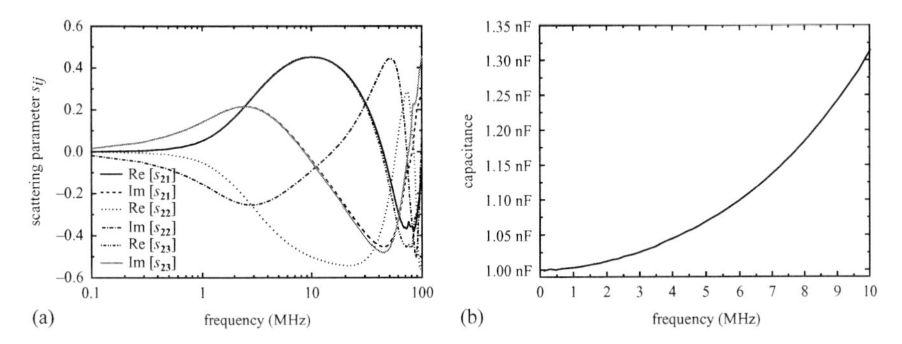

Figure 9.30 (a) Scattering parameter measurements of a 1-nF capacitor. (b) The frequency dependence calculated from these measurements

Despite the convenience and elegance of this method, evaluation of the uncertainty is difficult and is still the subject of debate by RF and microwave experts. In particular, it is not clear how to evaluate the uncertainties of the input quantities in (9.32) and how to address their propagation through this equation because it involves complex functions and strong correlations. Callegaro [14] has suggested performing a numerical uncertainty evaluation with propagation being accounted for by the Monte Carlo method.

9.4.8 Electronic four-terminal-pair impedance-measuring instruments

Commercial automated LCR and impedance analyser instruments are commercially available, which claim to offer moderate accuracies, operate in milliseconds and cover a wide frequency range from 20 Hz to 110 MHz and a wide impedance range from 1 mΩ to 1 GΩ. In these precision instruments, an attempt is also made to implement the four-terminal-pair definition of impedances Z to be measured in order to eliminate the effect of contact resistances and stray capacitances and mutual inductances in the connections to the impedance. The maximum AC voltage that these instruments supply for impedance measurement is usually 1 V_{rms}, but they can also provide up to 40 V DC bias for various applications. The claimed measurement uncertainty varies depending on the impedance being measured and the measurement frequency but is typically between 0.1% over most of their range up to 10% at the boundaries of the instrument's measuring capability. Their applications are wide ranging, from electrical and electronic component and materials evaluation to chemical and biological process measurements in a number of industries such as aerospace, biomedical, general research and development and telecommunications.

Most employ an auto-balancing method in which high-input-impedance amplifiers and feedback techniques create a virtual potential reference point P ('earth') at the low-potential port of the instrument, as shown in Figure 9.31. The instruments usually have four selectable range resistors, R_{ref}, across which a voltage

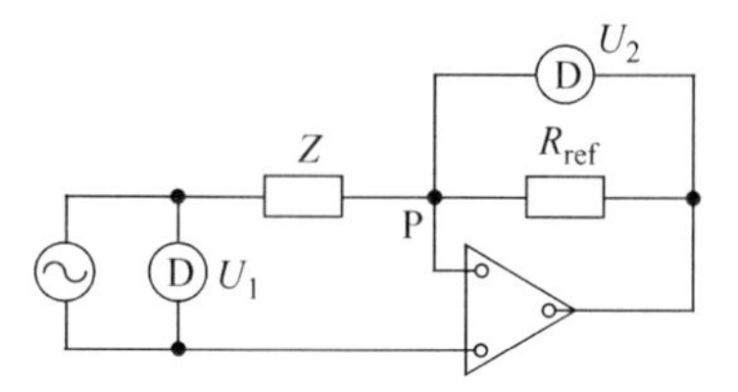

Figure 9.31 The auto-balancing technique typically employed in commercial instruments to measure a four-terminal-pair impedance Z

U_2 is detected to determine the current flowing through Z. Often an assumption is made that the high-frequency value of these resistors is the same as their DC value, which is not entirely true for accurate measurements. These, and other imperfections such as inexact obeying of the four-terminal-pair defining conditions, are corrected by internal computer software to obtain the expected results for known calibrating impedances. Unfortunately, the connections to calibrating impedances may not mimic sufficiently exactly connections to other devices to be measured and there are other effects such as shunt capacitive loading at the high- and low-potential ports, inadequate input impedance of the voltage detectors U_1 and U_2 over the whole frequency range, inequality of the currents in the inner and outer conductors of connecting coaxial cables and incorrect accounting for cable corrections (see section 5.4).

In quality control applications where pass/fail measurements are performed on passive components relatively quickly, instrument manufacturers supply a number of probes and adaptors. A common example is 'Kelvin clips' in which four coaxial cables are connected to the impedance analyser at one end and their inners are connected to two crocodile clips and their outers are shorted together at the other. The two inner surfaces of each crocodile clip are insulated from each other, and one makes the potential connection, and the other, the corresponding high- or low-current connection. A similar connection scheme is employed in various four-terminal-pair to four-terminal adaptors for measuring chip surface-mounted components.

For measurements at frequencies higher that 100 MHz up to at least 110 GHz, two-port 50-Ω matched impedance or vector network analysers are usually employed.

At the less expensive end of the market, there are numerous low-accuracy two-terminal hand-held LCR meters. These typically measure a wide range of impedances at one or two fixed frequencies, 120 Hz and 1 kHz being the most common, but offer at best only about 1% accuracy.

References

1. Cutkosky R.D. 'Techniques for comparing four terminal-pair admittance standards'. *J. Res. NBS* 1970;**740**:63–78

2. Kibble B.P., Rayner G.H. *Coaxial AC Bridges*. Bristol: Adam Hilger Ltd.; 1984. (Presently available from NPL, Teddington, TW11 0LW, U.K.)
3. Awan S.A., Jones R.G., Kibble B.P. 'Evaluation of coaxial bridge systems for accurate determination of the SI farad from the DC quantum Hall effect'. *Metrologia*. 2003;**40** (5):264–70
4. Hanke R., Kölling A., Melcher J. 'Inductance calibration in the frequency range 50 Hz to 1 MHz at PTB'. *CPEM 2002 Digest*. 2002;186–87
5. Côté M. 'A four-terminal coaxial-pair Maxwell-Wien bridge for the measurement of self-inductance'. *IEEE Trans*. 2009;**IM58 No. 4**:962–66
6. Sedlacek R., Bohacek J. 'Bridges for calibrating four terminal-pair standards of self-inductance at frequencies up to 10 kHz'. *Meas. Sci. Technol.* 2009; **20**:25–105
7. British Standards Institution BS1041-3. BSI 389 Chiswick High Road, London, W4 4AL U.K.
8. Callegaro L. The metrology of electrical impedance at high frequency: a review. *Meas. Sci. Technol.* 2009;**20**:1–14
9. Awan S.A., Kibble B.P., Robinson I.A. 'Calibration of IVD's at frequencies up to 1MHz by permuting capacitors'. *IEE Proc. Sci. Meas. Technol.* 2000;**147** (4):193–95
10. Awan S.A., Kibble B.P. 'A universal geometry for calculable frequency response coefficient of LCR standards and new 10 MHz resistance and 1.6 MHz quadrature bridge systems'. *IEEE Trans. Instrum. Meas.* 2007;**56** (2):221–25
11. Awan S.A., Callegaro L., Kibble B.P. 'Resonance frequency of four-terminal-pair air-dielectric capacitance standards and closing the metrological impedance triangle'. *Meas. Sci. Technol.* 2004;**15**:969–72
12. Suzuki K. 'A new universal calibration method for four-terminal-pair admittance standard'. *IEEE Trans. Instrum. Meas.* 1991;**40** (2):420–22
13. Koffman A.D., Avramov-Zamurovic S., Waltrip B.C., Oldham N.M. 'Uncertainty analysis for four terminal-pair capacitance and dissipation factor characterization at 1 and 10 MHz'. *IEEE Trans. Instrum. Meas.* 2000;**49** (2):346–48
14. Callegaro L., Durbiano F. 'Four-terminal-pair impedances and scattering parameters'. *Meas. Sci. Technol.* 2003;**14**:1–7

Chapter 10

Application of interference-free circuitry to other measurements

10.1 Resistance thermometry (DC and low-frequency AC)

Resistance thermometry is the most precise method of measuring temperature in the range from 30 to 1000 K. The sensor is usually a compact resistor of platinum (or platinum-rhodium alloy for higher temperatures), whose value at ambient temperatures is about 25 or 100 Ω. The change of resistance with temperature is about 0.4% per Kelvin, so that measurement of the resistance to 1 ppm yields temperature with a resolution of about 0.25 mK. A given thermometer can be calibrated at the various 'fixed points', which are the melting or triple points of water and various metals, and a temperature scale derived from interpolation between these.

10.1.1 DC resistance thermometry

Measuring the resistance of a thermometer by DC resistance comparison bridges is the older technique and is still in use. A plethora of schemes of increasing complexity for connection to a Wheatstone bridge (see section 5.1.1) or a Kelvin double bridge (see section 5.6.1) have been devised to overcome to a more-or-less degree the problem of the resistance of connecting wires affecting the measurement (see, for example, Reference 4). We would advocate the simple and elegant use of just two coaxial cables devised by Cutkosky and described in section 5.3.5. Because one cable carries the measuring current to and from the thermometer and the other connects its two potential contacts through which there is no current, both cables are inherently current-balanced with the outer conductors acting as screens, this connection scheme is interference free. Any bridge or instrument that satisfies the defining conditions of a four-terminal resistance will yield a correct measurement. The enclosing shields of the thermometer element and its attached cables should not be electrically connected to anything other than the measuring instrument or circuit to ensure that it is isolated. The incomplete isolation of any commercial instrument that measures its resistance should then not cause a problem.

10.1.2 AC resistance thermometry

The connection scheme of section 5.3.5 is also suitable for low-frequency AC methods of measuring thermometer resistance. The capacitance of the cables shunts the thermometer and gives it a quadrature component, so that the measuring bridge or instrument must be capable of ignoring this and reporting only the in-phase resistance component. The dielectric loss in the cable is also a potential cause of error, but present-day materials have a sufficiently low loss, for lengths of cables of the order of tens of metres and for the usual measurement frequencies of 25 or 75 Hz, the effect of dielectric loss is negligible. Measuring instruments are commonly based on the IVD ratio bridge described in section 9.1.3. Often the IVDs are switched by relays under the control of an internal computer so that the bridge is self-balancing, and the computer reports the thermometer resistance directly.

10.2 Superconducting cryogenic current comparator

10.2.1 Determining the DC ratio of two resistances R_1/R_2

This elegant instrument is the product of superconducting technology applied to the DC comparison of two resistances. In particular, one resistance can be a quantum Hall device, and therefore, the instrument can set up a resistance scale in terms of the defined value of the von Klitzing constant, which represents the SI defined ohm. Here, we only describe briefly the salient features of the instrument needed to use it as an example of the particular problems encountered in DC measurements in general. Figure 10.1 is a schematic diagram of a cryogenic current comparator (CCC) circuit. More details can be found in Reference 5.

A CCC produces a near-perfect flux linkage between its ratio windings carrying currents I_1 and I_2 and a detector winding. A fixed proportion of any residual flux imbalance produced by the ratio windings threads the detector winding and is transferred to the coil of the Superconducting Quantum Interference Device (SQUID) sensor whose output, via electronics and an integrator, is applied as feedback to adjust I_2 so as to keep the residual flux imbalance constant.

The SQUID response cycles repeatedly with changing current, and therefore, it is necessary that the current sources track one another quite precisely during reversal to ensure that the SQUID remains locked within one particular cycle.

Assuming for the moment that the residual flux is zero and that the voltage difference detector D is nulled by adjusting R_A,

$$\frac{R_1}{R_2} = \frac{N_1}{[N_2 - N_A R_B/(R_A + R_B)]} \tag{10.1}$$

A non-zero residual flux and any offset of the voltage null detector are eliminated by reversing both current directions and taking the mean of the two results.

The perfection of flux linkage ensures the accuracy of (10.1) in terms of the ratio if integer numbers of turns can be experimentally verified by a build-up procedure in which successive larger number of turns are compared with the sum of lesser numbers, starting with single turns. Relative accuracies of better than 1 part in 10^9 are routine, but the stability of laboratory resistors may not be good enough to make this meaningful.

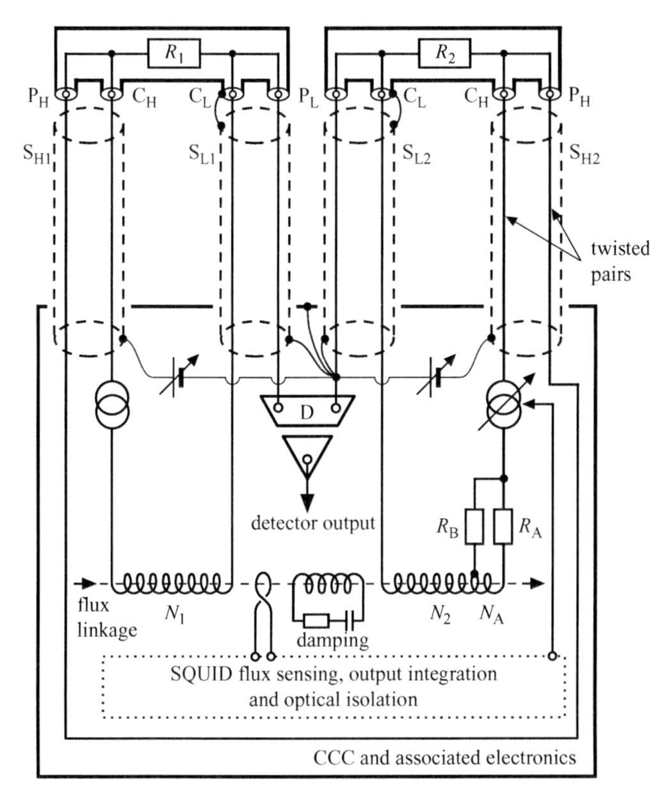

Figure 10.1 Schematic diagram of a CCC for comparing the DC values of two resistors

The overall values of the two current sources are governed by two independent digital-to-analogue convertors interfaced by optical fibre links to a controlling computer. DC power is supplied to them and any other electronics either by batteries within the enclosing screen or an isolating mains power supply (see section 1.1.5).

The particular problems posed by this instrument include the following:

(i) The extreme sensitivity of the SQUID null detector to wide-band interference (up to and beyond 100 MHz). Therefore, isolation and screening must be very carefully and perfectly carried out, and the circuit so designed as to keep the open areas of circuit meshes as small as possible by routing go-and return-current conductors together, and preferably twisting them (see section 1.1.1).

(ii) Thermoelectric effects in the room temperature circuitry (see section 1.1.11). Reasonably rapid reversal of the current sources allows for the time constant of these thermoelectric effects; rebalancing and taking the mean of the two results to eliminate them is desirable.

(iii) Unfortunately, the desirability of rapid reversal is contrary to the requirement for a settling period to allow any dielectric storage in any insulation, which experiences a change of voltage across it to die away. Storage of charge in a dielectric acts somewhat as a current source of high internal impedance, and its effects are therefore removed with a longer time constant than that of the inner-to-outer cable capacitance combined with a shunting resistance. The current from the decay of stored charge discharges through the resistors but avoids the comparator windings and therefore causes an error. To alleviate this problem, the insulation material should be one that exhibits minimal dielectric storage. PTFE is suitable. Also, the bridge sources are traditionally current sources as shown, but there seems to be no reason why voltage sources of negligible internal impedance should not be employed instead. This would create an additional low-impedance discharge path through the comparator windings, which might be undesirable however. As a further measure, the screens S_{H1} and S_{H2} in Figure 10.1 can be driven with a voltage source, which tracks the voltage of the conductors within them. A further advantage of the screens is that they intercept leakage currents between 'Hi' and 'Lo' and divert them harmlessly back to a source, and this greatly reduces the insulation resistance required for accuracy.

(iv) The cables to each resistor should be tightly twisted whether they are screened twisted pairs as shown or individually screened (see section 1.1.1).

A quantum Hall device measured with this instrument can have a single connection made either simply to each of its two potential terminals on either side of the Hall current or to the junction point of a multiple-series connection scheme. The junction point can either be adjacent to the device or remote from it, near to the CCC. An advantage of the first connection is that the measurement can be made using various pairs of potential terminals, which are not directly opposite to each other, as a check to see whether the device satisfies the important criterion of zero longitudinal resistance, as prescribed in the DC guidelines. An advantage of a multiple-series connection is that it also greatly reduces the value required for the insulation resistance between the 'Hi' and 'Lo' conductors needed to ensure an accurate result.

10.3 Josephson voltage sources and accurate voltage measurement

Accurate comparison of two nominally equal voltage standards only requires that they be connected back to back and that the difference voltage be measured with a calibrated voltmeter (Figure 10.2). Because the output impedance of voltage

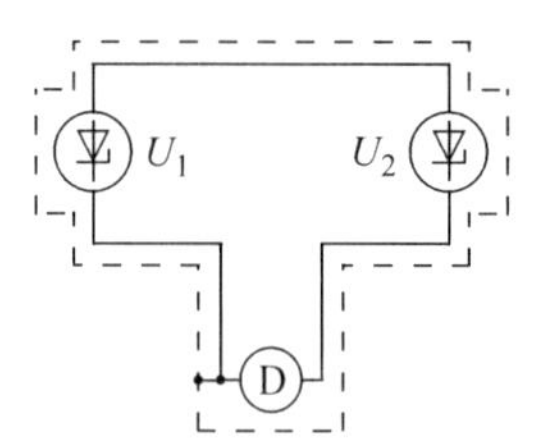

Figure 10.2 Comparing voltage sources, for example, Zener standards

sources, even including the connection conductors in series with their output, is very small, typically a few milliohms, the requirements for insulation resistance are not demanding. To eliminate interference, the power to their components and their temperature control needs careful isolation and should be provided by a totally isolated power supply (see section 1.1.5). The conductors to and from the standards should encompass as small an area as possible. Ideally, they should be a twisted conductor pair or coaxial and be surrounded by a screen, as in the diagram. If this is done, the detector need not be isolated. If one of the sources is a Josephson voltage standard, the same considerations apply.

Thermoelectric emfs are a problem, albeit less so now that 10-V Zener diode voltage standards and 10-V Josephson arrays are usual. To eliminate thermoelectric emfs, the measurement can be repeated with the polarity of the sources reversed. This can be done properly for a Josephson array by reversing its bias current, but for a Zener standard, the best way would be if the standard had an internal reversing relay located within the same temperature-controlled enclosure as the Zener itself. Since these standards are not usually constructed this way, it is necessary to regard their voltage as being defined by that found between two unstrained pure copper conductors lightly clamped under their output terminals. Then an external means of reversing the polarity needs to be provided.

One successful way of making reversing relay having very low thermoelectric emfs is to take a relay having massive silver contacts (because it is intended to switch heavy currents) and drill holes into the side of the contacts, which are of the diameter of a fine pure copper wire. After the wire is inserted into the hole, a sharp blow with a centre punch applied clear of the contact area can swage the contact and copper wire together to make a sound pressure joint. The wires can be routed from the contact to low thermoelectric emf terminals, or better, terminals can be avoided altogether if the wires are left long enough to go individually into insulating sleeves directly to their final destination. The thermoelectric voltages in a relay of this kind can die away in as little as 1 s to less than 1 nV following a reversal. From time to time, the contacts need cleaning by gentle rubbing with alcohol-soaked paper, so they would be even better if they were gold plated.

Lee *et al.* [6] have demonstrated an example of comparing two resistances supplied from two independent programmable Josephson sources generating an AC waveform at the same frequency (Figure 10.3).

Because the Josephson sources are each isolated, the resistances can be compared as defined as two-terminal-pair components with equal go-and-return currents without the need for current equalisers. Isolation is accomplished by powering each source with an isolated supply (see section 1.1.5) and sending the programming instructions over an optical fibre data link. Sources, resistors and the detector are connected by coaxial cables. The measured ratio of the two nominally equal 10-kΩ resistances agreed with the result from a conventional AC coaxial bridge based on a voltage ratio transformer (see section 9.1.5) over the 0.2–5 kHz frequency range of the latter, but the advantage of the programmable Josephson junction technique is that its frequency range extends from DC to 10 kHz at present.

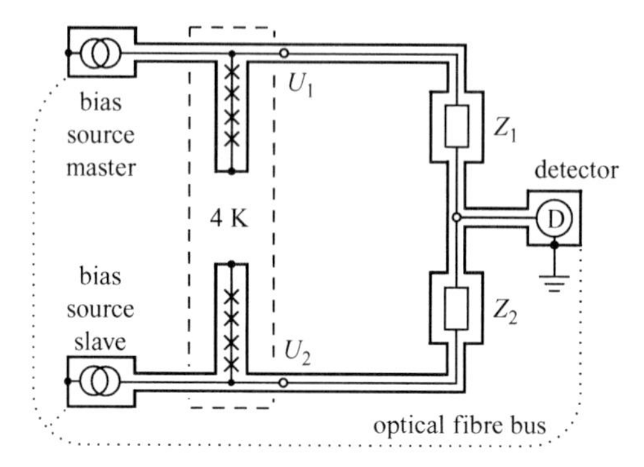

Figure 10.3 Comparing two-terminal-pair resistances using two AC independent programmable Josephson sources

There should be little difficulty in extending the technique to other values and combinations of resistance, capacitance or inductance with another since it is possible to generate a 90° or 180° phase change of one source with respect to the other. Extension to four-terminal-pair definition of the impedances may also be possible.

10.4 Future directions

There are many new and exciting possibilities for applying the higher-frequency techniques described in the previous chapters, and we now indicate only a few examples of the key future directions of this technology.

10.4.1 Higher-frequency measurements of quantum Hall resistance

As noted in section 6.7, quantum Hall resistance has been firmly established as an impedance standard between DC and 5 kHz. The principle limitation on the attainable accuracy is a residual linear frequency dependence, which has experimentally been shown to be less than 10^{-9}/kHz, a limit that is more reliable and less than that offered by low-frequency resistors of calculable frequency dependence. It seems possible that the frequency range might be extendable towards 1, or even 10 MHz, with accuracy better than 0.1% at the higher frequencies. Doing so would reveal whether quadratic or even higher-order frequency dependence coefficients might be present and, if so, the cause investigated. A higher-frequency quantum resistance standard would be important for the contribution of Johnson noise thermometry to the thermodynamic temperature scale.

10.4.2 Comparing calculable resistance standards up to 100 MHz with finite-element models

The frequency dependence of the high-frequency calculable resistance standards described in section 6.5.3 can be analytically calculated from Maxwell's equations to around 0.1% uncertainty at 100 MHz. Since the coaxial geometry of a HF-calculable resistance standard lends itself quite easily to accurate simulation by finite-element analysis, it would be very useful in the future to make robust comparisons between analytically calculated with simulated and measured values of calculable resistance standards. This would make possible direct traceability, for example, for 50- or 75-Ω radiofrequency and microwave impedance standards to DC electrical standards as complementary to, or even as an alternative to, their present traceability to length standards.

10.4.3 Radiofrequency and microwave measurements of carbon nanotubes and graphene

Carbon nanotubes have been extensively studied with potential applications in analogue and digital electronics [1] to replace silicon-based field effect transistors and copper interconnections in microprocessors because of their much greater current-carrying capacity and for bio/chemical sensors. Establishing the transport properties at higher frequencies is therefore important, and DC to 100-MHz measurements of the impedance of a carbon nanotube with an accuracy of 0.1% seems possible. These results might then be extrapolated to the microwave region (up to about 10 GHz) encountered in microprocessors. But for the moment, microwave transport measurements on individual carbon nanotubes are not possible due to the fundamental mismatch problem between 50-Ω instrumentation and the quantum contact resistance ($\approx h/4e^2 \approx 6.5$ kΩ) of any nanowire. A number of techniques are currently being researched to overcome this mismatch problem.

Carbon nanotubes may be viewed as rolled-up sheets of graphene, an atomically thin two-dimensional allotrope of carbon with a single plane of honeycomb crystal structure, extracted from graphite [2,3]. Graphene displays a whole plethora of unique physical properties with potential applications in many diverse fields of science and technology, particularly in nanoelectronics and photonics. For nanoelectronics applications, it is vital to develop a robust understanding of the DC and dyanamic transport properties of Dirac fermions in graphene (as well as developing wafer-scale, defect free fabrication processes). The dynamic transport properties of graphene can be probed using a precision, calibrated LCR meter over the range DC to 100 MHz. However, for RF and microwave transport properties of graphene, where many electronic applications such as transistors, amplifiers, mixers, interconnects etc., may be expected in the near future, the impedance mis-match problem is still present to some extent. Since the intrinsic conductance of graphene is at most $4e^2/h$ then the resitance of a realistic exfoliated sample (100 μm^2) can be of the order of a few kilo-ohms (even with quality ohmic contacts). Nevertheless, due to the two-dimensional nature of graphene this aspect may largely be circumvented by deploying device sizes having certain axial and lateral dimensions as well as impedance-matching networks which could enable accurate characterisation of graphene devices at RF and microwave frequencies.

Much work remains to be done in the areas we have briefly detailed above, but the rate of progress is rapid. In particular, since the RF/microwave properties of these novel materials are experimentally unexplored, potential remains for new physics, especially if significant differences emerge in comparing experimental results with the numerous theoretical models that have recently appeared in the literature. This should provide the stimulus for future investigations into precision electromagnetic measurements of carbon nanotubes and graphene ranging from DC to microwave frequencies.

References

1. Rutherglen C., Jain D., Burke P. 'Nanotube electronics for radiofrequency applications'. *Nature Nanotech.* 2009;**4**:811–19
2. Novoselov K.S., Geim A.K., Morozov S.V., Jiang D., Katsnelson M.I., Grigorieva I.V., et al. 'Two-dimensional gas of massless Dirac fermions in graphene'. *Nature.* 2005;**438**:197–200
3. Ferrari A.C., Meyer J.C., Scardaci V., Casiraghi C., Lazzeri M., Mauri F., et al. 'Raman spectrum of graphene and graphene layers'. *Phys. Rev. Lett.* 2006;**97**:187401
4. British Standards Institution BS1041-3. BSI 389 Chiswick High Road, London, W4 4AL U.K.
5. Hartland A. 'The quantum Hall effect and resistance standards'. *Metrologia.* 1992;**29**:175–91
6. Riddle J.L., Funikawa G.T., Plumt H.H. Platinum Resistance Thermometry, 1973. National Bureau of Standards Monograph 126

Appendix 1

T–Δ transformations. The notation is as in Figure A1.1.

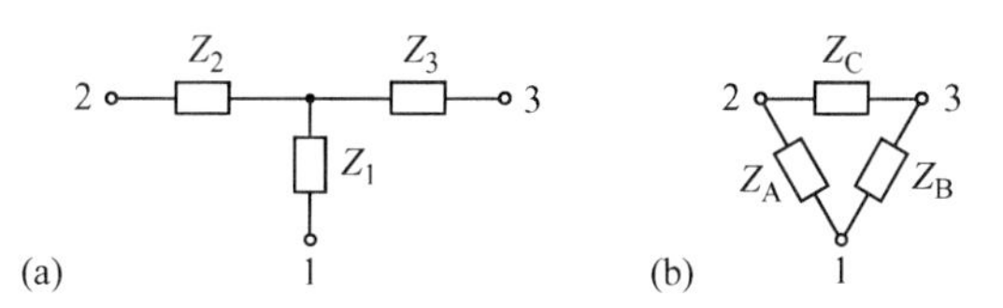

Figure A1.1 T–Δ transformations

$$Z_1 = \frac{Z_A Z_B}{Z_A + Z_B + Z_C}$$

The inverse is

$$Z_C = Z_2 + Z_3 + \frac{Z_2 Z_3}{Z_1}$$

Because of the rotational symmetry of the diagrams, the transformations for the other impedances are obtained by cyclic permutation of the subscripts.

Appendix 2

Theorem. In a network having m meshes, l links (cables) and n nodes, $m = l - n + 1$.

Proof. Suppose the theorem is true for some definite value of l, m and n.

There are just five ways in which another link can be added to the network. An unattached end of an added link is a new node.

1. One end of a new link is attached to an existing node of the network.

$$m \to m, \quad l \to l + 1, \quad n \to n + 1$$

2. One end of a new link is attached to a newly created node in the network. The new node divides the link it is attached to into two links.

$$m \to m, \quad l \to l + 2, \quad n \to n + 2$$

3. A new link is attached between two existing nodes.

$$m \to m + 1, \quad l \to l + 1, \quad n \to n$$

4. A new link is attached between an existing node of the network and a newly created one.

$$m \to m + 1, \quad l \to l + 2, \quad n \to n + 1$$

5. A link is attached between two newly created nodes in the network.

$$m \to m + 1, \quad l \to l + 2, \quad n \to n + 1$$

In each case, the theorem is still true. But the theorem is true for a network of just one node, $m = 0$, $l = 0$, $n = 1$. Hence, the theorem is universally true.

Appendix 3

Theorem. There is exactly one cable without an equaliser joining any two nodes of a network, that is, all the nodes of a network are joined by a simply connected linkage of cables without equalisers. Applying the theorem of Appendix 2 to this linkage of l_0 cables, since it has $m = 0$, $l_0 = n - 1$.

Proof. If there were to be more than one such cable, a mesh without an equaliser would be created.

Appendix 4

From section 6.5.12, the parameters α and β are as follows:

$$\begin{aligned}\alpha &= \left(\frac{I_{\text{pk}}}{2\pi bc}\right)\frac{bK_1(\eta b)+cK_1(\eta c)}{[I_1(\eta b)K_1(\eta c)-I_1(\eta c)K_1(\eta b)]} \\ &= \left(\frac{I_{\text{pk}}}{2\pi bc}\right)\left[\frac{(p+q)+j(s+t)}{u}\right]\end{aligned} \tag{A4.1}$$

where $\eta = \sqrt{j\omega\mu_0\sigma} = (1+j)/\delta$ and the Kelvin functions

$$K_1(\eta r) = K_1\left(r\zeta\sqrt{j}\right) = j[Ker_1(r\zeta) + jKei_1(r\zeta)] \tag{A4.2}$$

and

$$I_1(\eta r) = I_1\left(r\zeta\sqrt{j}\right) = -j[Ber_1(r\zeta) + jBei_1(r\zeta)] \tag{A4.3}$$

such that

$$p = -p_0[bKei_1(b\zeta) + cKei_1(c\zeta)] \tag{A4.4}$$

$$p_0 = \begin{bmatrix} Ber_1(b\zeta)Ker_1(c\zeta) - Bei_1(b\zeta)Kei_1(c\zeta) \\ -Ber_1(c\zeta)Ker_1(b\zeta) + Bei_1(c\zeta)Kei_1(b\zeta) \end{bmatrix} \tag{A4.5}$$

$$q = q_0[bKer_1(b\zeta) + cKer_1(c\zeta)] \tag{A4.6}$$

$$q_0 = \begin{bmatrix} Ber_1(b\zeta)Kei_1(c\zeta) + Bei_1(b\zeta)Ker_1(c\zeta) \\ -Ber_1(c\zeta)Kei_1(b\zeta) - Bei_1(c\zeta)Ker_1(b\zeta) \end{bmatrix} \tag{A4.7}$$

$$s = s_0[bKer_1(b\zeta) + cKer_1(c\zeta)] \tag{A4.8}$$

$$s_0 = \begin{bmatrix} Ber_1(b\zeta)Ker_1(c\zeta) - Bei_1(b\zeta)Kei_1(c\zeta) \\ -Ber_1(c\zeta)Ker_1(b\zeta) + Bei_1(c\zeta)Kei_1(b\zeta) \end{bmatrix} \tag{A4.9}$$

$$t = -t_0[bKei_1(b\zeta) + cKei_1(c\zeta)] \tag{A4.10}$$

$$t_0 = \begin{bmatrix} Ber_1(b\zeta)Kei_1(c\zeta) + Bei_1(b\zeta)Ker_1(c\zeta) \\ -Ber_1(c\zeta)Kei_1(b\zeta) - Bei_1(c\zeta)Ker_1(b\zeta) \end{bmatrix} \tag{A4.11}$$

$$u = \left\{ \begin{array}{l} [Ber_1(b\zeta)Ker_1(c\zeta) - Bei_1(b\zeta)Kei_1(c\zeta) - Ber_1(c\zeta)Ker_1(b\zeta) \\ \quad + Bei_1(c\zeta)Kei_1(b\zeta)]^2 \\ + [Ber_1(b\zeta)Kei_1(c\zeta) + Bei_1(b\zeta)Kei_1(c\zeta) - Ber_1(c\zeta)Kei_1(b\zeta) \\ \quad - Bei_1(c\zeta)Ker_1(b\zeta)]^2 \end{array} \right\} \tag{A4.12}$$

and

$$\begin{aligned} \beta &= -\left(\frac{I_{pk}}{2\pi bc}\right) \frac{bI_1(\eta b) + cI_1(\eta c)}{[I_1(\eta b)K_1(\eta c) - I_1(\eta c)K_1(\eta b)]} \\ &= -\left(\frac{I_{pk}}{2\pi bc}\right) \left[\frac{(p' + q') + j(s' + t')}{u^2}\right] \end{aligned} \tag{A4.13}$$

where

$$p' = p_0[bBei_1(b\zeta) + cBei_1(c\zeta)] \tag{A4.14}$$

$$q' = -q_0[bBer_1(b\zeta) + cBer_1(c\zeta)] \tag{A4.15}$$

$$s' = -s_0[bBer_1(b\zeta) + cBer_1(c\zeta)] \tag{A4.16}$$

$$t' = t_0[bBei_1(b\zeta) + cBei_1(c\zeta)] \tag{A4.17}$$

Index